Applied Probability and Statistics

Mario Lefebvre

Applied Probability and Statistics

With 58 illustrations

 Springer

Mario Lefebvre
Département de mathématiques
  et de génie industriel
École Polytechnique de Montréal, Québec
C.P. 6070, succ. Centre-ville
Montréal H3C 3A7
Canada
mlefebvre@polymtl.ca

Cover design by Mary Burgess.

Mathematics Subject Classification (2000): 60-01, 62-01

ISBN 978-1-4419-2106-2   e-ISBN 978-0-387-28505-4

Printed on acid-free paper.

Printed in the United States of America.   (TXQ/EB)

9 8 7 6 5 4 3 2 1

springer.com

To my parents

Preface

This book is based mainly on the lecture notes that I have been using since 1993 for a course on applied probability for engineers that I teach at the École Polytechnique de Montréal. This course is given to electrical, computer and physics engineering students, and is normally taken during the second or third year of their curriculum. Therefore, we assume that the reader has acquired a basic knowledge of differential and integral calculus.

The main objective of this textbook is to provide a reference that covers the topics that every student in pure or applied sciences, such as physics, computer science, engineering, etc., should learn in probability theory, in addition to the basic notions of stochastic processes and statistics. It is not easy to find a single work on all these topics that is both succinct and also accessible to non-mathematicians.

Because the students, who for the most part have never taken a course on probability theory, must do a lot of exercises in order to master the material presented, I included a very large number of problems in the book, some of which are solved in detail. Most of the exercises proposed after each chapter are problems written especially for examinations over the years. They are not, in general, routine problems, like the ones found in numerous textbooks.

The exercises that can be done after a given section is read are listed in Appendix C. The reader will also find, in Appendix D, the answers to all the multiple choice questions proposed in the manual. (Of course, the student is recommended to first try to solve an exercise before looking at the answer.) Appendix E provides the answers to selected supplementary exercises included in Chapters 6 and 7.

The book contains a few biographical notes on nearly all the mathematicians mentioned in the text. The reader interested in learning more about these great mathematicians can consult the various books or Web sites dedicated to the biographies of scientists.

Most of the figures in this book were realized with the help of a software program that enables one to *draw* curves or diagrams. When the figures involved mathematical functions, such as the exponential function, a mathematical software was used, when possible, to obtain precise curves.

I wish to express my gratitude to my colleagues who taught the course on probability theory for engineers with me during the last ten years. They contributed in providing interesting exercises that are now part of this manual. I am also grateful to Jean-Luc Guilbault, who helped me by typing most of the exercises found in the book. This was made possible by a grant from the Service pédagogique of the École Polytechnique, which I thank as well.

Mario Lefebvre

Montréal, August 2005

Contents

List of Tables

List of Figures

1

Introduction

1.1 The Beginnings of Probability

We often hear that the theory of probability started in the seventeenth century, when a French nobleman, the Chevalier de Méré, proposed the following problem in 1654 to his friend Pascal: Why is one more likely to obtain a "6" in four throws of a die than to obtain a double "6" in 24 throws of two dice? This problem is known as *de Méré's paradox*. We use the word *paradox*, because, based on the fact that there are 6 possible results when we roll a die and 36 possible results when we roll two dice, some people thought that the two *events* above should have the same *probability*. Indeed, notice that the number of throws, divided by the number of possible results, is equal to 2/3 in both cases (4/6 = 24/36 = 2/3). Nowadays, we can easily compute the probability of each event. We find that the probability of obtaining at least one "6" in four rolls of a (fair or non-biased) die is $1 - (5/6)^4 = 671/1296 \simeq 0.5177$, while the probability of getting at least a double "6" in throwing two dice 24 times is $1 - (35/36)^{24} \simeq 0.4914$. We can deduce that the Chevalier de Méré must have spent a lot of time throwing dice to discover such a small difference!

According to some historians, this problem was not proposed by de Méré. William Feller, who wrote two books on probability theory which are considered as true classics, mentions that the problem was in fact first treated by Cardano[1] in the preceding century. Gerolamo Cardano, or Jerome Cardan as he is called in English, was a colorful character who, in addition to being a mathematician, was also a doctor and an astrologer. He was an inveterate gambler, which caused him many problems. He also analyzed dice games and a card game similar to poker. Furthermore, he made astrological predictions. It is said that he had predicted the day of his death. Since he was in good health when the day in question arrived, he would

---

[1] Gerolamo Cardano, 1501–1576, was born and died in Italy. He became interested in the domain of probability to gain an advantage over his opponents in card and dice games. He also worked on algebraic equations. He gave the resolution method for third- and fourth-degree equations. The formula for the solution of third-degree equations had been previously obtained by the Italian mathematician Tartaglia.

certainly have committed suicide so as not to lose face! Note that another version of the story says that he managed to die of hunger on that day.

Remark. In order to pay homage to the great mathematicians who left their mark on the history of probability, we have included some biographical notes on almost every one whose name appears in the book, for example, Poisson, who gave his name to a random variable and an important stochastic process.

One thing is certain, Pascal[2] exchanged correspondence with Fermat[3] concerning the above-mentioned problem (see reference [5], p. 128, where an excerpt of a letter is reproduced). They also exchanged letters about other games of chance, including one known as the *problem of points*, which contributed greatly to the development of the domain of probability.

The first complete treatise on the calculus of probabilities was written by Huygens[4] in the seventeenth century. It is however James (or Jacob) Bernoulli (see p. 70), in posthumous works published in 1713, who really founded the calculus of probabilities. Afterward, Laplace (see p. 85), with his book *Théorie analytique des probabilités* written between the years 1812 and 1820, developed the theory of probability in a more rigorous way. Finally, in the twentieth century, Kolmogorov (see p. 227) gave the domain of probability its modern formulation.

1.2 Examples of Applications

If the conditions under which an experiment is carried out determine the result of this experiment, then it is said to be *deterministic*. For example, suppose that we observe an object moving in the sky along a decreasing exponential trajectory. Suppose also that we can control this object from a distance. Let x be the height of the object with respect to the ground at time 0. Then, if there are no perturbations, we can use the following model to determine the value of $x(t)$, the height of the object at time t:

---

[2] Blaise Pascal, 1623–1662, was born and died in France. His father, Étienne, was his professor. He invented a *calculating machine* to help his father with his work as a tax collector. Along with Fermat, he is one of the founders of the theory of probability. He was also interested in geometry and physics. From 1654 on, he became deeply religious and published books on philosophy and theology.

[3] Pierre de Fermat, 1601–1665, was born and died in France. He is especially known for his work on number theory, in particular his famous "last theorem." He has been a precursor in the domains of probability, differential calculus and analytic geometry. He was also a lawyer and, in addition to doing his research in mathematics, he was a councillor for the parliament at Toulouse.

[4] Christiaan Huygens, 1629–1695, was born and died in the Netherlands. He studied law and mathematics at the university of Leiden. Descartes showed interest in his mathematical progress. He worked, in particular, in astronomy, mechanics and optics. In 1655, using instruments he had made himself, he discovered Titan, the largest satellite of Saturn. The theory of pendulum motion is also due to him.

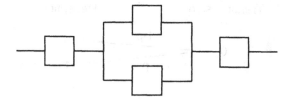

Figure 1.1. Example of a system.

$$dx(t) = -ax(t)\, dt + bu(t)\, dt, \tag{1.1}$$
$$x(0) = x,$$

where a (> 0) and b are constants and the variable $u(t)$ is called the control.

Whittle (see [24]) and the author used the equations above in research papers as a rudimentary model for the displacement of an airplane that is preparing to land.

When we cannot predict the result of an experiment repeated under the same conditions, we say that it is a *random* experiment. We can then make a list of all possible outcomes and try to compute the *probability* of each of these possible outcomes. In this textbook, we are only interested in random experiments.

Probability is used in practically every pure or applied science. Examples in engineering, particularly in electrical engineering, where we must resort to the calculus of probabilities are the following.

a) Many systems may be represented by a number of components placed in series or in parallel. For example, consider the system described by the diagram in Fig. 1.1. In **reliability**, we must be able to calculate the probability that such a system will function during a certain period of time, or at a given instant. In the latter case, we must then know, for each component, the probability that it functions at this instant and take into account the fact that the components perhaps do not operate independently from one another. In the case of reliability during a given period of time, we must know the distribution of the lifetime of each component.

b) In **communication**, we must often take into account the "noise" present in a system. For example, suppose that a system transmits either a 0 or a 1, and that there is a risk p that the number transmitted will not be received correctly (see Fig. 1.2). We may be interested in computing the probability that a 0 has been transmitted, given that a 0 has been received, or that a transmission error has occurred, etc.

c) In **automatic control**, if we take the random perturbations into account, then the model (1.1) above becomes

$$dx(t) = -ax(t)\, dt + bu(t)\, dt + dW(t), \tag{1.2}$$

where $W(t)$ is called a Brownian motion (or a Wiener process). We also say that $dW(t)/dt$ is a Gaussian white noise. Equation (1.2) is an example of a *stochastic differential equation*.

d) In **computer science**, probability is used, in particular, to help us make decisions in expert systems. We also make use of probability in simulation and in artificial intelligence, as well as in the field of queueing theory.

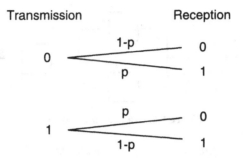

Figure 1.2. Example of a communication system.

e) In **physics**, the domain known as *statistical physics* requires some knowledge of the theory of probability, and so does that of *quantum mechanics*. In this last case, we must be familiar with diffusion processes, such as the Wiener process.

1.3 Relative Frequencies

To obtain the probability of one of the possible outcomes when we perform a random experiment, we repeat this experiment a large number of times. Let

$$f_k(n) := \frac{N_k(n)}{n},\qquad(1.3)$$

where $N_k(n)$ is the number of times that the possible outcome k has occurred during n repetitions of the experiment. The quantity $f_k(n)$ is called the **relative frequency** of outcome k. If there are K possible outcomes, which we denote by $1, 2, \ldots, K$, then we may write that

$$0 \le f_k(n) \le 1 \quad \text{for } k = 1, 2, \ldots, K \qquad(1.4)$$

and that

$$\sum_{k=1}^{K} f_k(n) = 1. \qquad(1.5)$$

Indeed we have, of course: $N_k(n) \in \{0, 1, \ldots, n\}$, so that $f_k(n) \in [0, 1]$, and

$$\sum_{k=1}^{K} f_k(n) = \frac{N_1(n) + \cdots + N_K(n)}{n} = \frac{n}{n} = 1.$$

Moreover, if A is a set that contains two possible outcomes, j and k, then

$$f_A(n) = f_j(n) + f_k(n), \qquad(1.6)$$

because j and k cannot occur on the *same* repetition of the experiment.

For instance, if we consider the random experiment that consists in observing the outcome of the roll of a die, then there are $K = 6$ possible results: $1, 2, \ldots, 6$. Let A be the set $\{1, 6\}$. Because we cannot obtain both "1" and "6" on the same roll, we may write that $f_A(n) = f_1(n) + f_6(n)$.

Finally, the **probability** of the outcome k is obtained by taking the limit of $f_k(n)$ as the number n of repetitions tends to infinity:

$$P[\{k\}] := \lim_{n \to \infty} f_k(n). \tag{1.7}$$

2

Elementary Probabilities

2.1 Basic Concepts

Definition 2.1.1. *An experiment that can be repeated under the same conditions and whose outcome cannot be predicted with certainty is called a* **random** *experiment.*

Example 2.1.1 A box contains 10 brand *A* transistors and 10 brand *B* transistors. We consider the following four random experiments:
E_1: three transistors are taken, at random and with replacement, and the number of brand *A* transistors (among the three selected) is counted.

Remark. Sampling *with* (respectively *without*) replacement means that the object that has been selected *is* (resp. *is not*) replaced in the box before taking the next one. Therefore, in the case of sampling with (resp. without) replacement, the same object can (resp. cannot) be selected more than once.

E_2: three transistors are taken, at random and with replacement, and the brand of each transistor is noted.

E_3: transistors are taken one at a time, at random and with replacement, until a brand *A* transistor has been obtained; the number of brand *B* transistors taken before obtaining a brand *A* transistor is counted.

E_4: a transistor is taken at random and its lifetime is measured (in hours).

Definition 2.1.2. *The set S of all possible outcomes of a random experiment is called the* **sample space** *of this experiment. Each possible outcome is also called an* **elementary event**.

When a repetition of a random experiment is performed, one and only one of the elementary events occurs. That is, the elementary events are **incompatible** (or mutually exclusive) and **exhaustive**.

Example 2.1.1 (continued) Corresponding to the random experiments above, we have the following sample spaces:

$S_1 = \{0, 1, 2, 3\}$.
$S_2 = \{AAA, AAB, ABA, BAA, ABB, BAB, BBA, BBB\}$.
$S_3 = \{0, 1, \ldots\}$.
$S_4 = [0, \infty)$.

The number of elementary events in a sample space may be finite (S_1 and S_2), denumerably infinite (S_3), or non-denumerably infinite (S_4).

Remark. A set is called denumerably (or countably) infinite if we can establish a one-to-one relationship between its elements and the positive integers.

A sample space that is finite or denumerably infinite is said to be **discrete**, whereas if S is non-denumerably infinite, we say that it is **continuous**.

Remark. In this chapter, we will not consider the case when the sample space S would be the union of a finite or denumerably infinite set of points and a non-denumerably infinite set of points. For instance, let E be the following random experiment: first a coin is tossed; if we get "tails," then the point 0 is chosen, otherwise a point is taken at random in the interval $[1, 2]$. In this case, we would have $S = \{0\} \cup [1, 2]$. An example like this will be called a "mixed" type in Chapter 3.

Definition 2.1.3. *An* **event** *is a subset of the sample space S. Thus, the empty set $\emptyset$ and the sample space S itself are events.*

Remarks. i) The empty set $\emptyset$ is called the **impossible** (or null) event and S is the **certain** event.
ii) There are 2^n events that can be defined with n elementary events.
iii) We generally use capital letters like A, B, etc., to denote events.

Example 2.1.1 (continued) Events defined with respect to the sample spaces associated with the random experiments above are the following:
A_1: exactly one brand A transistor is obtained; that is, $A_1 = \{1\}$.
A_2: one brand A transistor and two brand B transistors are obtained; that is, $A_2 = \{ABB, BAB, BBA\}$.
A_3: five or six brand B transistors are picked before a (first) brand A transistor is obtained; that is, $A_3 = \{5, 6\}$.
A_4: the selected transistor lasts more than 200 hours; that is, $A_4 = (200, \infty)$.

Operations with Sets (see Fig. 2.1, p. 9)

Union: $A \cup B$ denotes the set of outcomes that belong to A **or** to B (or to both). Similarly, in general, for n events we write:

$$A_1 \cup A_2 \cup \ldots \cup A_n = \bigcup_{i=1}^{n} A_i.$$

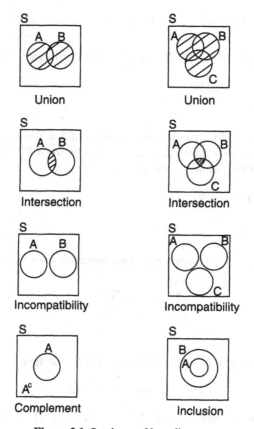

Figure 2.1. Set theory: Venn diagrams.

Intersection: $A \cap B$ denotes the set of outcomes that belong to both A **and** B. In general, we have:

$$A_1 \cap A_2 \cap \ldots \cap A_n = \bigcap_{i=1}^{n} A_i.$$

If two events are incompatible, then we write: $A \cap B = \emptyset$.

Complement: The complement of an event A is the set of outcomes that do not belong to A; it is denoted by A^c.

Inclusion: If all the outcomes that belong to event A also belong to event B, then we say that A is included in B and we write: $A \subset B$.

Equality: Two events are said to be equal if they contain the same outcomes; we then write: $A = B$.

We can easily show the following relationships:

1) $A \cup B = B \cup A$ and $A \cap B = B \cap A$ (Commutativity).

2) $A \cup (B \cup C) = (A \cup B) \cup C$ and $A \cap (B \cap C) = (A \cap B) \cap C$ (Associativity).

3) $A \cup (B \cap C) = (A \cup B) \cap (A \cup C)$ and $A \cap (B \cup C) = (A \cap B) \cup (A \cap C)$ (Distributivity).

4) $(A \cap B)^c = A^c \cup B^c$ and $(A \cup B)^c = A^c \cap B^c$ (De Morgan's laws).

2.2 Probability

Definition 2.2.1. *Let E be a random experiment and S a sample space associated with E. To each event A in S we assign a real number noted $P[A]$, called the **probability** of A, so that the following properties are satisfied:*

Axiom I: $P[A] \geq 0 \ \forall \ A \subset S$;

Axiom II: $P[S] = 1$;

Axiom III: *If $A_1, A_2, \ldots$ is a sequence of incompatible events, then*

$$P\left[\bigcup_{k=1}^{\infty} A_k\right] = \sum_{k=1}^{\infty} P[A_k]. \tag{2.1}$$

Remarks. i) We deduce from Axiom III that if $A \cap B = \emptyset$, then

$$P[A \cup B] = P[A] + P[B]. \tag{2.2}$$

Indeed, we only have to take $A_1 = A$, $A_2 = B$ and $A_k = \emptyset$ for $k \geq 3$, because $P[\emptyset] = 0$ (see Proposition 2.2.1).

ii) The function P is called a *probability measure*; it is a function from S into the interval $[0, 1]$.

With the help of the three axioms above, we easily show the following proposition.

Proposition 2.2.1. *We have:*

1) $P[A^c] = 1 - P[A]$.

2) $P[A] \leq 1$.

3) $P[\emptyset] = 0$.

4)

$$P\left[\bigcup_{k=1}^{n} A_k\right] = \sum_{k=1}^{n} P[A_k] - \sum_{j<k} P[A_j \cap A_k] + \cdots + (-1)^{n+1} P\left[\bigcap_{k=1}^{n} A_k\right].$$

5) *If $A \subset B$, then $P[A] \leq P[B]$.*

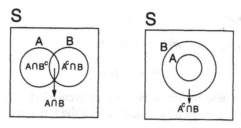

Figure 2.2. Figures for the proof of parts 4) and 5) of Proposition 2.2.1.

Proof.

1) Since $A^c \cup A = S$ and $A^c \cap A = \emptyset$, we deduce from Axioms II and III that

$$P[A^c \cup A] \overset{\text{II}}{=} 1 \quad \text{and} \quad P[A^c \cup A] \overset{\text{III}}{=} P[A^c] + P[A] \Rightarrow P[A^c] = 1 - P[A].$$

2) $P[A] = 1 - P[A^c] \le 1$ because $P[A^c] \ge 0$ (by Axiom I).

3) We have: $S^c = \emptyset$ and $P[S^c] = 1 - P[S] \overset{\text{II}}{=} 1 - 1 = 0$.

4) For $n = 2$, let us write $A_1 = A$ and $A_2 = B$; we then have (see Fig. 2.2):

$$P[A \cup B] = P[A \cap B^c] + P[A \cap B] + P[A^c \cap B],$$

because the three events are incompatible. Moreover,

$$P[A \cap B^c] = P[A] - P[A \cap B] \quad \text{and} \quad P[A^c \cap B] = P[B] - P[A \cap B].$$

So, we have:

$$P[A \cup B] = P[A] + P[B] - P[A \cap B]. \tag{2.3}$$

Next, let $D := B \cup C$; then, we may write that

$$P[A \cup B \cup C] = P[A \cup D] = P[A] + P[D] - P[A \cap D]$$
$$= P[A] + P[B] + P[C] - P[B \cap C] - P[A \cap (B \cup C)]$$
$$= P[A] + P[B] + P[C] - P[B \cap C] - P[(A \cap B) \cup (A \cap C)].$$

Hence, we obtain the following formula:

$$P[A \cup B \cup C] = P[A] + P[B] + P[C] - P[A \cap B]$$
$$- P[A \cap C] - P[B \cap C] + P[A \cap B \cap C]. \tag{2.4}$$

To prove the formula in the general case, we proceed by induction (that is, we assume that the formula is true for the case of n events and we try to show that it is then also valid for $n + 1$ events).

5) If $A \subset B$, we may write (see Fig. 2.2) that

$$P[A] = P[B] - P[A^c \cap B] \Rightarrow P[A] \le P[B]. \qquad \square$$

Discrete Sample Spaces

If S is a discrete sample space, we may write that

$$S = \{e_1, e_2, \ldots\},$$

where e_k is a possible outcome (or an elementary event). Let $A \subset S$; then the probability of event A can be obtained by making use of the following formula:

$$P[A] = \sum_{k: e_k \in A} P[\{e_k\}]. \tag{2.5}$$

If the number n of elementary events e_k is *finite* and if the e_k's are *equiprobable* (or equally likely), so that $P[\{e_k\}] = 1/n \; \forall k$, then we may write that

$$P[A] = n(A)/n, \tag{2.6}$$

where $n(A)$ is the number of elementary events in A.

Example 2.2.1 i) In the case of the sample space S_1 in Example 2.1.1, the four elementary events are *not* equiprobable. If we denote the probability $P[\{0\}]$ by p, then we may write that

$$P[\{1\}] = P[\{2\}] = 3p \quad \text{and} \quad P[\{3\}] = p.$$

Moreover,

$$\sum_{k=0}^{3} P[\{k\}] = 1 \Rightarrow p = 1/8.$$

Hence, we find that

$$P[A_1] = P[\{1\}] = 3/8 \quad (\neq 1/4).$$

On the other hand, the e_k's are *equally likely* in the case of S_2 and we may write, directly, that

$$P[A_2] = P[\{ABB, BAB, BBA\}] = 3/8.$$

Note that the probabilities $P[A_1]$ and $P[A_2]$ must be equal because the events A_1 and A_2 correspond to the same outcome of the three draws in the random experiments E_1 and E_2. Thus, to calculate the probability of getting exactly one brand A transistor in three draws at random and with replacement from a box containing 10 brand A and 10 brand B transistors, it is simpler to consider the sample space S_2 for which the outcomes are equally likely.

ii) Since the sample space S_3 is *denumerably infinite*, the e_k's cannot be equiprobable. We can show that

$$P[A_3] = P[\{5, 6\}] = P[\{5\}] + P[\{6\}] = (10/20)^5(10/20) + (1/2)^6(1/2).$$

The second equality above is obtained by incompatibility of the events $\{5\}$ and $\{6\}$, while the third equality results from the notion of independence (which will be discussed in Section 2.5).

Continuous Sample Spaces

When the number of possible outcomes is non-denumerably infinite, each e_k has a *zero* probability of occurring. We must then give a formula that enables us to calculate the probability that the outcome of the random experiment will be located in any interval $[a, b]$.

Example 2.2.2 Suppose that

$$P[(a, b]] = e^{-a/100} - e^{-b/100},$$

where $0 \leq a < b < \infty$, is the probability that the lifetime of the transistor taken at random in the random experiment E_4, in Example 2.1.1, belongs to the interval $(a, b]$. Then, we may write that

$$P[A_4] = P[(200, \infty)] = e^{-2} - e^{-\infty} \simeq 0.1353.$$

2.3 Combinatorial Analysis

1) **Tree diagrams**. In order to more easily make the list of all possible outcomes when we perform a random experiment, we sometimes use a *tree diagram*. For example, suppose that we perform the experiment E_2 in Section 2.1. The eight possible outcomes are obtained by following each branch or path in the diagram of Fig. 2.3.

2) **Principle of multiplication**. Suppose that we perform k consecutive random experiments E_i. Suppose also that the outcome of a given random experiment does *not* influence the subsequent experiments. Let n_i be the number of possible outcomes in the sample space S_i of the experiment E_i, for $i = 1, 2, \ldots, k$. Then there are

$$n_1 \times n_2 \times \cdots \times n_k$$

possible results in all. Indeed, this corresponds to the number of ways of choosing an element from a set that contains n_1 elements, then an element from another set that contains $n_2, \ldots$ and finally an element from a set that contains n_k.

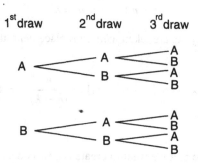

Figure 2.3. Example of a tree diagram.

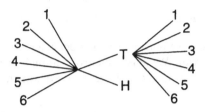

Figure 2.4. Tree diagram drawn compactly.

For example, if we want to buy a computer system made up of three components: the computer itself, a screen and a printer, and if we can choose between three brands of computers, two brands of screens and four brands of printers, then we can buy $3 \times 2 \times 4 = 24$ different systems.

Similarly, if a cafeteria is offering for lunch a choice of two soups, three main courses and four desserts, then the person working at the cash register can come across $3 \times 4 \times 5 - 1 = 59$ different trays (since a customer does not have to buy a dessert, for instance, but he or she must buy something).

Remark. It is important that the outcome of a given experiment does not influence the subsequent experiments. For example, consider the following random experiment: a die is thrown, then a coin; if "tails" is obtained, then the die is thrown a second time. In this case, the sample space of this experiment is $S = \{1H, \ldots, 6H, 1T1, \ldots, 6T6\}$ (see Fig. 2.4). So, there are $6 + 36 = 42$ elementary events in S, and not $6 \times 2 \times 6 = 72$. The first six elementary events are equiprobable among themselves if the die is non-biased, and so are the last 36. In fact, using the notion of independence (see Section 2.5), we immediately find that the first six elementary events have a probability of $1/12$, and the others have a probability of $1/72$ (if the coin also is non-biased).

3) Permutations. Suppose that we possess n *distinct* objects and that we pick k at random. If the *order* in which the objects are chosen matters and if the objects are taken *with* replacement, then the number of different *permutations* (or arrangements) of these objects that we can obtain is given by

$$n \times n \times \cdots \times n = n^k; \tag{2.7}$$

in the case when the objects are taken *without* replacement, the number of different permutations is

$$n \times (n-1) \times \cdots \times [n-(k-1)] = \frac{n!}{(n-k)!} := (n)_k := P_k^n, \tag{2.8}$$

where $k \in \{(0,)1, \ldots, n\}$.

Example 2.3.1 Suppose that we want to create access codes, to a computer system, consisting of five letters taken among the 26 letters of the alphabet. Then, we can create 26^5 different codes if the same letter can be used more than once. However, if repetitions are not allowed, the number of different codes is given by

$$26 \times 25 \times 24 \times 23 \times 22 = 7,893,600.$$

Remarks. i) If $k = n$, we have: $P_k^n = n!$.

ii) **Stirling's formula**: we can show that

$$n! \sim (2\pi n)^{1/2} n^n e^{-n}.$$

That is, the ratio $n!/(2\pi n)^{1/2} n^n e^{-n}$ tends to 1 when n tends to infinity.

4) **Combinations**. Suppose now that we take k objects, at random and *without* replacement, among n *distinct* objects and that the order in which the objects are chosen does *not* matter. Then, the number of different *combinations* (or samples) that we can obtain is given, for $k \in \{(0,)1, \ldots, n\}$, by

$$\frac{n \times (n-1) \times \cdots \times [n-(k-1)]}{k!} = \frac{n!}{k!(n-k)!} := \binom{n}{k} := C_k^n. \qquad (2.9)$$

Remarks. i) Each combination of k objects enables us to form $k!$ different permutations.

ii) We can easily check that $C_k^n = C_{n-k}^n$.

iii) We can show that if the objects are taken *with* replacement, then the number of different combinations is C_k^{n+k-1}. In this case, there are no restrictions on k.

iv) We have: $C_k^n = 0$ if $k > n$ or $k < 0$.

Example 2.3.2 Suppose that we have ten diskettes in a box. If we pick three, at random and without replacement, then there are

$$C_3^{10} = \frac{10!}{3!(10-3)!} = 120$$

different groups of three diskettes that we can obtain. In practice, it is not easy to take each diskette in a really random way, so that the first diskette that we pick had 1 chance in 10 of being chosen, the second one 1 chance in 9, and the third one 1 chance in 8. To do so, we can use a random numbers generator. For instance, if we have access to a computer that generates random numbers between 0 and 1, we can generate three numbers (or more, if necessary) and take first diskette no. 2 if the first number generated is located in the subinterval $[1/10, 2/10)$, etc. We can also use tables of random numbers.

5) **Permutations of non-distinguishable objects**. Suppose finally that among the n objects that we have, there are n_i of type i, where $i = 1, 2, \ldots, k$ and $n_1 + n_2 +$

$\cdots + n_k = n$. Then, the number of different permutations of the n objects taken *all at once* is given by

$$\frac{n!}{n_1!n_2!\cdots n_k!}. \tag{2.10}$$

The numbers obtained for varying values of n_i in this expression are called the *multinomial coefficients*.

Example 2.3.3 With the letters a, a, b, b, b and c, we can form

$$\frac{6!}{2!3!1!} = 60$$

different code words (or access words).

We will complete this section with other examples of combinatorial analysis problems.

Example 2.3.4 If a letter (or two) may be repeated (only once) in Example 2.3.1, then there are

$$P_5^{26} + C_4^{26}C_1^4\frac{5!}{2!} + C_3^{26}C_2^3\frac{5!}{2!2!} = 7,893,600 + 3,588,000 + 234,000 = 11,715,600$$

different codes. The first term above corresponds to the case when there are no repetitions, the second term to the case when there is exactly one repetition and the third one to the case when there are exactly two repetitions. Moreover, we may write that

$$P_5^{26} = C_5^{26} \cdot 5!.$$

Example 2.3.5 With the letters in the word *essays*, the number of distinct words containing four letters that we can form is given by

$$4! \text{ (with 1 s)} + C_2^3\frac{4!}{2!} \text{ (with 2 s's)} + C_1^3\frac{4!}{3!} \text{ (with 3 s's)} = 24 + 36 + 12 = 72.$$

Example 2.3.6 If we have one one-dollar coin, two quarters and three nickels, then we can pay $2 \times 3 \times 4 - 1 = 23$ different sums exactly. Note that we can employ the principle of multiplication here, because whatever the number of coins of a certain type that we use to pay, the sum will always be different. The problem would be more complicated if we also had a dime, for instance.

Example 2.3.7 If there are n persons in a class, then the probability that at least two of these n persons have the same birthday is given (disregarding leap years) by

$$1 - \frac{P_n^{365}}{365^n} \quad \text{for } n = 2, 3, \ldots, 365.$$

Note that the term P_n^{365} in the numerator corresponds to the number of permutations of n objects taken at random and *without* replacement among 365 distinct objects, while 365^n is the number of permutations *with* replacement of n objects taken among 365 distinct objects. We might think that this formula is equivalent to that obtained by computing the number of *combinations* with and without replacement, that is,

$$1 - \frac{C_n^{365}}{C_n^{365+n-1}} \quad \text{for } n = 2, 3, \ldots, 365.$$

However, this last formula is wrong, because the combinations with replacement are *not* equally likely, so that we cannot simply divide the number of favorable cases by the total number of cases, like we did with the permutations.

Example 2.3.8 We have 20 components of type I, 5 of which are defective, and 30 components of type II, 15 of which are defective.

a) We want to build a system made up of ten components of type I and five components of type II placed in series. What is the probability that the system functions if the components are taken at random?

b) How many distinct systems made up of four components placed in series, with at least two components of type I, can be constructed if the order of the components is taken into account?

Solution. a) The total number of distinct systems that can be built is given by

$$C_{10}^{20} \times C_5^{30}.$$

The number of distinct systems that function is

$$C_0^5 \times C_{10}^{15} \times C_0^{15} \times C_5^{15}.$$

Then, by *equiprobability*, we can write that the required probability is

$$\frac{C_{10}^{15} \times C_5^{15}}{C_{10}^{20} \times C_5^{30}} \simeq 0.00034.$$

b) The total number of distinct systems that we can construct is P_4^{50}; among those, there are P_4^{30} that contain no components of type I, and

$$20 \times 4 \times P_3^{30}$$

that contain exactly one. Then, the required number is

$$P_4^{50} - P_4^{30} - 80 \times P_3^{30}.$$

Remark. If we assume that we cannot distinguish between two components of the same type, then the number of distinct systems that we can construct is given by 6 (with 2 components of type I) + 4 (with 3 components of type I) + 1 (with 4 components of type I) = 11.

2.4 Conditional Probability

Let A and B be two events defined with respect to a sample space S associated with a random experiment E. Suppose that we perform the experiment E and that the event B occurs. Then B becomes the new sample space, for this trial, and in order that A too has occurred, $A \cap B$ must now have occurred.

Notation. The expression $P[A|B]$ denotes the probability of the event A, *given that* the event B has occurred.

Definition 2.4.1. *We write:*

$$P[A|B] = \frac{P[A \cap B]}{P[B]} \quad \text{if } P[B] > 0. \tag{2.11}$$

Remarks. i) The conditional probabilities satisfy the three axioms in the definition of probability (see p. 10). In fact, every probability is a conditional probability since, for any $A \subset S$,

$$P[A|S] = \frac{P[A \cap S]}{P[S]} = \frac{P[A]}{1} = P[A].$$

It follows that Proposition 2.2.1 is still valid for conditional probabilities. For example, we may write that

$$P[A^c|B] = 1 - P[A|B].$$

However, in general,

$$P[A|B^c] \neq 1 - P[A|B].$$

Moreover,

$$P[A \cup B|C] = P[A|C] + P[B|C] - P[A \cap B|C],$$

if $P[C] > 0$, etc.

ii) If $A \cap B = \emptyset$, then $P[A|B] = 0$.

iii) If $B \subset A$, then $P[A|B] = 1$. Furthermore, if $A \subset B$, then

$$P[A|B] = \frac{P[A]}{P[B]}.$$

iv) We may have: $P[A|B] < P[A]$, $P[A|B] > P[A]$, or $P[A|B] = P[A]$. That is, there is no relationship between conditional probabilities and the corresponding *marginal* probabilities.

From Definition 2.4.1, we deduce at once the following proposition.

Proposition 2.4.1. (Multiplication rule) *We have:*

$$P[A \cap B] = P[A|B] \times P[B] \quad \text{if } P[B] > 0, \qquad (2.12a)$$

$$P[A \cap B] = P[B|A] \times P[A] \quad \text{if } P[A] > 0. \qquad (2.12b)$$

In general,

$$P[A_1 \cap A_2 \ldots \cap A_n] = P[A_1] \times P[A_2|A_1] \times P[A_3|A_1 \cap A_2] \times \cdots$$
$$\times P[A_n|A_1 \cap A_2 \cap \ldots \cap A_{n-1}] \qquad (2.13)$$

if $P[A_1 \cap A_2 \cap \ldots \cap A_{n-1}] > 0.$

Example 2.4.1 Two components are taken at random and *without* replacement in a box containing ten brand A and ten brand B components. What is the probability of getting a) two brand A components? b) two components of the same brand? c) two components of different brands?

Solution. Let A_k = a brand A component is obtained on the kth draw.
a) We want

$$P[A_1 \cap A_2] = P[A_2|A_1]P[A_1] = 9/19 \times 10/20 = 9/38.$$

Note that, according to the multiplication rule, we may also write that

$$P[A_1 \cap A_2] = P[A_1|A_2]P[A_2].$$

However, this formula does not enable us to directly find the probability $P[A_1 \cap A_2]$.

b) By symmetry (since there are as many brand A components as brand B components), we deduce from the preceding result that the required probability is given by $9/38 + 9/38 = 9/19$.

c) We now deduce from b) that the required probability is $1 - 9/19 = 10/19$.

Proposition 2.4.2. *Let A and B be two events such that $P[A] \times P[B] > 0$. Then,*

$$P[A|B] = \frac{P[B|A]P[A]}{P[B]}. \qquad (2.14)$$

Proof. We only have to make use of the definition of $P[A|B]$ and the multiplication rule. $\qquad \square$

Remark. This result is sometimes called *Bayes' formula.*[1]

---

[1] The Reverend Thomas Bayes, 1702–1761, was born and died in England. He was first educated by tutors. After his ordination, he worked with his father, who was also a pastor. His works on probability theory were published in a posthumous scientific article in 1764. He wanted to find a method by which "we might judge concerning the probability that an event has to happen, in given circumstances, upon supposition that we know nothing concerning it but that, under the same circumstances, it has happened a certain number of times, and failed a certain other number of times." Nowadays, a statistical school of thought is called Bayesian.

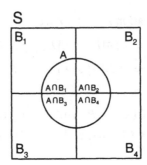

Figure 2.5. Example of a partition of a sample space with $n = 4$.

Definition 2.4.2. *Let B_1, B_2, ... , B_n be events such that*

i) $B_i \cap B_j = \emptyset \; \forall i \neq j$;

ii) $\bigcup_{k=1}^{n} B_k = S$.

*We say that the events B_k form a **partition** of the sample space S.*

Remarks. i) In probability, it is natural to impose in the preceding definition the additional condition: iii) $P[B_k] > 0$ for $k = 1, 2, \ldots , n$.

ii) A and A^c constitute a partition of S. Similarly, $A \cap B$, $A \cap B^c$, $A^c \cap B$ and $A^c \cap B^c$ constitute a partition of S.

Now, let $A \subset S$. We can split up A into n disjoint and exhaustive parts as follows (see Fig. 2.5):

$$A = (A \cap B_1) \cup (A \cap B_2) \cup \ldots \cup (A \cap B_n).$$

Hence, using Axiom III in the definition of $P[A]$, we obtain the following proposition.

Proposition 2.4.3. (Total probability rule) *Let B_1, B_2, ... , B_n be a partition of S. Then, for any $A \subset S$ we have:*

$$P[A] = \sum_{k=1}^{n} P[A \cap B_k] = \sum_{k=1}^{n} P[A|B_k]P[B_k] \quad \text{if } P[B_k] > 0 \; \forall k. \qquad (2.15)$$

Remark. The second equality above follows from the multiplication rule.

Finally, making use of Proposition 2.4.2 and the total probability rule, we obtain *Bayes' rule.*

Proposition 2.4.4. (Bayes' rule) *Let $A \subset S$ and B_1, B_2, ... , B_n be a partition of S such that $P[B_k] > 0$ for $k = 1, 2, \ldots , n$. Then,*

$$P[B_j|A] = \frac{P[A|B_j]P[B_j]}{\sum_{k=1}^{n} P[A|B_k]P[B_k]} \quad \text{for } j = 1, 2, \ldots , n. \qquad (2.16)$$

Example 2.4.2 We consider a communication system that transmits either a 0 or a 1. Because of the "noise," the signal transmitted is sometimes received incorrectly. We define the events

$$E_i = i \text{ is transmitted} \quad \text{and} \quad R_i = i \text{ is received}$$

for $i = 0$ and 1.

We assume that $P[R_0|E_0] = 0.7$, $P[R_1|E_1] = 0.8$ and that a 0 is transmitted 60% of the time.

a) Calculate $P[E_0|R_1]$.

b) Find the probability of a transmission error.

Solution. a) We have:

$$P[E_0|R_1] = \frac{P[R_1|E_0]P[E_0]}{P[R_1|E_0]P[E_0] + P[R_1|E_1]P[E_1]}$$
$$= \frac{(1-0.7)(0.6)}{(1-0.7)(0.6) + (0.8)(0.4)} = 0.36.$$

b)
$$P[\text{Transmission error}] = P[E_0 \cap R_1] + P[E_1 \cap R_0]$$
$$= P[R_1|E_0]P[E_0] + P[R_0|E_1]P[E_1]$$
$$= (1-0.7)(0.6) + (1-0.8)(0.4)$$
$$= 0.26.$$

Remark. The events E_0 and E_1 constitute a partition of S, and so do R_0 and R_1.

2.5 Independence

Definition 2.5.1. *Let A and B be two events. We say that A and B are* **independent** *if*

$$P[A \cap B] = P[A]P[B]. \tag{2.17}$$

Remarks. i) Two independent events may or may not be incompatible. However, if two events A are B are incompatible and independent, then $P[A]$ or $P[B]$ (or both) must be equal to zero. Indeed, we then have:

$$P[A \cap B] \overset{inc.}{=} P[\emptyset] = 0 \text{ and } P[A \cap B] \overset{ind.}{=} P[A]P[B] \Rightarrow P[A] = 0 \text{ or } P[B] = 0.$$

ii) We also define **conditional independence** as follows: A and B are said to be *conditionally independent with respect to C if*

$$P[A \cap B|C] = P[A|C]P[B|C]. \tag{2.18}$$

Equation (2.18) may be satisfied, whether the events A are B are independent or not. Moreover, two independent events may be conditionally dependent with respect to an event C. For instance, suppose that we throw a (non-biased) coin twice, independently. Let us define the following events:

A = "heads" is obtained on the first throw,

B = "heads" is obtained on the second throw,

C = exactly one "heads" is obtained in all.

Then A and B (as well as A and B^c) are independent, but

$$(0 =) \; P[A \cap B|C] \neq P[A|C]P[B|C] > 0.$$

In fact, we have:

$$P[A|C] = \frac{P[A \cap C]}{P[C]} = \frac{P[A \cap B^c]}{P[C]} \stackrel{\text{ind.}}{=} \frac{(1/2)(1/2)}{(1/2)^2 + (1/2)^2} = 1/2$$

and $P[B|C] = P[A|C]$, by symmetry.

Proposition 2.5.1. *Two events, A and B, such that $P[A] \times P[B] > 0$, are independent if and only if*

$$P[A|B] = P[A] \quad or \quad P[B|A] = P[B]. \tag{2.19}$$

Proof. We have: $P[A \cap B] = P[A|B]P[B]$. Then, if A and B are independent, we obtain that

$$P[A]P[B] = P[A|B]P[B] \Rightarrow P[A|B] = P[A].$$

Similarly, since $P[A \cap B] = P[B|A]P[A]$, we also have: $P[B|A] = P[B]$.

Conversely, if $P[A|B] = P[A]$, then

$$P[A] = \frac{P[A \cap B]}{P[B]} \Rightarrow P[A]P[B] = P[A \cap B].$$

In a similar way, we show that if $P[B|A] = P[B]$, then $P[A \cap B] = P[A]P[B]$. $\square$

Proposition 2.5.2. *If A and B are independent, then so are A^c and B.*

Proof. Let us consider the case when $P[A] \times P[B] > 0$. We have: $P[A^c|B] = 1 - P[A|B] \stackrel{\text{ind.}}{=} 1 - P[A] = P[A^c]$. $\square$

Remark. Likewise, we show that A and B^c, as well as A^c and B^c, are independent events.

Definition 2.5.2. *The events $A_1, A_2, \ldots, A_n$ are said to be* **pairwise independent** *if*

$$P[A_i \cap A_j] = P[A_i]P[A_j] \quad \forall i \neq j, \tag{2.20}$$

where $i, j = 1, 2, \ldots, n$.

Remark. There are C_2^n conditions to check.

Definition 2.5.3. *The events $A_1, A_2, \ldots, A_n$ are said to be* **(globally) independent** *if for any $k \leq n$, whatever the events $A_{i(1)}, \ldots, A_{i(k)}$, we may write that*

$$P[A_{i(1)} \cap A_{i(2)} \cap \ldots \cap A_{i(k)}] = P[A_{i(1)}]P[A_{i(2)}] \cdots P[A_{i(k)}], \tag{2.21}$$

where $i(j) \neq i(m)$ if $j \neq m$.

Remarks. i) This means that the events can be taken $2, 3, \ldots, n$ at a time and the probability of the intersection is always equal to the product of the marginal probabilities, that is, of the probabilities of the individual events.

ii) This time, the number of conditions that we must check is

$$\sum_{k=2}^{n} C_k^n = \sum_{k=0}^{n} C_k^n - C_1^n - C_0^n = 2^n - n - 1,$$

by Newton's binomial theorem.[2]

Example 2.5.1 We consider a system made up of three components that operate independently from one another. We suppose that the system functions if at least two of its components are working. A system of this type may be represented by the diagram in Fig. 2.6, p. 24. Let the events be as follows:

$$F = \text{the system is functioning at time } t$$

and

$$F_i = \text{the component } i \text{ is functioning at time } t, \text{ for } i = 1, 2 \text{ and } 3.$$

Then, if we assume that $P[F_i] = 0.9$ for all i, we can write (by symmetry) that

$$P[F] = 3 \times P[F_1 \cap F_2 \cap F_3^c] + P[F_1 \cap F_2 \cap F_3]$$

$$\stackrel{\text{ind.}}{=} 3 \times (0.9)(0.9)(0.1) + (0.9)^3 = 0.9720.$$

---

[2] Sir Isaac Newton, 1643–1727, was born and died in England. He is famous for his contributions to the fields of mechanics, optics and astronomy. He and Leibniz (and, according to some, Fermat) invented differential calculus. Actually there was a great controversy between the two scientists about that, each one claiming to be the inventor of infinitesimal calculus. He also worked on alchemy and wrote theological books.

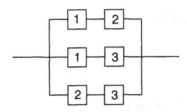

Figure 2.6. Graphical representation of a two-out-of-three system.

Example 2.5.2 A box contains 10 brand A diodes and 20 brand B diodes. Ten diodes are taken, at random and with replacement, and the brand of each one is noted. Compute the probability of getting at least one brand A diode.

Solution. Let the events be as follows:

$$F = \text{at least one brand } A \text{ diode is obtained}$$

and

$$A_k = \text{the } k\text{th diode selected is a brand } A \text{ diode.}$$

We want

$$P[F] = P[A_1 \cup A_2 \cup \ldots \cup A_{10}] = 1 - P[A_1^c \cap A_2^c \cap \ldots \cap A_{10}^c].$$

Now, because the diodes are taken with replacement, the events A_k are *(globally) independent*. Moreover, $P[A_k] = 10/30 = 1/3$ for any k. Then, we have:

$$P[F] = 1 - (2/3)^{10} \simeq 0.9827.$$

Example 2.5.3 (see [14]) In a certain factory, 96% of the computers manufactured conform to the norms. Each computer produced is subjected to two independent quality control operations. We assume that each operation classifies as "good" 98% of the computers that effectively conform to the norms, and 6% of the computers that actually do not conform to the norms. Calculate the probability that a computer sold effectively conforms to the norms.

Solution. Let the events be as follows:

$$A = \text{the computer has been classified as "good" twice}$$

and

$$B = \text{the computer conforms to the norms.}$$

We want

$$P[B|A] = \frac{P[A|B]P[B]}{P[A|B]P[B] + P[A|B^c]P[B^c]}$$

$$\stackrel{\text{ind.}}{=} \frac{(0.98)^2(0.96)}{(0.98)^2(0.96) + (0.06)^2(0.04)} \simeq 0.9998.$$

Remarks. i) We know that $B_1 := B$ and $B_2 := B^c$ constitute a partition of S.

ii) Let A_k = the computer has been classified as "good" by the kth control operation; then, we may write that $A = A_1 \cap A_2$. Furthermore, A_1 and A_2 are conditionally independent with respect to B (and B^c), but are not independent.

Example 2.5.4 ("The Monty Hall problem") A man takes part in a television game show. At the end, he is presented with three doors and is asked to choose one among them. The grand prize is hidden, at random, behind one of the doors, while there is nothing behind the other two doors. The game show host knows where the grand prize has been hidden. Suppose that the man has chosen door no. 1 and that the host tells him that he did well in not choosing door no. 3, because there was nothing behind it. He then offers the man the opportunity to change his choice and, therefore, to select door no. 2 instead. What is the probability that the man will win the grand prize if he decides to stick with door no. 1?

Solution. Let A_k = the grand prize is hidden behind door no. k, for $k = 1, 2, 3$, and let F = the game show host has eliminated door no. 3. Assume, in all logic, that if the man has chosen the right door, then the host will eliminate door no. 3 with probability 1/2. In this case,

$$P[F] = P[F|A_1]P[A_1] + P[F|A_2]P[A_2] + P[F|A_3]P[A_3] = \frac{1}{2} \cdot \frac{1}{3} + 1 \cdot \frac{1}{3} + 0 = \frac{1}{2}$$

and then

$$P[A_1|F] = \frac{P[F|A_1]P[A_1]}{P[F]} = \frac{1/6}{1/2} = 1/3.$$

Therefore, the man has a probability of 2/3 of winning the grand prize if he decides to switch doors. In general, if there are n doors and if the host eliminates $n - 2$ among them (he naturally cannot eliminate the door chosen by the participant), then the probability that the grand prize is hidden behind the only remaining door, among the $n - 1$ doors in play, is equal to $(n - 1)/n$.

Example 2.5.5 (The liars problem) A says that B told him that C has lied. If the three persons tell the truth and lie with probability $p \in (0, 1)$, independently from one another, what is the probability that C has indeed lied?

Solution. Let the events be as follows:

$$F = A \text{ says that } B \text{ told him that } C \text{ has lied}$$

and

$$F_I = I \text{ lied, for } I = A, B, C.$$

We want

$$P[F_C|F] = \frac{P[F|F_C]P[F_C]}{P[F|F_C]P[F_C] + P[F|F_C^c]P[F_C^c]}.$$

Now, we have:

$$P[F|F_C] = P[F_A^c \cap F_B^c] + P[F_A \cap F_B] \stackrel{\text{ind.}}{=} p^2 + (1-p)^2$$

and

$$P[F|F_C^c] = P[F_A^c \cap F_B] + P[F_A \cap F_B^c] \stackrel{\text{ind.}}{=} 2p(1-p).$$

Hence,

$$P[F_C|F] = \frac{[p^2 + (1-p)^2](1-p)}{[p^2 + (1-p)^2](1-p) + [2p(1-p)]p} = \frac{p^2 + (1-p)^2}{3p^2 + (1-p)^2}.$$

Note that if $p = 1/2$, then $P[F_C|F] = 1/2$, which is logical.

2.6 Exercises, Problems, and Multiple Choice Questions

Solved Exercises

Exercise no. 1 (2.3)[3]
License plates are made up of three letters followed by four digits. We assume that the letters I and O are never used and that no license plates end with 0000.

a) How many distinct license plates can there be?

b) What is the answer in a) if, in addition, no plates bear either three identical letters or four identical digits?

Solution

a) The number of possible distinct plates is

$$(24 \times 24 \times 24) \times \left(10^4 - 1\right) = 138, 226, 176.$$

b) In this case, the total number of possible plates is given by

$$\left(24^3 - 24\right) \times \left(10^4 - 10\right) = (13, 800)\,(9990) = 137, 862, 000.$$

Exercise no. 2 (2.3)
How many distinct code words made up of four letters can be formed by using (without replacement) the letters of the word ESSAY?

---

[3] The material that must have been read to be able to solve each exercise is indicated in the parentheses.

Solution

If we take a single s, then we can form 4! = 24 distinct code words. If we use the two s's, then the number of distinct code words that can be formed is given by $\binom{3}{2}\frac{4!}{2!1!1!} = 3\frac{24}{2} = 36$. Hence, the total number is 60 code words.

Exercise no. 3 (2.3)
A case holds two boxes containing ten objects each.

a) Suppose that, among the 20 objects, there are exactly five that are defective. What is the probability that these five defective objects are all in the same box?

b) If each box is made up of two rows of five objects each and if the 20 objects (that are distinguishable) are placed at random into the two boxes, how many distinct arrangements of the ten objects inside a box are there?

Solution

a)
$$P\left[\text{All defective objects in the same box}\right]$$
$$= 2 \times \frac{\binom{5}{5}\binom{15}{5}}{\binom{20}{10}} = 2 \times \frac{15!}{5!10!}\frac{10!10!}{20!} \simeq 0.0325.$$

b) The number of distinct arrangements is given by $\binom{20}{10}10! = \frac{20!}{10!}$.

Exercise no. 4 (2.3)
A certain computer language uses the 26 letters of the alphabet and 10 special characters ($, #, etc.). We use these 36 characters to generate (at random) access codes, to a computer, made up of four characters.

a) Let E be the random experiment that consists in counting the number of special characters in a code taken at random.
    i) Write the sample space S for this experiment.
    ii) Compute the probability of each elementary event.

b) We consider a code taken at random. We define the events
    A = the given code contains at least one letter
and
    B = the given code contains exactly one letter.
Are the events A and B^c incompatible? Justify your answer.

Solution

a) i) $S = \{0, 1, 2, 3, 4\}$.

    ii)
$$P\left[\{0\}\right] = \left(\frac{26}{36}\right)^4 \simeq 0.2721;$$

$$P\left[\{1\}\right] = \binom{4}{1}\left(\frac{10}{36}\right)\left(\frac{26}{36}\right)^3 \simeq 0.4186;$$

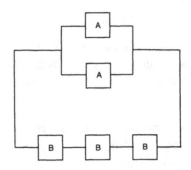

Figure 2.7. System in Exercise no. 5.

$$P[\{2\}] = \binom{4}{2}\left(\frac{10}{36}\right)^2\left(\frac{26}{36}\right)^2 \simeq 0.2415;$$

$$P[\{3\}] = \binom{4}{3}\left(\frac{10}{36}\right)^3\left(\frac{26}{36}\right) \simeq 0.0619;$$

$$P[\{4\}] = \left(\frac{10}{36}\right)^4 \simeq 0.0060.$$

b) $P\left[A \cap B^c\right] = P[\{0\} \cup \{1\} \cup \{2\}] > 0$, which implies that A and B^c are not incompatible.

Exercise no. 5 (2.5)

We have 10 brand A components, denoted by $A_1, \ldots, A_{10}$, and 20 brand B components, denoted by $B_1, \ldots, B_{20}$. We want to build a system made up of two subsystems: three brand B components placed in series and two brand A components placed in parallel. Moreover, the two subsystems are placed in parallel (see Fig. 2.7).

a) How many distinct systems can be built?

Remark. The disposition of the components inside the subsystems does not matter.

b) Assume that two brand A components are defective, while five brand B components are defective. What is the probability that the system functions if the components are taken at random and operate independently from one another?

Solution

a) There are $\binom{20}{3} = 1140$ possibilities for the first subsystem and $\binom{10}{2} = 45$ possibilities for the second one. Then, by the principle of multiplication, there are $1140 \times 45 = 51,300$ distinct systems that can be built.

b) We have:

$$P\left[\text{Subsystem } B \text{ functions}\right] = \frac{15 \times 14 \times 13}{20 \times 19 \times 18} = \frac{91}{228}$$

and

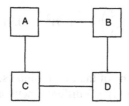

Figure 2.8. Figure for Exercise no. 6.

$$P\left[\text{Subsystem } A \text{ functions}\right] = 1 - \frac{2 \times 1}{10 \times 9} = \frac{88}{90}$$

$$\Rightarrow \quad P\left[\text{System functions}\right] = 1 - \left(\frac{137}{228}\right)\left(\frac{1}{45}\right) \simeq 0.9866.$$

Exercise no. 6 (2.5)

A network is constituted of four terminals and four links, as shown in the diagram in Fig. 2.8. Thus, there are two paths that connect any pair of terminals. Moreover, a terminal that transmits information to another terminal does so by sending this information in both directions, independently. The transmission is considered to be successful if the information is received along one path or the other (or both). Finally, we assume that the links fail, independently from one another, with probability 0.1.

a) What is the probability that the information transmitted by terminal A to terminal B will be successful?

b) What is the probability that the information transmitted by a given terminal to another terminal will be successful?

c) In how many different ways can the four terminals be disposed inside the network if we suppose that two networks are identical when each terminal has the same neighbors in both networks?

Solution

Let F = the transmission is successful, and F_{AB} = the information transmitted in the direction AB is well received, etc.

a)

$$P[F] = P[F_{AB} \cup F_{ACDB}] \overset{\text{ind.}}{=} P[F_{AB}] + P[F_{ACDB}] - P[F_{AB}] P[F_{ACDB}]$$
$$\overset{\text{ind.}}{=} (0.9) + (0.9)^3 - (0.9)^4 = 0.9729.$$

b) Let D_k = the minimum number of links between the two terminals is equal to k. We have: $k = 1$ or 2. Then,

$$P[F] = P[F \mid D_1] P[D_1] + P[F \mid D_2] P[D_2]$$
$$\overset{\text{a)}}{=} (0.9729)\left(\frac{2}{3}\right) + P[F_{ABD} \cup F_{ACD}]\left(\frac{1}{3}\right)$$

$$\stackrel{\text{ind.}}{=} (0.9729) \left(\frac{2}{3}\right) + \left[(0.9)^2 + (0.9)^2 - (0.9)^4\right] \left(\frac{1}{3}\right)$$

$$= 0.9699.$$

c) First we fix the position of one of the four terminals, and next we move the other three terminals. There are then $\binom{3}{2} = 3$ different networks (because the order of the neighbors does not matter):

$$
\begin{array}{ccc}
A\ B & A\ B & A\ C \\
C\ D & D\ C & D\ B
\end{array}
$$

Exercise no. 7 (2.5)

A system is made up of three components that operate independently from one another. For the system to function, at least two of its components must function. We suppose that the reliability of component no. 1 is equal to 0.95, that of component no. 2 to 0.9, and that of component no. 3 to 0.8.

a) What is the probability that the system functions?

b) Given that the system functions, what is the probability that exactly two components function?

c) Given that the system does not function, what is the probability that no components function?

d) Given that component no. 1 functions, what is the probability that the system functions?

Solution

Let F = the system functions, and F_i = component i functions, for $i = 1, 2, 3$.

a) $P[F] = P\left[F_1 \cap F_2 \cap F_3^c\right] + P\left[F_1 \cap F_2^c \cap F_3\right] + P\left[F_1^c \cap F_2 \cap F_3\right]$

$$+ P[F_1 \cap F_2 \cap F_3]$$

$$\stackrel{\text{ind.}}{=} (0.95)\,(0.9)\,(0.2) + (0.95)\,(0.1)\,(0.8) + (0.05)\,(0.9)\,(0.8)$$

$$+ (0.95)\,(0.9)\,(0.8) = 0.283 + 0.684 = 0.967.$$

b) Let G = exactly two components function.

$$P[G \mid F] = \frac{P[F \mid G]\,P[G]}{P[F]} \stackrel{\text{a)}}{=} \frac{1\,(0.283)}{0.967} \simeq 0.2927.$$

c) Let H = no components function.

$$P\left[H \mid F^c\right] = \frac{P\left[F^c \mid H\right]P[H]}{P[F^c]} \stackrel{\text{a)}}{=} \frac{1\,(0.05)\,(0.1)\,(0.2)}{1 - 0.967} = 0,\overline{03}.$$

d) $P\left[F \mid F_1\right] \stackrel{\text{ind.}}{=} P[F_2 \cup F_3]$

$$= P[F_2] + P[F_3] - P[F_2 \cap F_3]$$

$$\stackrel{\text{ind.}}{=} 0.9 + 0.8 - (0.9)\,(0.8) = 0.98.$$

Unsolved Problems

Problem no. 1

Two dice are thrown simultaneously. If a sum equal to six or ten is obtained, then a coin is tossed.

a) How many elementary events (of the form (die 1, die 2) or (die 1, die 2, coin)) are there in the sample space S?

b) Assume that the dice and the coin are non-biased (or well-balanced). Given that the coin has been tossed, what is the probability that a "1" has been rolled with the second die?

Problem no. 2

Let A and B be events for which $P[A \cap B] = P[A^c \cap B] = P[A \cap B^c] = p$ (see Fig. 2.9). Calculate a) $P[A^c \cap B^c]$; b) $P[A^c \cup B^c]$.

Problem no. 3

In a certain lottery, five balls are picked, at random and without replacement, among 25 balls numbered from 1 to 25. We win the grand prize if the five balls that we have chosen are selected in the same order as they appear on our ticket.

a) What is the probability of winning the grand prize?

b) What is the probability of not winning the grand prize because of a single ball?

Problem no. 4

We have nine electronic components, including two defective ones. Four components are taken at random to construct a system in series.

a) What is the probability that the system does not function?

b) If a fifth component (taken among the remaining five) is placed in parallel with the first four, what is the probability that the system will function?

Problem no. 5

Let $P[A \mid B] = \frac{1}{2}$, $P[B^c] = \frac{1}{3}$ and $P[A \cap B^c] = \frac{1}{6}$. Calculate a) $P[A]$; b) $P[A^c \cap B]$.

Problem no. 6

Let A and B be independent events such that $P[A] < P[B]$ and

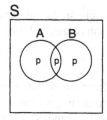

Figure 2.9. Figure for Problem no. 2.

$$P[A \mid B] + P[B \mid A] = 1.$$

a) Calculate $P[A]$ if $P[A \cap B] = \frac{4}{25}$.

b) Calculate $P[B]$ if $P[A \cup B] = \frac{19}{25}$.

Problem no. 7

A communication system transmits three signals: s_1, s_2 and s_3, with equal probability. The reception is sometimes erroneous, because of the "noise." It was found, experimentally, that the probability p_{ij} of receiving the signal s_j, given that the signal s_i has been transmitted, is given by the following table:

		Reception		
		s_1	s_2	s_3
	s_1	0.8	0.1	0.1
Transmission	s_2	0.05	0.90	0.05
	s_3	0.02	0.08	0.90

a) Calculate the probability that the signal s_1 has been transmitted, given that the signal s_2 has been received.

b) If we assume that the transmissions are independent, what is the probability of receiving two consecutive s_3 signals?

Problem no. 8

A box contains five brand A, three brand B and two brand C transistors. The transistors are all distinguishable. In how many ways can we pick, at random and without replacement, four transistors if we want to get at least one of each brand?

Remark. The order in which the transistors are selected does not matter.

Problem no. 9

According to the data collected, 40% of all human beings have type A blood, 10% have type B, 45% type O and 5% type AB. Moreover, we think that 90% of people who have type O blood are incorrectly classified, whereas 3% of people with type B, 10% with type AB and 2% with type A blood are classified as having type O blood.

a) What is the probability that a person classified as having type O blood really has this type of blood?

b) If we assume that each person is classified independently from the others, what is the probability that two given persons classified with type O blood do not have this type of blood?

Problem no. 10

A diskette is taken at random from a box containing 10 brand A, 15 brand B and 25 brand C diskettes. We assume that, in general, 95% of the brand A diskettes are perfect. This percentage is 97% for the brand B diskettes and 99% for those of brand C.

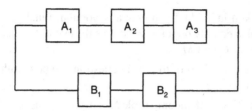

Figure 2.10. System in Problem no. 12.

a) Given that the diskette taken at random is not perfect, what is the probability that it is of brand A?

b) If we repeat the random experiment described above ten times, with replacement (so that the trials are independent), what is the probability of obtaining at least one imperfect diskette?

Problem no. 11

Four calls were directed at random to a telephone central during a one-hour period. What is the probability that one call was received during each quarter of an hour of the hour considered?

Problem no. 12

A system is made up of subsystems A and B placed in parallel. The subsystem A contains three components, A_1, A_2 and A_3, placed in series, while the components B_1 and B_2 placed in series constitute the subsystem B (see Fig. 2.10). We suppose that the probability that the component A_i functions at the end of one year is equal to 0.9, for $i = 1, 2, 3$, independently of the other components. We also suppose that the components B_1 and B_2 operate independently of each other. Moreover, we assume that the probability that the component B_k, $k = 1, 2$, functions at the end of one year is equal to 0.95 if the subsystem A functions at this moment, and to 0.80 otherwise. Let us define the following events:

$$F = \text{the system functions at the end of one year}$$

and

$$F_X = \text{the subsystem } X \text{ functions at the end of one year, for } X = A, B.$$

Calculate a) $P[F_B]$; b) $P[F]$; c) $P[F_B \mid F]$; d) $P\left[F_A^c \cap F\right]$.

Problem no. 13

A person buying a certain make of car can choose one or several of the following options:

$A = $ an automatic transmission, $C = $ air-conditioning,   and $M = $ a V6 engine.

Based on the data collected so far, 90% of the customers choose at least one of the three options, 75% choose at least two and 45% choose the three options. Moreover, the three options taken individually are equally popular. Finally, all possible groups of two options are also equally popular. We define the event A = the customer chooses option A; similarly for C and M.

a) How many different versions of the car in question can be bought if at least one of the three options is chosen?

b) Are the events A, C and M incompatible? Justify your answer.

c) Are the events A, C and M (globally) independent? Justify.

d) Calculate the probability that a person buying this car chooses exactly one option.

e) Calculate the probability that a person buying this car has chosen the three options, given that he/she has chosen at least one.

Hint. Use a Venn diagram.

Problem no. 14

An electricity distribution network is made up of three stations: A, B and C. We know, by experience, that during a heat wave the probability that an overload occurs at station A is equal to 0.4; this probability is equal to 0.3 for station B and to 0.2 for station C. Moreover, we assume that an overload at station A causes a breakdown of the entire network in 2% of the cases. An overload at station B brings about a breakdown of the entire network 3% of the time, and an overload at station C causes a breakdown of the network 4% of the time. We define the events:

$$D = \text{a breakdown of the network occurred during a certain heat wave}$$

and

$$A \text{ (respectively } B, C) = \text{an overload occurred at station } A \text{ (resp. } B, C).$$

Remark. We assume that a network breakdown can only occur when there is an overload at one of the three stations, and that two overloads cannot happen exactly at the same moment (so that the events A, B and C are incompatible for a given breakdown).

Calculate a) $P[D]$; b) $P[A \mid D]$; c) $P[D \mid B \cup C]$.

d) Suppose that the breakdowns are independent of one another. Compute the probability that two consecutive breakdowns have been caused by an overload at the same station.

Problem no. 15

A respirator used during surgical operations breaks down with probability p (> 0). A breakdown of the respirator entails that the patient will die, unless a monitor detects the breakdown and warns the surgeon. In this case, there is an 80% chance that the surgeon will be able to save the patient. The monitor fails with probability 0.005, independently of the respirator. We define the events:

$$F = \text{a patient dies because of a breakdown of the respirator,}$$
$$R = \text{the respirator fails}$$

and

M = the monitor breaks down.

a) The respirator is made up of three components placed in parallel that operate independently from one another. If the reliability of each component is equal to 0.9, what is the probability p that the respirator will fail?

b) The monitor is constituted of two brand A and two brand B components. The four components are placed in series. We have 10 brand A and 20 brand B components to build monitors. How many different devices can we construct if we assume that the order of the four components placed in series matters and that the 10 brand A components and the 20 brand B components are not distinguishable among themselves?

c) What is the answer in b) if we assume that the 30 components that we have are numbered from 1 to 30?

d) Are the events R and M incompatible? Justify your answer.

Answer the following questions assuming that $p = 0.01$.
Calculate e) $P[F]$; f) $P[M \mid F]$; g) $P[F \mid R \cup M]$.

h) We consider two patients that died because of respirator failures. What is the probability that the monitor worked in one case but failed in the other case?

Remark. We assume that the monitor breakdowns are independent.

Problem no. 16

We divide the length of telephone calls into three categories: those lasting less than one minute (I), those lasting between one and three minutes (II), and the ones lasting more than three minutes (III). Moreover, we suppose that 60% of all calls are personal and the rest are business calls. Finally, we suppose that 10%, 20% and 70% of the personal calls are of category I, II and III, respectively. In the case of the business calls, these percentages are equal to 20%, 40% and 40%, respectively.

a) Two calls are made during a certain period. If we record the total number of personal calls and the number of personal calls of each category, how many elementary events (like 0000, for example) are there in the sample space S?

b) We consider three independent business calls. What is the probability that there is one of each category?

c) What is the probability that a given call lasts less than one minute?

Problem no. 17

You are waiting for a friend who entered a shop. At the time of paying, he goes at random to one of the three cash registers: A, B or C. The probability that the service is slow at A is equal to 0.7; this probability is equal to 0.6 at B and to 0.5 at C. When he arrives, your friend tells you that the service has been slow at the cash register.

a) Calculate the probability that your friend went to cash register C.

b) Calculate the probability that, on two other independent visits to this shop, your friend did not go to cash register C, given that the service has been slow on each occasion.

Problem no. 18

In a certain city, 60% of the calls to the emergency number require the help of the police department, 40% that of the ambulance service, and 20% ask for the fire department. Moreover, 40%, 20% and 5% of the calls require only the help of the police department, the ambulance service and the fire department, respectively. Finally, 5% of the calls require the help of the three services and 10% do not ask for any of these three services.

a) Calculate the probability that a given call requires the help of the police department and of the ambulance service, but not of the fire department.

Hint. Use a Venn diagram.

b) Let the events be as follows:

$$A = \text{a certain call requires the help of the police department}$$

and

$$B = \text{a certain call requires the help of the ambulance service.}$$

Are the events $A \cup B$ and $A^c \cup B^c$ incompatible (or mutually exclusive)? Justify your answer by a numerical computation.

c) We consider ten calls to the emergency number. Among these calls, there are exactly seven that required the help of the police department. If four calls are taken at random (and without replacement) among the ten calls considered, what is the probability that exactly two of these four calls required the help of the police department?

Problem no. 19

Computers A and B exchange information within a network. The probability that the information sent from A to B is incorrectly transmitted is equal to 0.01, while this probability is equal to 0.005 in the case when the information is sent from B to A. We consider 10 messages transmitted from A to B and 15 messages from B to A. We assume that all transmissions are independent.

a) What is the probability that a message taken at random among the 25 messages considered has been incorrectly transmitted?

b) Two messages are taken at random and without replacement among the 25 messages considered. Given that both messages have been incorrectly transmitted, what is the probability that they were transmitted by the same computer?

c) Suppose that exactly two of the 25 messages considered have been incorrectly transmitted. What is the probability that the two erroneous messages were transmitted by the same computer?

Problem no. 20

An engineer responsible for the control of the quality in the company where she works receives a batch of 200 parts used in the computers that they build. She decides to pick 10, at random and without replacement, and to test them. Let E be the random experiment that consists in counting the number of defective parts among the 10 parts tested. Answer the following questions by assuming that there are in fact two defective parts among the 200 parts received.

a) Write the sample space S.

b) Calculate the probability of all the elementary events.

c) We define the event A_k = there are exactly k defective parts among the 10 parts tested, for $k = 0, 1, \ldots$. Calculate $P[A_0 \cup (A_1^c \cap A_2)]$.

Problem no. 21

A small company has two telephone lines: a local number (line 1) and an 800 number for people outside the city (line 2). We suppose that the probability that line 1 is busy during working hours is equal to 0.1; this probability is equal to 0.05 in the case of line 2. Moreover, we suppose that the probability that a customer who calls for the first time (and finds the line busy) is lost is equal to 0.3 when the call is local and to 0.4 when the call comes from outside the city. Finally, we estimate that 60% of the calls are local.

a) Calculate the probability that the next customer who calls the company for the first time will be lost.

b) Given that a customer has been lost after having called only once, what is the probability that this customer has called the 800 number?

c) Among the next 10 customers who will call for the first time to the local number, what is the probability that at least two of them will be lost?

Remark. We assume that the customers are independent of one another.

Problem no. 22

a) We consider four viruses (V_1, V_2, V_3 and V_4) that can infect the computers of the computer network of a certain institution. Let F_i = virus V_i has infected the network, for $i = 1, 2, 3, 4$. We assume that the events F_1 and F_2 are independent, as well as the events F_3 and F_4. Moreover, the events F_1 and F_3 are incompatible, and so are the events F_1 and F_4, F_2 and F_3, and F_2 and F_4. Finally, we suppose that $P[F_1] = P[F_2] = 0.05$ and that $P[F_3] = P[F_4] = 0.01$.

i) Calculate the probability that the network will be infected by any of the viruses.

ii) Can we say that the event F_1 is included in the event F_2? Justify your answer.

iii) Are the four intersections of three distinct events equally likely? Justify your answer.

b) We consider five other viruses that can infect the computers of another computer network. How many conditions must we check (at most) to determine whether the five viruses are (globally) independent? Justify your answer.

Problem no. 23

A sample of five empty CDs is taken at random from a batch of 100 CDs. We consider the random experiment E that consists in counting the number of CDs that conform to the norms. Assume that the batch of 100 CDs contains in fact exactly four defective CDs.

a) Write the sample space for this random experiment if
   i) the CDs are taken one at a time and with replacement;
   ii) the five CDs are taken all at once.

b) Calculate the probability that the sample of five CDs contains at least one defective in cases i) and ii) above.

c) Let the events be as follows:

$$A = \text{there is at least one CD that conforms to the norms among}$$
$$\text{the five taken at random}$$

and

$$B = \text{there is at least one defective CD among the five taken at random.}$$

In case i) above, are the events A^c and B^c incompatible? Are they independent? Justify your answers with numerical calculations.

Problem no. 24
   A system is made up of three components placed in parallel that operate simultaneously. The probability that component no. 1 functions is equal to 0.9. In the case of component no. 2, this probability is equal to 0.95 if component no. 1 functions, and to 0.8 otherwise. Finally, the probability that component no. 3 functions is equal to 0.99 if components nos. 1 and 2 function, to 0.75 if components nos. 1 and 2 do not function, and to 0.8 otherwise.

a) Calculate the probability that the system functions.

b) Given that the system functions, what is the probability that component no. 2 functions?

c) Calculate the probability that component no. 3 functions.

Multiple Choice Questions

Question no. 1
   A box contains 100 objects, namely 40 brand A and 60 brand B objects. We take 50 objects at random and without replacement and we count the number of brand A objects obtained. Write the sample space corresponding to this random experiment.
a) $\{0, \dots, 40\}$ b) $\{0, \dots, 50\}$ c) $\{10, \dots, 40\}$ d) $\{10, \dots, 50\}$
e) none of these answers

Question no. 2
   A certain store has 5 brand A, 10 brand B and 20 brand C items in stock. How many different orders of these three brands can it satisfy?
a) 35   b) 999   c) 1000   d) 1385   e) 6545

Question no. 3

We have 20 components, of which two are defective. Three components are taken at random to construct a machine made up first of two components placed in parallel, and then a component placed in series. We suppose that all the components operate at the same time, and independently from one another. Calculate the probability that the machine functions.

a) 0.8895   b) 0.8947   c) 0.9   d) 0.9942   e) 1

Question no. 4

Let A and B be independent events such that $P[A] = 1/4$ and $P[B] = 1/2$. Moreover, let C be an event such that $P[C|A^c \cap B] = 1/2$ and $P[C|A^c \cap B^c] = 1$. Finally, A and C are incompatible. Calculate $P[A^c \cap B^c|C]$.

a) 0   b) 1/3   c) 1/2   d) 2/3   e) 1

Question no. 5

A class is made up of five female students and 45 male students. Among the five girls, four are second-year students, while 30 out of the 45 boys are second-year students. Two students are taken at random, and without replacement, among the 50 students in this class. Knowing that in both cases the person chosen was a second-year student, what is the probability that a male and a female students were chosen?

a) 8/289   b) 30/289   c) 60/289   d) 229/289   e) 259/289

Question no. 6

Let E be the following random experiment: two dice are tossed; if the two numbers obtained are equal, then the two dice are tossed again (only once). How many elementary events are there in the sample space S?

a) 72   b) 216   c) 246   d) 252   e) 1296

Question no. 7

In a class, there are 20 students. What is the probability that exactly two of them share the same birthday?

a) 0.3232   b) 0.4114   c) 0.5886   d) 0.6768   e) none of these answers

Question no. 8

A number x is taken at random in the interval $[0, 3]$ so that the probability that the number chosen is in the interval $[k, k + 1]$ is equal to $(k + 1)p$, for $k = 0, 1, 2$. Calculate $P[A]$, where $A := \{x \in [0, 3] : |x - \frac{3}{2}| > 1\}$.

a) 1/12   b) 1/4   c) 1/3   d) 1/2   e) 2/3

Question no. 9

In a certain factory, 80% of the parts fabricated conform to the norms. Every part fabricated in this factory is subjected to three independent quality control operations. We suppose that each of these operations classifies as non-defective 95% of the parts that effectively conform to the norms, and 10% of the parts that are in fact defective. Calculate the probability that a part that has been sold conforms to the norms.

a) 0.80   b) 0.9250   c) 0.95   d) 0.9744   e) 0.9997

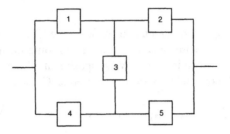

Figure 2.11. System in Multiple choice question no. 10.

Question no. 10
We consider a system constituted of five identical components that operate independently from one another and that are connected as in the diagram in Fig. 2.11. Knowing that exactly three of the five components work, what is the probability that the system functions?

a) 0.6   b) 0.7   c) 0.8   d) 0.9   e) we cannot compute it

Question no. 11
The license plates in a certain country bear six characters taken at random among the 26 letters of the alphabet and the ten digits $\{0, 1, \ldots, 9\}$. What is the probability that a given license plate bears at least one digit?

a) 0.0546   b) 0.1419   c) 0.3275   d) 0.8581   e) 0.9995

Question no. 12
A student gets up at time X and goes to bed at time Y, where $0 < X < Y \leq 24$. Let $S = \{(x, y) : 0 < x < y \leq 24\}$. Write in mathematical form the event $A = $ the student is up at least three hours more than he lies in bed.

a) $\{(x, y) \in S : |y - x| \geq 3\}$　　　　b) $\{(x, y) \in S : |y - x| \leq 13.5\}$
c) $\{(x, y) \in S : y - x \geq 13.5\}$　　　　d) $\{(x, y) \in S : y - x \leq 15\}$
e) $\{(x, y) \in S : y - x \geq 15\}$

Question no. 13
Let A, B and C be events such that $A \subset B$, A and C are incompatible, $P[(B \cup C)^c] = 1/10$, $P[B \cap C] = 3/10$, $P[A^c \cap B] = 1/2$ and $P[C] = 13/20$. Calculate $P[A]$.

a) 0   b) 0.05   c) 0.35   d) 0.5   e) 0.9

Question no. 14
Two men and two women sit at a circular table. In how many different ways can they be seated if the two men must be opposite each other?

Remark. Two ways are equivalent if each person has the same neighbors to his/her left **and** to his/her right.

a) 2   b) 3   c) 4   d) 6   e) 24

Question no. 15

How many distinct license plates made up of three letters and three digits can there be if the three letters appear either at the beginning or at the end of the plate?

a) $C_3^{10}C_3^{26}$ b) $2(C_3^{10}C_3^{26})$ c) $10^3 26^3$ d) $2(10^3 26^3)$ e) $C_3^6\, 10^3 26^3$

Question no. 16

Let A, B and C be events such that $P[A] = P[B] = P[C] = 1/3$, $P[C \mid A \cap B] = 1/2$, and A and B are independent. Calculate $P[A \cap B|C]$.

a) 0   b) 1/18   c) 1/6   d) 1/3   e) 1/2

Question no. 17

The events A and B form a partition of the sample space S. Calculate $P[A|B] + P[A|B^c]$.

a) 0   b) 1/4   c) 1/2   d) 1   e) none of these answers

Question no. 18

In a certain computer store, a study revealed that the three most popular options when people buy a new computer are the following: A = a faster processor, B = a larger hard disk and C = a wider screen. According to the data collected so far, 40% of the customers have chosen at least one of these three options, and 10% chose only option C. Moreover, all the customers who chose option A also chose option B, and vice versa. Finally, options A and B taken together have been twice as popular as option C.

In the following questions, only the options A, B and C are considered.

A) How many different computers have been sold up to now by this computer store?

a) 2   b) 3   c) 4   d) 6   e) 8

B) How many different computers can be sold to future customers?

a) 4   b) 5   c) 6   d) 7   e) 8

C) What proportion of the customers have chosen the three options so far?

a) 0   b) 0.05   c) 0.1   d) 0.15   e) 0.2

D) Up to now, 30% of the customers have been women. Moreover, 80% of these women have chosen none of the options considered. What is the probability that a man taken at random has chosen at least one of the three options?

a) 2/5   b) 17/35   c) 18/35   d) 3/5   e) none of these answers

E) Under the same hypotheses as in D), what is the probability that a customer that has chosen none of the three options is a woman?

a) 0.2   b) 0.3   c) 0.4   d) 0.6   e) 0.8

F) We consider five persons who each bought a computer. Assuming that the customers are independent, what is the probability that at least two of them chose none of the three options?

a) 0.0870   b) 0.2304   c) 0.7696   d) 0.9130   e) 0.9898

Question no. 19

A transistor is taken at random and its lifetime is measured. The sample space for this random experiment is given by $S = [0, \infty)$. We consider the events $A = [0, 1]$, $B = [0, 2]$ and $C = [1, \infty)$. Give the intervals that correspond to the following events:

A) $D = A \cup (B \cap C)$;

a) [0, 1]   b) [0, 2]   c) [1, 2]   d) [2, ∞)   e) [1, ∞)

B) $F = [A \cap (B^c \cup C^c)]^c$.

a) [0, 1]   b) [0, 2]   c) [1, 2]   d) [2, ∞)   e) [1, ∞)

Question no. 20

Two dice are thrown simultaneously. One of the dice has four red faces and two white faces, whereas the other one has two red faces and four white faces. Let R = at least one red face is obtained, and B = at least one white face is obtained. Calculate, assuming that all faces have an equal probability of coming up (namely, one chance in six)

A) $P[R]$;

a) 1/3   b) 4/9   c) 5/9   d) 7/9   e) 8/9

B) $P[R \cap B]$.

a) 1/3   b) 4/9   c) 5/9   d) 2/3   e) 7/9

Question no. 21

The combination of a padlock is made up of three digits (taken in the set $\{0, 1, \ldots, 9\}$). How many possible combinations are there if

A) each digit cannot be chosen more than once?

a) 360   b) 450   c) 495   d) 720   e) 990

B) each digit cannot be chosen more than twice?

a) 495   b) 720   c) 900   d) 950   e) 990

Question no. 22

An engineer subscribes to two independent electronic mail services. The probability that service no. 1 does not function on a given day is equal to 1/20, while the probability of a breakdown of service no. 2 is equal to only 1/100. Moreover, when service no. 1 functions, the probability that a message sent is received by its addressee is equal to 0.995. This probability is equal to 0.99 in the case of service no. 2. To be safer, the engineer wants to send an important message through both services.

A) What is the probability that the addressee receives this message?

a) 0.94525   b) 0.9627   c) 0.9801   d) 0.9989   e) 0.9999

B) Given that the addressee has received the message, what is the probability that only the message that the engineer wants to send through service no. 1 arrived at its destination?

a) 0.01883   b) 0.03627   c) 0.04569   d) 0.05372   e) 0.07255

Question no. 23

A factory recycles defective car parts. Among the last 50 parts that it recycled, five are still defective. A company buys 20 of these 50 parts.

A) What is the probability that the company receives at least one defective part?

a) 0.1216   b) 0.2587   c) 0.2702   d) 0.8784   e) 0.9327

B) Given that the company has received at least one part, what is the probability that it received exactly two?

a) 0.3246   b) 0.3641   c) 0.3903   d) 0.6097   e) 0.6754

Question no. 24

Ten candidates are interviewed to fill two posts in a company. In how many ways can the company fill these posts, if

A) both posts are identical?

a) 45   b) 50   c) 55   d) 90   e) 100

B) one post is permanent and the other temporary?

a) 50   b) 90   c) 100   d) 110   e) 200

Question no. 25

A) A system is made up of three subsystems placed in series. Each subsystem comprises two components placed in parallel (see system A in Fig. 2.12, p. 43). We suppose that all the components operate independently from one another and all have a 90% probability of working at a given time. Calculate the reliability of the system at that moment.

a) 0.2710   b) 0.7290   c) 0.9266   d) 0.9703   e) 0.9991

B) Under the same hypotheses as in A), what is the reliability of a system made up of two subsystems placed in parallel, if each subsystem comprises three components placed in series (see system B in Fig. 2.12)?

a) 0.9266   b) 0.9703   c) 0.9797   d) 0.9980   e) 0.9991

Question no. 26

Among the 100 passengers on board a plane landing at an airport, 80 arrive at their destination, whereas the others are in transit. Moreover, among those arriving at destination, 70% started their journey inside the country. This percentage is equal to 40% in the case of the passengers in transit. Finally, we assume that the probability that a given passenger travels for business is equal to 10%, independently of what

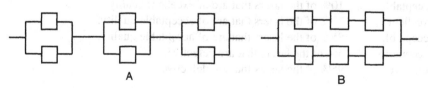

A   B

Figure 2.12. Systems for Multiple choice question no. 25.

precedes. A passenger traveling on board this plane is taken at random. What is the probability that the passenger

A) arrives at destination or started his/her journey outside the country?

a) 0.12   b) 0.24   c) 0.375   d) 0.56   e) 0.875

B) travels for business or started his/her journey inside the country?

a) 0.064   b) 0.424   c) 0.676   d) 0.74   e) 0.964

Question no. 27

Customers arrive at a store, at random and independently from one another, from 9:00 a.m.

A) Suppose that there have been two customers between 9:00 a.m. and 10:00 a.m.

i) Calculate the probability that the second customer arrived before 9:30, given that the first one arrived at 9:15.

a) 1/4   b) 1/3   c) 7/15   d) 29/60   e) 1/2

ii) What is the probability that the two customers arrived inside a 15-minute interval (that can be located anywhere inside the hour considered)?

a) 1/4   b) 1/3   c) 7/16   d) 1/2   e) 9/16

B) Suppose now that there have been four customers between 10:00 a.m. and 11:00 a.m.

i) What is the probability that exactly three of these four customers arrived between 10:00 and 10:30?

a) 1/16   b) 1/8   c) 1/4   d) 1/2   e) 3/4

ii) What is the probability that there has been one customer in each quarter of an hour of the hour considered?

a) 0.00391   b) 0.00586   c) 0.04167   d) 0.06250   e) 0.09375

Question no. 28

A maker of lasers estimates that 20% of the lasers the company produces are of excellent quality, 75% are of acceptable quality, and 5% are defective. Moreover, we assume that the quality of a given laser does not depend on the other lasers. To improve the reliability of the lasers the company sells (those they think are non-defective), they submit their products to a quality control procedure. Suppose that this procedure classifies as

excellent   90% of the lasers that are of excellent quality,
acceptable   10% of the lasers that are of excellent quality,
excellent   10% of the lasers that are of acceptable quality,
acceptable   85% of the lasers that are of acceptable quality,
acceptable   5% of the lasers that are defective,
defective   95% of the lasers that are defective.

A) Answer the following questions based on the maker's *estimations*.

i) What is the probability that a batch of ten lasers contains at least one laser of excellent quality or at least one defective laser?

a) 0.9137   b) 0.9237   c) 0.9337   d) 0.9437   e) 0.9537

ii) What is the probability that a batch of ten lasers contains at least one laser of excellent quality and at least one defective laser?

a) 0.3502   b) 0.3602   c) 0.3702   d) 0.3802   e) 0.3902

B) i) Calculate the probability that a given laser has been incorrectly classified by the procedure.

a) 0.135   b) 0.15   c) 0.235   d) 0.25   e) 0.3

ii) Calculate the probability that a laser sold is defective.

a) 0   b) 0.0017   c) 0.0027   d) 0.0037   e) 0.0047

Question no. 29
We consider the 10^7 telephone numbers made up of seven digits. For security reasons, the numbers starting with 0 or with 911, or ending with 911, are not used. We define the events

A = a number starts with 0,
B = a number starts with 911, and
C = a number ends with 911.

A) Which pairs of events are incompatible?

a) A and B only   b) A and C only   c) B and C only
d) A and B, and B and C only   e) the three pairs of events

B) Calculate the number of telephone numbers that are not used. That is, calculate the number of elementary events in $D := A \cup B \cup C$.

a) 810,190   b) 998,190   c) 1,018,990   d) 1,431,190   e) 2,100,010

C) Let $F := A \cup (B \cap C)$. How many elementary events are there in F?

Suggestion. Use a Venn diagram.

a) 1,000,000   b) 1,000,010   c) 1,010,000   d) 1,100,000   e) 1,110,000

Question no. 30
In an interview for a job as a computer analyst, first at most five questions are asked to every candidate. As soon as the candidate has answered four questions correctly, he (or she) is hired. If the candidate does not succeed in this first stage of the interview, he is asked a last question. In this case, he is hired if and only if he answers this question correctly. Suppose that each time the candidate answers a question, he has a probability of 1/2 of responding correctly, independently from a question to another.

A) What is the probability that the candidate is hired in the first stage of the interview?

a) 1/16   b) 3/32   c) 1/8   d) 5/32   e) 3/16

B) Let p be the answer in A). What is the probability that the candidate is hired?

a) $\frac{1}{2} - \frac{p}{2}$ b) $\frac{1}{2} + \frac{p}{2}$ c) $\frac{1}{2} + p$ d) $1 - p$ e) $1 - \frac{p}{2}$

C) Suppose now that, in fact, the candidate has a probability of 1/2 of responding correctly to the first question, but that his self-confidence increases after a right answer, so that the probability of responding correctly after a right answer is equal to 3/4. Conversely, the probability that the candidate responds correctly after a wrong answer is equal to only 1/4. What is then the probability that the candidate responds correctly to exactly one of the first two questions?

a) 1/4   b) 3/8   c) 1/2   d) 5/8   e) 3/4

Question no. 31

A box contains ten components, of which two are defective. The components are taken one at a time, at random and without replacement, and are tested until the two defective components have been identified.

A) What is the probability that the two defective components are identified as early as after the second test?

a) 1/100   b) 1/95   c) 1/90   d) 1/50   e) 1/45

B) What is the probability that at least one defective component has been identified after three tests (if necessary)?

a) 0.4667   b) 0.4880   c) 0.5120   d) 0.5333   e) 0.5600

C) Calculate the probability that more than eight components have to be tested in order to identify the two defective ones.

a) 4/15   b) 14/45   c) 16/45   d) 6/15   e) 4/9

Question no. 32

Let A, B and C be three events. Write in mathematical form the event $F = $ exactly one of the events A, B, C does not occur.

a) $(A \cap B) \cup (B \cap C) \cup (A \cap C)$
b) $(A \cap B \cap C^c) \cup (A \cap B^c \cap C) \cup (A^c \cap B \cap C)$
c) $(A^c \cap B^c \cap C) \cup (A^c \cap B \cap C^c) \cup (A \cap B^c \cap C^c)$
d) $A^c \cup B^c \cup C^c$
e) $(A \cup B \cup C)^c$

Question no. 33

Let A and B be two events such that $P[A] = 1/2$ and $P[B] = 3/10$. Which of the following statements can, under certain conditions, be true?

i) $P[A \cup B] = 4/5$;   ii) $P[A \cup B] = 13/20$;   iii) $P[A \cup B] = 2/5$.

a) i) only   b) i) and ii) only   c) i) and iii) only   d) ii) and iii) only
e) all

Question no. 34

Each of the tools $O_1, \ldots, O_5$ has a particular pocket in a certain tool belt. After use, you must replace each tool in its proper pocket. Because of lack of time, you put at random (equiprobability) a tool in each pocket.

A) What is the probability that the second tool that you put into the tool belt is actually in its proper pocket?

a) 1/20   b) 1/12   c) 1/10   d) 1/5   e) 1/4

B) Calculate the probability that each of the tools O_1, O_2 and O_3 are put into the right pockets?

a) 1/125   b) 1/120   c) 1/60   d) 1/40   e) 1/20

Question no. 35

A company buys electrical components in batches of ten components. Upon reception of each batch, two components are taken at random and without replacement and are then tested. The company accepts the batch if and only if the two components tested are non-defective. Based on previous data, we estimate that the probability that a batch of ten components contains no defectives is equal to 7/10, the probability that it contains exactly one defective is 1/5, and the probability that it contains exactly two defectives is 1/10. We then find that the probability that a batch is accepted is approximately equal to 0.9222. Use, if needed, this result to calculate the probability that

A) a batch contains no defectives and is accepted;

a) 0.3   b) 0.6222   c) 0.7   d) 0.7222   e) 0.9222

B) a batch contains exactly two defectives or is accepted;

a) 0.92   b) 0.93   c) 0.94   d) 0.95   e) 0.96

C) a batch contains exactly one defective, given that it has been rejected;

a) 0.5141   b) 0.5541   c) 0.6141   d) 0.6541   e) 0.7141

D) three consecutive (independent) batches are rejected.

a) 0.00047   b) 0.0014   c) 0.0778   d) 0.2157   e) 0.2333

Question no. 36

Let A, B and C be events such that $P[A] = P[B] = P[C] = 1/4$, $A \subset B$, and A and C are incompatible.

A) Calculate $P[A \cup B \cup C]$.

a) 1/4   b) 3/8   c) 1/2   d) 5/8   e) 3/4

B) Calculate $P[A^c \cap (B \cup C)]$.

a) 1/4   b) 3/8   c) 1/2   d) 5/8   e) 3/4

Question no. 37

In physics, many systems, for example Einstein's and Debye's models for simple solids, are made up of subsystems that are independent of one another. Consider a system of this type, made up of three particles. Each particle, independently from the other two particles, can have energy level ϵ_1, ϵ_2 or ϵ_3. The energy level of the system is defined by the vector (n_1, n_2, n_3), where n_i is the number of particles having energy level ϵ_i, for $i = 1, 2, 3$.

A) How many distinct energy levels can this system have?

a) 6   b) 7   c) 8   d) 9   e) 10

B) In A), how many distinct arrangements of the energy levels of the particles corre-spond to the energy level $(2, 1, 0)$ of the system or to level $(1, 1, 1)$?

a) 9   b) 10   c) 12   d) 14   e) 15

C) i) Let P_i be the probability that a particle has energy level ϵ_i, for $i = 1, 2, 3$. Suppose that $P_1 = 1/4$, $P_2 = 1/2$ and $P_3 = 1/4$. What is the probability that the three particles all have a different energy level?

a) 1/16   b) 1/8   c) 3/16   d) 1/4   e) 3/8

ii) Given that the three particles have the same energy level, what is the probabil-ity that this energy level is ϵ_1 or ϵ_2?

a) 2/3   b) 7/10   c) 4/5   d) 5/6   e) 9/10

Question no. 38
A) A number is taken at random in the set $\{1, 2, \ldots, 8\}$, so that $P[\{k\}] = 1/8$ for $k = 1, 2, \ldots, 8$. We define the events

$$A = \{1, 2, 3, 4\}, \quad B = \{2, 3, 4, 5\} \quad \text{and} \quad C = \{4, 5, 6, 7\}.$$

Which pairs of events are independent?

a) (A, B) only   b) (B, C) only   c) (A, B) and (B, C) only

d) no pairs   e) all the pairs

B) Suppose that in A), we have: $P[\{k\}] = c_1$ if $k = 1, 3, 5, 7$ and $P[\{k\}] = c_2$ if $k = 2, 4, 6, 8$. For what values of c_1 and c_2 are the events $D = \{1, 2\}$ and $F = \{2, 3\}$ independent?

a) $c_1 = c_2 = 1/16$ b) $c_1 = c_2 = 3/16$ c) $c_1 = 1/16; c_2 = 3/16$

d) $c_1 = 3/16; c_2 = 1/16$ e) no values of c_1 and c_2

Question no. 39
    The probabilities of the events A, B and C are $P[A] = 1/2$, $P[B] = 1/3$ and $P[C] = 1/4$. Moreover, A and C are incompatible, B and C are incompatible, and $P[A \cap B] = P[A^c \cap B]$.

A) Calculate $P[A^c \cap B^c]$.

a) 1/12   b) 1/6   c) 1/4   d) 1/3   e) 1/2

B) Calculate $P[A| B \cup C^c]$.

a) 1/4   b) 1/3   c) 1/2   d) 7/12   e) 2/3

Question no. 40
    A box contains five brand A components, five components of brand B and five of brand C. Five components are taken at random and without replacement.

A) What is the probability that the five components taken at random are of the same brand?

a) $\frac{1}{3003}$ b) $\frac{1}{3000}$ c) $\frac{1}{1001}$ d) $\frac{1}{1000}$ e) $\frac{1}{500}$

B) What is the probability that the five components are of the same brand, given that at least four of the five components are of the same brand?

a) $\frac{1}{101}$ b) $\frac{1}{51}$ c) $\frac{1}{50}$ d) $\frac{1}{11}$ e) $\frac{1}{10}$

Question no. 41

We have three boxes containing only brand A and brand B components. The first box comprises 10 brand A and 10 brand B components, the second box contains 10 brand A and 20 brand B components, and the third one has 20 brand A and 10 brand B components. Let E be the random experiment that consists in taking first, independently, a component at random in each of the first two boxes (starting with the first box). If the two components are of the same brand, then the random experiment is over; otherwise a component is taken at random from the third box. The *brand* of each component taken at random in this random experiment is noted.

We consider the events

A_k = a brand A component is obtained on the kth draw, for $k = 1, 2, 3$,
B_k = a brand B component is obtained on the kth draw, for $k = 1, 2, 3$,
F = three components have to be taken at random to complete the random experiment.

A) How many elementary events are there in the sample space S associated with E if we take the order into account?

a) 5   b) 6   c) 7   d) 8   e) 9

B) Which groups of events, among the following, are incompatible?

I) A_1, B_2 II) A_1, B_1 III) A_1, A_2, A_3 IV) A_1, A_2, B_3 V) A_1, B_2, B_3
a) II, III and IV only   b) I and IV only   c) II and III only
d) I, IV and V only   e) II only

C) Calculate $P[B_3]$.

a) 1/18   b) 1/9   c) 1/6   d) 1/3   e) 1/2

D) Calculate $P[A_1 \cap (A_2 \cup A_3)|F]$.

a) 1/9   b) 2/9   c) 1/3   d) 4/9   e) 5/9

Question no. 42

A point is taken at random in the interval $[0, 1]$, so that the probability of the event F_I = the point chosen is in the interval $I = [a, b]$ is given by $P[F_I] = b - a$ for any $I \subset [0, 1]$. We consider the events F_A, F_B, F_C and F_D, where

$$A = [0, \tfrac{1}{2}], \quad B = [\tfrac{1}{4}, \tfrac{3}{4}], \quad C = [\tfrac{5}{8}, \tfrac{3}{4}] \quad \text{and} \quad D = [0, 1].$$

A) Which are the only pairs of (distinct) events that are incompatible (or mutually exclusive)?

a) no pairs   b) (F_A, F_B) c) (F_A, F_C) d) (F_A, F_B) and (F_A, F_C)
e) $(F_A, F_B), (F_A, F_D), (F_B, F_D)$ and (F_C, F_D)

B) Which are the only pairs of (distinct) events that are independent?

a) no pairs   b) (F_A, F_B) c) (F_A, F_C) d) (F_A, F_B) and (F_A, F_C)
e) $(F_A, F_B), (F_A, F_D), (F_B, F_D)$ and (F_C, F_D)

Question no. 43

A particle is at the origin at the initial time and is then moving on the positive integers as follows: at each time unit, a coin for which the probability of getting "tails" is equal to 1/3 (and $P[\{Heads\}] = 2/3$) is thrown (independently). If "tails" is obtained, then the particle moves one integer to the right, whereas if "heads" is obtained, the particle moves two integers to the right.

A) What is the probability that the particle does not visit the point 3?

a) 2/9   b) 8/27   c) 10/27   d) 4/9   e) 14/27

B) In how many different ways can the particle go from 0 to 10 without visiting the point 2?

a) 21   b) 22   c) 23   d) 24   e) 25

Question no. 44

Two defective fuses have been mixed by mistake with three good fuses. The fuses are taken one at a time, at random and without replacement, and tested until the two defective fuses can be identified.

A) What is the probability that the first two fuses tested are two defective or two good fuses?

a) 0.25   b) 0.3   c) 0.35   d) 0.4   e) 0.5

B) What is the probability that the last fuse that needs to be tested to identify the two defective fuses is a good one?

a) 0.25   b) 0.3   c) 0.35   d) 0.4   e) 0.5

Question no. 45

In a certain course, 45% of the students are in electrical engineering, 40% are in computer engineering, and 15% are in physics engineering. Furthermore, 60% of the electrical engineering students registered in this course are fourth-term students. In the case of the computer engineering students this percentage is equal to 70%, while it is equal to 90% in the case of the physics engineering students.

A) Two students are taken, at random and with replacement, among the students in the course considered. What is the probability that they are both fourth-term students?

a) 0.4492   b) 0.4692   c) 0.4892   d) 0.5092   e) 0.5292

B) Let p_A be the answer in A). Knowing that the two students taken at random in A) are fourth-term students, what is the probability that they are both in electrical engineering?

a) $\dfrac{0.0729}{p_A}$ b) $0.0729\, p_A$ c) $\dfrac{0.27}{p_A}$ d) $0.5625\, p_A$ e) $\dfrac{0.5625}{p_A}$

Question no. 46

The students in a certain department of a university must choose exactly two courses among three optional courses: A, B and C. We find that the number of students registered for course A is 150% greater than the number of students registered for course C. Moreover, the number of students registered for course B is 50% greater than the number of students registered for course C. A student in this department is taken at random. Let F_A = the student taken at random has chosen course A, etc.

A) Calculate the probability $P\,[(F_A \cap F_B) \cup (F_A \cap F_C) \cup (F_B \cap F_C)]$.

a) 0.6   b) 0.7   c) 0.8   d) 0.9   e) 1

B) What is the probability that the student has chosen courses A and B?

a) 0.4   b) 0.5   c) 0.6   d) 0.7   e) 0.8

Question no. 47

The probability that a certain person spends an amount of time (in minutes) comprised in the interval $[a, b]$ surfing the Internet, on a weekday, is given by

$$\frac{b - a}{20} \quad \text{for } 10 \le a < b \le 30.$$

We assume that the days are independent. We consider a time period of five weekdays, namely from Monday to Friday.

A) What is the probability that the shortest period of time spent surfing the Internet, during that week, was that on Monday, and the longest period that on Friday?

a) 1/120   b) 1/20   c) 1/5   d) 1/4   e) 2/5

B) What is the probability that the length of the surfing sessions has increased each day?

a) 1/720   b) 1/120   c) 1/60   d) 1/40   e) 1/20

Question no. 48

We consider the system shown in Fig. 2.13. Each component functions with probability 1/2, independently from the other three.

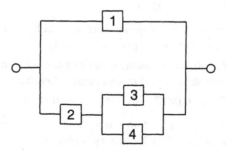

Figure 2.13. System for Multiple choice question no. 48.

A) What is the probability that the system functions?

a) 7/16   b) 1/2   c) 9/16   d) 5/8   e) 11/16

B) Let p be the answer in A). Given that the system functions, what is the probability that component no. 1 functions?

a) $\frac{1}{2p}$ b) $\frac{1}{3p}$ c) $\frac{1}{4p}$ d) $\frac{p}{2}$ e) $\frac{p}{4}$

Question no. 49

A hydroelectric station is subject to receive three types of shocks. The probability of a shock of type I is equal to 30%, that of a shock of type II is equal to 50%, and that of a shock of type III is equal to 20%. Moreover, the various types of shocks are incompatible.

The probability that a shock of type I causes a power failure is equal to 40%. In the case of type II and type III shocks, this probability is equal to 20% and to 90%, respectively.

A) What is the probability that the station receives a type II shock before a type III shock, from a given time instant?

a) 3/7   b) 1/2   c) 4/7   d) 5/7   e) 6/7

B) Given that a power failure (caused by a shock) has occurred, what is the probability that a type II shock has caused this failure?

a) 1/7   b) 1/6   c) 1/5   d) 1/4   e) 1/3

Question no. 50

Manufactured objects are classified into three categories: compliant with the norms (I), containing one or more minor defects only (II), or containing one or more major defects (III).

A) Four objects will be taken, at random and with replacement, among those that will be manufactured tomorrow. In how many different ways will it be possible to classify them if

i) the order in which the objects will be taken matters?

a) 64   b) 81   c) 100   d) 121   e) 196

ii) the order in which the objects will be taken does not matter?

a) 12   b) 13   c) 14   d) 15   e) 16

B) Suppose that in a batch of 100 objects, there are 70 objects that comply with the norms and 23 that have one or more minor defects only.

i) A sample of three objects is taken, at random and without replacement. What is the probability that the sample contains an object of each category?

a) 0.0697   b) 0.0797   c) 0.0897   d) 0.0997   e) 0.1097

ii) Four objects are taken, at random and without replacement. What is the probability that at least one object of each category is obtained?

a) 0.0952   b) 0.1052   c) 0.1152   d) 0.1252   e) 0.1352

Question no. 51

Two tests based on the DNA of an individual have been developed in order to identify the culprits in criminal cases. The tests work by comparing two samples: one taken on the victim and the other on the suspect. In the case of test A, the probability of concluding that the suspect is the person wanted, given that he is effectively the culprit, is equal to 99.5%. Moreover, the probability of arriving at the same conclusion, given that he is not the person wanted, is equal to 1.5%. In the case of test B, these probabilities are equal to 99.7% and to 2%, respectively. Finally, we assume that the two tests are conditionally independent with respect to the event $C =$ the suspect is the culprit, and with respect to C^c.

A) A suspect in a particular case has been arrested. Suppose that the probability that he is the culprit is equal to 80%.

i) Given that test A clears the suspect, what is the probability that he is in fact guilty?

a) 0.0179   b) 0.0189   c) 0.0199   d) 0.0209   e) 0.0219

ii) Given that the two tests concluded that the suspect is guilty, what is the probability that he is really guilty?

a) 0.999914   b) 0.999924   c) 0.999934   d) 0.999944   e) 0.999954

iii) If the suspect is the criminal wanted, what is the probability that one test concludes that he is guilty and the other that he is innocent?

a) 0.00497   b) 0.00597   c) 0.00697   d) 0.00797   e) 0.00897

B) Suppose that ten suspects have been arrested and that one of them is the culprit. What is the probability that only the culprit is found guilty by test A?
Remark. We suppose that the tests are independent from one person to another.

a) 0.4685   b) 0.5685   c) 0.6685   d) 0.7685   e) 0.8685

3

Random Variables

3.1 Introduction

Definition 3.1.1. *Let S be a sample space associated with a random experiment E. A function X that associates a real number $X(s) = x$ with each outcome $s \in S$ is called a **random variable**. The set of all possible values of X will be denoted by S_X (see Fig. 3.1).*

Example 3.1.1 i) Let E_i be the following random experiment: a transistor is taken at random in a box containing 10 brand A and 20 brand B transistors, and its brand is noted. Then, $S = \{A, B\}$. Let X be the random variable that takes the value 1 if the transistor is of brand A and the value 0 otherwise. We have: $S_X = \{0, 1\}$. Here, the random variable X is in fact the *indicator function* of the event $F = $ a brand A transistor is picked.

ii) Let E_{ii} be the random experiment that consists in taking a number at random in the interval $[0, 10)$. Let X be the random variable that takes the value of the number obtained and Y be the random variable that is equal to the integer part of the number obtained. We have:

$$S = [0, 10), \quad S_X \equiv S \quad \text{and} \quad S_Y = \{0, 1, \dots, 9\}.$$

In this case, the random variable X is the identity function. Note that if the number is taken at random in the interval $[0, 10]$, then we can write that

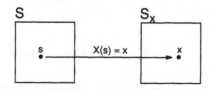

Figure 3.1. Notion of random variable.

$$S_Y = \{0, 1, \ldots, 10\}.$$

However, the probability that Y takes on the value 10 is equal to zero, because the number taken at random in the interval $[0, 10]$ should be the number 10 (see Definition 3.1.2 below), and $S = [0, 10]$ is a *continuous* sample space.

Remark. Since a random variable is a real-valued *function* defined on S, the set S_X cannot have more elements than S. Therefore, if S is a discrete sample space, then S_X is discrete. However, if S is continuous, then S_X can be either discrete or continuous (as in Example 3.1.1 ii). Note that if Z is the random variable that is equal to twice the number taken at random in the interval $[0, 10)$ in the preceding example, that is, if we define $Z = 2X$, then we have:

$$S_Z = [0, 20).$$

However, $S (= S_X)$ and S_Z are both non-denumerably infinite.

Definition 3.1.2. An **event** *with respect to* S_X *is a subset of* S_X. *Moreover, let* $A \subset S$ *and* $B \subset S_X$. *We say that* A *and* B *are* **equivalent events** *if*

$$A = \{s \in S : X(s) \in B\}. \tag{3.1}$$

Then, we write that $P[B] = P[A]$.

Remark. We could denote the probability $P[B]$ of the event B by $P_X[B]$, or $P[X \in B]$.

Example 3.1.2 A box contains two brand M_1, one brand M_2 and one brand M_3 components. Two components are taken at random and without replacement and the brand of the components is noted. Then, if the order in which the components are chosen does not matter, we can write that

$$S = \{M_1 M_1, M_1 M_2, M_1 M_3, M_2 M_3\}.$$

Let X be the number of brand M_1 components obtained. We have:

$$S_X = \{0, 1, 2\}.$$

Now, let $B = \{1\}$. The event equivalent to B is

$$A = \{M_1 M_2, M_1 M_3\}.$$

It follows (by symmetry) that

$$P[X \in \{1\}] \equiv P[X = 1] = P[B] = P[A] = 2 \times (2)(2/4)(1/3) = 2/3.$$

Remarks. i) If we identify the two brand M_1 components as M_{11} and M_{12}, then the sample space S can be rewritten as follows:

$$S = \{M_{11}M_{12}, M_{11}M_2, M_{12}M_2, M_{11}M_3, M_{12}M_3, M_2M_3\}.$$

In this case, we obtain:

$$A = \{M_{11}M_2, M_{12}M_2, M_{11}M_3, M_{12}M_3\}$$

and, given that the elementary events are now equiprobable, we find directly that

$$P[X = 1] = P[B] = P[A] = 4/6 = 2/3.$$

ii) We can also define the events F = exactly one brand M_1 component is obtained, and F_1 = a brand M_1 component is obtained on the first draw. Then, by the total probability rule, we can write that

$$P[F] = P[F \mid F_1]P[F_1] + P[F \mid F_1^c]P[F_1^c] = \frac{2}{3} \cdot \frac{2}{4} + \frac{2}{3} \cdot \frac{2}{4} = \frac{2}{3},$$

or

$$P[F] = \frac{\binom{2}{1}\binom{2}{1}}{\binom{4}{2}} = \frac{2 \times 2}{6} = \frac{2}{3}.$$

3.2 The Distribution Function

Definition 3.2.1. *The **distribution function** of the random variable X is defined, for any real number x, by*

$$F_X(x) = P[X \le x]. \tag{3.2}$$

Remark. More precisely, we may write that

$$F_X(x) = P\left[\{s \in S: X(s) \le x\}\right]. \tag{3.2a}$$

Properties. i) $0 \le F_X(x) \le 1$ (since $F_X(x)$ is a probability).

ii) $\lim_{x \to \infty} F_X(x) = 1$ (since the event $\{X < \infty\}$ is certain).

iii) $\lim_{x \to -\infty} F_X(x) = 0$ (because the event $\{X \le -\infty\}$ is possible).

iv) The function F_X is *non-decreasing*; that is, if $x_0 < x_1$, then $F_X(x_0) \le F_X(x_1)$ (this follows from the fact that the event $\{X \le x_0\}$ is included in $\{X \le x_1\}$).

v) We can show that the function F_X is *right-continuous*; that is, $F_X(x) = F_X(x^+)$, where

$$F_X(x^+) := \lim_{\delta \downarrow 0} F_X(x + \delta).$$

Proposition 3.2.1. *We have:* $P[a < X \le b] = F_X(b) - F_X(a)$.

Proof. We only have to notice that the event $\{X \leq b\}$ can be divided into two disjoint (and exhaustive) parts:

$$\{X \leq b\} = \{X \leq a\} \cup \{a < X \leq b\}.$$

Since the two events in the right-hand member of the equation are *incompatible*, it follows that

$$F_X(b) = F_X(a) + P[a < X \leq b].\qquad\qquad\square$$

Corollary 3.2.1. *We have:* $P[X = x] = F_X(x) - F_X(x^-)$, *where*

$$F_X(x^-) := \lim_{\delta \downarrow 0} F_X(x - \delta).$$

Proof. We choose $a = x - \delta$ and $b = x$ in the preceding proposition, and we take the limit as δ decreases to 0. $\qquad\qquad\square$

Remark. We deduce from the corollary that if F_X is a continuous function, then the probability $P[X = x]$ is equal to zero for any real number x. It also follows that

$$P[a \leq X \leq b] = P[a < X \leq b] = P[a \leq X < b] = P[a < X < b]$$

when the distribution function of X is *continuous*.

Types of Random Variables

Definition 3.2.2. *Let X be a random variable that can take, at most, a denumerably infinite number of values. That is, S_X is of the form*

$$S_X = \{x_1, x_2, \ldots\}.$$

*Then, we say that X is a **discrete** random variable.*

Example 3.2.1 Suppose that we observe the lifetime of an electric light bulb that works. Suppose also that the probability that the lifetime T (in hours) of a light bulb of this type takes on a value in the interval $[a, b)$ is given by

$$P[a \leq T < b] = e^{-a/100} - e^{-b/100}, \quad \text{where } 0 \leq a < b \leq \infty.$$

Let X be the number of complete periods of 100 hours that the light bulb lasts. Then,

$$S_X = \{0, 1, 2, \ldots\}.$$

Therefore, X is a discrete random variable. Now, since X only takes on non-negative integer values, we have:

$$F_X(x) \equiv P[X \leq x] = 0 \quad \forall x < 0.$$

Remark. We may write that

$$F_X(x) = P[X \in \{x_k \in S_X : x_k \leq x\}].$$

Since there is not a single element of S_X that is inferior to zero, we have indeed: $F_X(x) = 0$ for all $x < 0$.

Next, for any real number in the interval $[0, 1)$, we can write that

$$F_X(x) = P[X \in \{0\}] \equiv P[X = 0] = P[T < 100] = 1 - e^{-1}.$$

Remark. Since we assume that the light bulb is working, we have:

$$P[T < 100] = P[0 < T < 100] = P[0 \leq T < 100],$$

because $P[T = 0] = 0$.

Similarly, for any real number in the interval $[1, 2)$, we obtain that

$$F_X(x) = P[X = 0] + P[X = 1] = 1 - e^{-1} + P[100 \leq T < 200]$$
$$= 1 - e^{-1} + (e^{-1} - e^{-2}) = 1 - e^{-2}.$$

In general, for any x in the interval $[n, n+1)$, where $n \in \{0, 1, \dots\}$, we find that

$$F_X(x) \equiv P[X \leq x] = P[T < (n+1)100] = 1 - e^{-n-1}.$$

For example, $F_X(2.5) = F_X(2) = P[X \in \{0, 1, 2\}] = P[T < 300] = 1 - e^{-3} = 1 - e^{-2-1}$.

Thus, we can write that

$$F_X(x) = \begin{cases} 0 & \text{if } x < 0, \\ 1 - e^{-\text{int}(x)-1} & \text{if } x \geq 0, \end{cases}$$

where $\text{int}(x)$ denotes the integer part of x. We say that F_X is a *staircase function* (see Fig. 3.2).

Definition 3.2.3. *Let X be a random variable that can take a non-denumerably infinite number of values. If F_X is a continuous function, then we say that X is a* **continuous** *random variable.*

Example 3.2.2 In the preceding example, we can write that $S_T = (0, \infty)$ and that

$$F_T(t) = P[0 < T \leq t] = P[0 \leq T < t] = \begin{cases} 0 & \text{if } t \leq 0, \\ 1 - e^{-t/100} & \text{if } t > 0, \end{cases}$$

since $P[T = 0] = 0$ (see above) and $P[T = t] = 0$ (because the sample space $S = (0, \infty)$ associated with the random experiment that consists in observing the lifetime of a light bulb that works is continuous). The function F_T being continuous (see Fig. 3.3), T is a continuous random variable.

Remark. $F_T(t)$ is differentiable everywhere, except at $t = 0$.

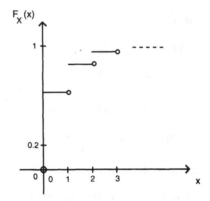

Figure 3.2. Distribution function of the random variable X in Example 3.2.1.

Definition 3.2.4. *Let X be a random variable that can take a non-denumerably in-finite number of values. If F_X is not a continuous function, then we say that X is a random variable of* **mixed** *type.*

Remark. A random variable of mixed type is in fact a function that is defined at the same time on a finite or denumerably infinite set and on one or more intervals. Moreover, we can write that

$$F_X(x) = p F_1(x) + (1 - p) F_2(x),$$

where $0 < p < 1$ is the probability that X is discrete and $F_1(x)$ (respectively $F_2(x)$) is the distribution function of a discrete random variable X_1 (resp. a continuous random variable X_2). In fact, we can also write that

$$X_1 = X \mid \{X \text{ is discrete}\} \quad \text{and} \quad X_2 = X \mid \{X \text{ is continuous}\}.$$

Example 3.2.3 In Example 3.2.1, suppose that the light bulb considered is taken at random and that the probability that it works is equal to 0.9. Let Y be the lifetime

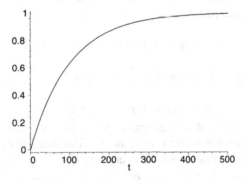

Figure 3.3. Distribution function of the random variable T in Example 3.2.2.

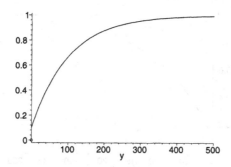

Figure 3.4. Distribution function of the random variable Y in Example 3.2.3.

of the light bulb and let us define the event F = the light bulb works. Then, we can write, by the total probability rule, that

$$F_Y(y) \equiv P[Y \le y] = P[Y \le y \mid F]P[F] + P[Y \le y \mid F^c]P[F^c]$$
$$= P[T \le y](0.9) + P[0 \le y](0.1) \quad \forall y \ge 0.$$

Remark. We can write that $Y \mid F \equiv T$ and $Y \mid F^c \equiv 0$. Furthermore, the probability $P[0 \le y]$ is equal to 1 if $y \ge 0$. In fact, the event $\{0 \le y\}$ is *deterministic*, so that its probability is either 0 or 1.

Using the preceding example, we find that

$$F_Y(y) = \begin{cases} 0 & \text{if } y < 0, \\ 0.1 & \text{if } y = 0, \\ 1 - 0.9e^{-y/100} & \text{if } y > 0. \end{cases}$$

Since $S_Y = [0, \infty)$ and F_Y is a discontinuous function (see Fig. 3.4), Y is of mixed type.

Remarks. i) We could in fact include the point $y = 0$ in the second interval.
ii) Here, we could write that

$$F_Y(y) = 0.1F_U(y) + 0.9F_T(y),$$

where

$$F_U(u) = \begin{cases} 0 \text{ if } u < 0, \\ 1 \text{ if } u \ge 0. \end{cases}$$

Actually, U is the constant 0 (a degenerate discrete random variable).

iii) Suppose, in addition, that we replace a light bulb that is still working after 200 hours, for preventive maintenance. Let W be the time that a light bulb is used. Then, we find that (see Fig. 3.5)

$$F_W(w) = \begin{cases} 0 & \text{if } w < 0, \\ 1 - 0.9e^{-w/100} & \text{if } 0 \le w < 200, \\ 1 & \text{if } w \ge 200. \end{cases}$$

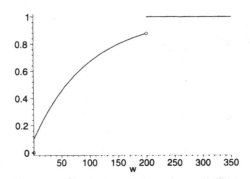

Figure 3.5. Distribution function of the random variable W in Example 3.2.3.

In this case, we can write that

$$F_W(w) = p F_{X_1}(w) + (1 - p) F_{X_2}(w),$$

where $p := P[W \text{ is discrete}] = 0.1 + 0.9 P[T \geq 200] = 0.1 + 0.9 e^{-2} \ (\Rightarrow 1 - p = 0.9(1 - e^{-2}))$,

$$F_{X_1}(w) = \begin{cases} 0 & \text{if } w < 0, \\ \frac{0.1}{p} & \text{if } 0 \leq w < 200, \\ 1 & \text{if } w \geq 200 \end{cases}$$

and

$$F_{X_2}(w) = \begin{cases} 0 & \text{if } w \leq 0, \\ \frac{1 - e^{-w/100}}{1 - e^{-2}} & \text{if } 0 < w < 200, \\ 1 & \text{if } w \geq 200. \end{cases}$$

Thus, we have:

$$X_1 = \begin{cases} 0 & \text{with probability } 0.1/p, \\ 200 & \text{with probability } 0.9 e^{-2}/p \end{cases}$$

and (see Section 3.3)

$$f_{X_2}(w) := \frac{d}{dw} F_{X_2}(w) = \begin{cases} \frac{(1/100)\, e^{-w/100}}{1 - e^{-2}} & \text{if } 0 < w < 200, \\ 0 & \text{elsewhere.} \end{cases}$$

Finally, we can also generalize the notion of distribution function as follows.

Definition 3.2.5. *Let X be a random variable and A an event. The* **conditional distribution function** *of X, given A, is defined by*

$$F_X(x \mid A) = \frac{P[\{X \leq x\} \cap A]}{P[A]} \quad \text{if } P[A] > 0. \tag{3.3}$$

Example 3.2.4 In Example 3.2.2, let

$$A = \text{the light bulb lasts less than } t_0 \text{ hours.}$$

Then, we have:

$$F_T(t \mid A) = F_T(t \mid T < t_0) = \frac{P[\{T \leq t\} \cap \{T < t_0\}]}{P[T < t_0]}$$

$$= \begin{cases} \dfrac{F_T(t)}{F_T(t_0)} & \text{if } t < t_0, \\ 1 & \text{if } t \geq t_0 \end{cases}$$

(since T is a continuous random variable). Using Example 3.2.2, we obtain that

$$F_T(t \mid T < t_0) = \begin{cases} 0 & \text{if } t \leq 0, \\ \dfrac{1 - e^{-t/100}}{1 - e^{-t_0/100}} & \text{if } 0 < t < t_0, \\ 1 & \text{if } t \geq t_0. \end{cases}$$

Remark. Without having to do any calculations, we could write, for instance,

$$P[T \leq 200 \mid T < 100] = 1.$$

However, in the case of the probability $P[T \leq 100 \mid T < 200]$, we must make use of the formula above.

Similarly, let

$$B = \text{the light bulb lasts more than } t_0 \text{ hours.}$$

Then,

$$F_T(t \mid B) = F_T(t \mid T > t_0) = \frac{P[\{T \leq t\} \cap \{T > t_0\}]}{P[T > t_0]}$$

$$= \begin{cases} 0 & \text{if } t \leq t_0, \\ \dfrac{P[t_0 < T \leq t]}{P[T > t_0]} & \text{if } t > t_0 \end{cases} = \begin{cases} 0 & \text{if } t \leq t_0, \\ \dfrac{F_T(t) - F_T(t_0)}{1 - F_T(t_0)} & \text{if } t > t_0. \end{cases}$$

Remark. In this case, we could write, directly, that

$$P[T \leq 100 \mid T > 200] = 0,$$

while the probability $P[T \leq 200 \mid T > 100]$ must be calculated.

3.3 The Probability Mass and Density Functions

Definition 3.3.1. *Let X be a discrete random variable. Then, we have: $S_X = \{x_1, x_2, \ldots\}$. The function p_X, defined by*

$$p_X(x_k) = P[X = x_k] \tag{3.4}$$

for $k = 1, 2, \ldots$, is called the **probability mass function** *of X.*

Remarks. i) The function p_X has the following properties:

  a) $p_X(x_k) \geq 0 \ \forall k$ (since it is a probability);

  b) $\sum_{k=1}^{\infty} p_X(x_k) = 1$ (because $P[X \in S_X] = P[S] = 1$).

ii) If we assume that $x_1 < x_2 < \cdots$, then we can write that

$$p_X(x_1) = F_X(x_1) \tag{3.5a}$$

and

$$p_X(x_k) = F_X(x_k) - F_X(x_{k-1}) \quad \text{for } k = 2, 3, \ldots \tag{3.5b}$$

We also have:

$$p_X(x) = F_X(x) - F_X(x^-) \quad \text{for any real number } x. \tag{3.6}$$

iii) Using Heaviside's function $u(x)$, defined by

$$u(x) = \begin{cases} 0 \text{ if } x < 0, \\ 1 \text{ if } x \geq 0, \end{cases} \tag{3.7}$$

we can express the distribution function of a discrete random variable X as follows:

$$F_X(x) = \sum_{k=1}^{\infty} p_X(x_k) u(x - x_k). \tag{3.8}$$

This formula is equivalent to

$$F_X(x) = \sum_{x_k \leq x} p_X(x_k). \tag{3.9}$$

Example 3.3.1 We deduce from the function $F_X(x)$ in Example 3.2.1 that

$$p_X(n) = F_X(n) - F_X(n^-) = e^{-n} - e^{-n-1} \quad \text{for } n = 0, 1, \ldots$$

(see Fig. 3.6, p. 65). To be more complete, we can add that $p_X(x) = 0$ if x is not a positive integer or zero.

Remarks. i) When S_X is finite, the function p_X is often presented as a table. For example, if

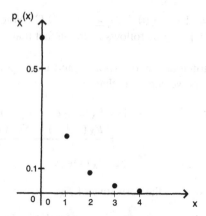

Figure 3.6. Probability mass function of the random variable X in Example 3.3.1.

$$p_X(x) = \begin{cases} 1/4 \text{ if } x = -1, \\ 1/2 \text{ if } x = 0, \\ 1/4 \text{ if } x = 1, \end{cases}$$

then we write:

x	−1 0 1
$p_X(x)$	1/4 1/2 1/4

Note that, in this particular case, we can also write the function p_X as a single formula as follows:

$$p_X(x) = \binom{2}{x+1}\frac{1}{4} \quad \text{for } x = -1, 0, 1,$$

which is not always possible. Moreover, the function F_X can likewise be presented in the form of a table:

x	−1 0 1
$F_X(x)$	1/4 3/4 1

However, we must be able to interpret these tables correctly. For example, we deduce from these two tables that $p_X(1/2) = 0$, while $F_X(1/2) = 3/4$.

ii) Since $p_X(x)$ is a probability, we can easily generalize Definition 3.3.1 by considering the **conditional probability mass function** of X, given an event A, defined by

$$p_X(x \mid A) = \frac{P[\{X = x\} \cap A]}{P[A]} \quad \text{if } P[A] > 0. \tag{3.10}$$

For example, let $A := \{X \geq 0\}$. Then we find, using the random variable X of part i) above, that

x	0 1
$p_X(x \mid X \geq 0)$	2/3 1/3

Indeed, we have:

$$p_X(0 \mid X \geq 0) = \frac{P[\{X = 0\} \cap \{X \geq 0\}]}{P[X \geq 0]} = \frac{P[X = 0]}{P[X \geq 0]} = \frac{1/2}{1/2 + 1/4} = 2/3$$

and $p_X(1 \mid X \geq 0) = 1 - p_X(0 \mid X \geq 0) = 1/3$. Note that the relationship $p_X(0 \mid X \geq 0) = 2p_X(1 \mid X \geq 0)$ follows from the fact that $p_X(0) = 2p_X(1)$.

Now, let X be a continuous random variable and x be a point where the function F_X is *differentiable*. Then, we can write that

$$P[x < X \leq x + \epsilon] = F_X(x + \epsilon) - F_X(x)$$
$$= \frac{F_X(x + \epsilon) - F_X(x)}{\epsilon} \epsilon$$
$$\simeq f_X(x)\epsilon,$$

where $\epsilon > 0$ is small and f_X is the derivative of F_X.

Definition 3.3.2. *Let X be a continuous random variable. The function f_X, defined by*

$$f_X(x) = \frac{d}{dx} F_X(x) \tag{3.11}$$

(if the derivative exists), is called the (**probability) density function** *of X.*

Remark. The function $f_X(x)$ is approximately equal to the probability that the random variable X takes a value in a small interval of length ϵ around the point x (from $x - (\epsilon/2)$ to $x + (\epsilon/2)$, for instance), divided by the length of this interval; $f_X(x)$ is *not* the probability that X takes on the value x (this probability is equal to zero, because X is a continuous random variable).

Properties. i) $f_X(x) \geq 0$ (by the preceding remark, or since F_X is a non-decreasing function).

ii) Integrating Equation (3.11) from $-\infty$ to x, we obtain:

$$\int_{-\infty}^{x} f_X(t)\, dt = \int_{-\infty}^{x} \frac{d}{dt} F_X(t)\, dt = F_X(t)|_{-\infty}^{x} = F_X(x) - 0 = F_X(x). \tag{3.12}$$

It follows that

$$\int_{-\infty}^{\infty} f_X(x)\, dx = F_X(\infty) = 1. \tag{3.13}$$

Actually, any non-negative piecewise continuous function that possesses the preceding property is a density function.

We also deduce from Equation (3.12) that

$$P[a < X \leq b] = F_X(b) - F_X(a) = \int_{a}^{b} f_X(x)\, dx. \tag{3.14}$$

Thus, the probability that X takes on a value in the interval $(a, b]$ is given by the area under the curve $f_X(x)$ from a to b.

Remarks. i) If, instead of Equation (3.13), we have:

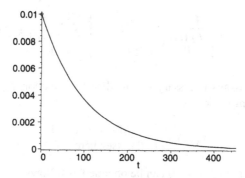

Figure 3.7. Probability density function of the random variable T in Example 3.3.2.

$$\int_{-\infty}^{\infty} f_X(x)\, dx = c,$$

where $0 < c < 1$, we say that the random variable X is *defective* or *improper*. In the general case when the integral is equal to $c > 0$, we only have to define $g_X(x) = f_X(x)/c$ to obtain a valid or proper random variable. Similarly, a discrete random variable can be defective.

ii) If S_X is an interval or a set of intervals, rather than the entire real line, we write that the function f_X is equal to zero outside S_X.

Example 3.3.2 Since the distribution function of the random variable T in Example 3.2.1 is differentiable $\forall t \neq 0$ (see Example 3.2.2), the density function of T is given by

$$f_T(t) = \begin{cases} \frac{1}{100} e^{-t/100} & \text{if } t > 0, \\ 0 & \text{elsewhere} \end{cases}$$

(Fig. 3.7). Notice that the function $f_T(t)$ is continuous everywhere, except at $t = 0$.

Now, as in the case of the distribution function, we can generalize the definition of the density function.

Definition 3.3.3. *Let X be a continuous random variable and A an event. We define the* **conditional (probability) density function** *of X, given A, by*

$$f_X(x \mid A) = \frac{d}{dx} F_X(x \mid A) \tag{3.15}$$

(when the derivative exists).

Example 3.3.3 The conditional (probability) density function of the random variable T in Example 3.2.1, given the event $A = \{T < t_0\}$, is given (see Example 3.2.4) by

$$f_T(t \mid T < t_0) = \frac{d}{dt} F_T(t \mid T < t_0)$$

$$= \begin{cases} \dfrac{f_T(t)}{F_T(t_0)} & \text{if } t < t_0, \\ 0 & \text{if } t \geq t_0 \end{cases} = \begin{cases} \dfrac{\frac{1}{100}e^{-t/100}}{1 - e^{-t_0/100}} & \text{if } 0 < t < t_0, \\ 0 & \text{elsewhere.} \end{cases}$$

Remark. It is not always necessary to calculate the function $F_X(x \mid A)$ to obtain $f_X(x \mid A)$. For example, let

$$f_X(x) = \begin{cases} 1/2 & \text{if } -1 < x < 1, \\ 0 & \text{elsewhere.} \end{cases}$$

Then, we find that $f_X(x \mid X < 0)$ can be obtained as follows:

$$f_X(x \mid X < 0) = \frac{f_X(x)}{P[X < 0]} \quad \text{if } -1 < x < 0$$

$$= \begin{cases} \frac{1/2}{1/2} = 1 & \text{if } -1 < x < 0, \\ 0 & \text{elsewhere.} \end{cases}$$

Finally, we would also like to generalize the notion of density function to the discrete case. To this end, we consider the *Dirac delta function*:

$$\delta(x) = \begin{cases} 0 & \text{if } x \neq 0, \\ \infty & \text{if } x = 0 \end{cases} \tag{3.16}$$

(with $\int_{-\infty}^{\infty} \delta(x)\, dx = 1$). We have the following relationship between the functions $u(x)$ and $\delta(x)$:

$$u(x) = \int_{-\infty}^{x} \delta(t)\, dt. \tag{3.17}$$

Using this relationship and Equation (3.8), we can define the density function of a discrete random variable by

$$f_X(x) = \sum_{k=1}^{\infty} p_X(x_k)\delta(x - x_k) \quad \forall x \in \mathbb{R}. \tag{3.18}$$

Example 3.3.4 Let

$$F_X(x) = \begin{cases} 0 & \text{if } x < 0, \\ 0.2 & \text{if } 0 \leq x < 1, \\ 0.7 & \text{if } 1 \leq x < 2, \\ 1 & \text{if } x \geq 2. \end{cases}$$

Then, we have:

x	0	1	2	Σ
$p_X(x)$	0.2	0.5	0.3	1

and

$$f_X(x) = 0.2\,\delta(x - 0) + 0.5\,\delta(x - 1) + 0.3\,\delta(x - 2) \quad \forall x \in \mathbb{R}.$$

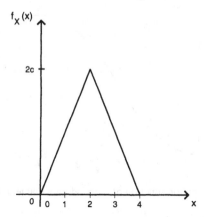

Figure 3.8. Probability density function of the random variable in Example 3.3.5.

Definition 3.3.4. *Let X be a continuous random variable with $S_X = [a, b]$. The number x_p for which*

$$F_X(x_p) = p, \tag{3.19}$$

*where $0 < p < 1$, is called the **quantile of order p** of X. In particular, $x_{1/2} \equiv x_m$ (or $\tilde{x}$) is called the **median** of X.*

Remarks. i) The number x_p is *unique*. However, if X is not a continuous random variable defined on $[a, b]$, then we must modify the foregoing definition. For example, we define the median x_m, in the general case, as follows:

$$P[X \leq x_m] \geq 1/2 \quad \text{and} \quad P[X \geq x_m] \geq 1/2. \tag{3.20}$$

If X is not a continuous random variable defined on a single interval $[a, b]$, then the median is not necessarily unique.

ii) If $100p = k$, an integer, x_p is also called the kth *percentile* of X. The terms *decile* (if k is a multiple of 10) and *quartile* are used as well.

Example 3.3.5 Let

$$f_X(x) = \begin{cases} cx & \text{if } 0 \leq x < 2, \\ c(4 - x) & \text{if } 2 \leq x \leq 4, \\ 0 & \text{elsewhere,} \end{cases}$$

where $c > 0$ (see Fig. 3.8). Calculate the constant c, the distribution function, the median and the 99th percentile of X.

Solution. We must have:

$$1 = \int_{-\infty}^{\infty} f_X(x)\, dx = \int_0^2 cx\, dx + \int_2^4 c(4 - x)\, dx = 4c \Rightarrow c = 1/4.$$

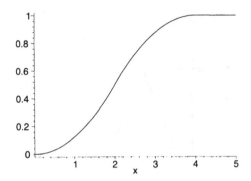

Figure 3.9. Distribution function of the random variable in Example 3.3.5.

Or we can use the fact that the graph of the function f_X is a triangle of area equal to $4c$, so that $c = 1/4$. Then, we can write that

$$
F_X(x) = \begin{cases}
0 & \text{if } x < 0, \\[2mm]
\displaystyle \int_{-\infty}^{0} 0\,dt + \int_{0}^{x} \frac{t}{4}\,dt = \frac{x^2}{8} & \text{if } 0 \le x < 2, \\[4mm]
\displaystyle \int_{0}^{2} \frac{t}{4}\,dt + \int_{2}^{x} \frac{1}{4}(4-t)\,dt = 1 - \frac{(4-x)^2}{8} & \text{if } 2 \le x \le 4, \\[3mm]
1 & \text{if } x > 4
\end{cases}
$$

(see Fig. 3.9). Making use of the function F_X, we find that $x_m = 2$ (because $F_X(2) = 1/2$) and, solving the equation

$$
1 - \frac{(4 - x_{0.99})^2}{8} = 0.99
$$

(with $2 \le x_{0.99} \le 4$), that $x_{0.99} \simeq 3.71$.

3.4 Important Discrete Random Variables

Bernoulli distribution.[1] Suppose that we perform a random experiment E. We consider a particular event A. Let X be the *indicator variable* of A; that is,

---

[1] Jacob (or Jacques) Bernoulli, 1654–1705, was born and died in Switzerland. He was a member of a famous family of mathematicians and physicists of Belgian origin that included his brother Johann and his nephew Daniel (Johann's son). He first studied philosophy and theology. He taught mathematics to his younger brother Johann, but later they quarreled and ended their relationship. He contributed greatly to algebra, infinitesimal calculus, mechanics and geometry. He solved what is now known as the *Bernoulli differential equation*. His important book on probability theory was published eight years after his death. In this book, we also find the *Bernoulli numbers*.

$$X = \begin{cases} 1 \text{ if } A \text{ occurs,} \\ 0 \text{ if } A^c \text{ occurs.} \end{cases} \tag{3.21}$$

We say that X has a **Bernoulli distribution** with **parameter** p, where $p := P[A]$ is called the probability of a **success**. We have:

x	0	1	Σ
$p_X(x)$	$1 - p$	p	1

This table can be replaced by

$$p_X(x) = p^x q^{1-x} \quad \text{for } x = 0 \text{ and } 1, \tag{3.22}$$

where $q := 1 - p$ is the probability of a **failure**.

Remark. A *parameter* is a constant (generally unknown, in practice) that appears in a probability mass function or a density function, and that can take any value in $\mathbb{R}$ or in a subset of $\mathbb{R}$, like $[0, 1]$, $(0, \infty)$, etc. We will see in Section 6.1 how to *estimate* unknown parameters.

Bernoulli trials. Suppose now that we repeat the random experiment E n times. We say that the repetitions $E_1, E_2, \ldots, E_n$ constitute a sequence of **Bernoulli trials** if

a) the repetitions are *independent*;

b) the probability of a success is *the same* for each repetition.

For example, suppose that a woman must cross n intersections with traffic lights when she goes to work. Let E_k be the random experiment that consists in observing if the woman must wait for the green light when she arrives at the kth intersection, for $k = 1, 2, \ldots, n$. Then the trials $E_1, \ldots, E_n$ are Bernoulli trials if and only if we may assume that the probability that the woman can cross a given intersection without having to stop is the same for each intersection and does not depend on what happened when she arrived at the previous intersections. In practice, unless the traffic lights are distant enough from one another, the assumption of independence of the trials is not (exactly) satisfied. Similarly, for the assumption b) to be satisfied, the length of the green light in the cycle of the traffic lights must be the same for each intersection, which is not true in general. However, to be able to estimate the probability that the woman has to stop at exactly five intersections in ten, for instance, it is almost necessary to make such simplifying assumptions.

Binomial distribution. Let X be the number of successes in n Bernoulli trials. We say that X has a **binomial distribution** with parameters n and p, where p is the probability of a success. We have: $S_X = \{0, 1, \ldots, n\}$. We write: $X \sim B(n, p)$.

Proposition 3.4.1. *The probability mass function of $X \sim B(n, p)$ is given by*

$$p_X(k) = \binom{n}{k} p^k q^{n-k} \quad \text{for } k = 0, 1, \ldots, n. \tag{3.23}$$

Proof. Let E be the random experiment that consists in observing, for each of the n Bernoulli trials, whether the event A occurred or not. Then, we have:

$$S = \{\overbrace{\underbrace{AA\ldots A}_{n \text{ times}}}^{X=n}, \ldots, \overbrace{\underbrace{A^c A^c \ldots A^c}_{n \text{ times}}}^{X=0}\}.$$

There are 2^n elementary events in all. Let s be one of the elementary events for which $X = k$. In particular, there is

$$s = \underbrace{AA\ldots A}_{k \text{ times}} \underbrace{A^c A^c \ldots A^c}_{(n-k) \text{ times}}.$$

By independence, we can write that $P[\{s\}] = p^k q^{n-k}$. Furthermore, there are $\binom{n}{k}$ different elementary events in S with exactly k successes and $(n-k)$ failures. Equation (3.23) then follows by equiprobability and incompatibility. $\qquad\square$

Properties. i) The function p_X defined in (3.23) is a proper probability mass function, because it is non-negative and

$$\sum_{k=0}^{n} p_X(k) = (p+q)^n = 1^n = 1,$$

by Newton's binomial theorem (see the example in Fig. 3.10).

ii) We can show that the function $p_X(k)$ reaches its maximum at $k_{max} = \text{int}[(n+1)p]$. Moreover, if $(n+1)p$ is an integer, then the maximum is also attained at $k_{max} - 1$.

iii) The distribution function of X, $F_X(x)$, is obtained by summing of the function $p_X(k)$, for k from 0 to $\text{int}(x)$:

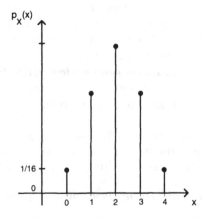

Figure 3.10. Probability mass function of a binomial random variable for which $n = 4$ and $p = 1/2$.

$$F_X(x) = \sum_{k=0}^{\text{int}(x)} p_X(k).$$

We can use a pocket calculator to evaluate this function. There are also tables of the functions p_X and F_X. Table 3.1 (p. 74) gives the value of the function $F_X(x)$ for a few values of $n \in \{2, 3, \ldots, 20\}$ and of $p \in [0.05, 0.50]$. Note that for values of p greater than 0.50, we can use the following relationship:

$$P[B(n, p) \le k] = 1 - P[B(n, 1 - p) \le n - k - 1] \quad \text{for } k \in \{0, 1, \ldots, n\}.$$

Remark. Suppose that n transistors are taken at random from a box containing m brand A and $N - m$ brand B transistors. Then, if the transistors are taken *with* replacement, the number X of brand A transistors obtained has a binomial distribution with parameters n and $p = m/N$. However, if the transistors are taken *without* replacement, then the trials are not independent. We can show that, in this case, we have:

$$p_X(k) = \frac{\binom{m}{k}\binom{N-m}{n-k}}{\binom{N}{n}} \quad \text{for } k = \max\{0, n - N + m\}, \ldots, \min\{n, m\}.$$

We say that X has a **hypergeometric distribution** with parameters N, n and m. We write: $X \sim \text{Hyp}(N, n, m)$.

Geometric distribution. Let X be the number of Bernoulli trials needed to obtain a first success. Then, $S_X = \{1, 2, \ldots\}$. We say that X has a **geometric distribution** with parameter p. We write: $X \sim \text{Geom}(p)$ (or $X \sim \text{Geo}(p)$).

Proposition 3.4.2. *Let $X \sim \text{Geom}(p)$. We have:*

$$p_X(k) = q^{k-1} p \quad \text{for } k = 1, 2, \ldots. \tag{3.24}$$

Proof. Let E be the random experiment in which the results of the Bernoulli trials are observed until event A occurs. We have:

$$S = \{ \overbrace{A}^{X=1}, \overbrace{A^c A}^{X=2}, \overbrace{A^c A^c A}^{X=3}, \ldots \}.$$

Then, we may write that

$$p_X(k) := P[X = k] = P[\{\underbrace{A^c A^c \ldots A^c}_{(k-1) \text{ times}} A\}] \overset{\text{ind.}}{=} q^{k-1} p.$$

Indeed, in order that $X = k$, there must first be $k - 1$ consecutive failures, and then a success. The above result is then obtained by independence. $\qquad\square$

Properties. i) The function p_X is non-negative and

Table 3.1. Distribution function of the binomial distribution.

		p					
n	x	0.05	0.10	0.20	0.25	0.40	0.50
2	0	0.9025	0.8100	0.6400	0.5625	0.3600	0.2500
	1	0.9975	0.9900	0.9600	0.9375	0.8400	0.7500
3	0	0.8574	0.7290	0.5120	0.4219	0.2160	0.1250
	1	0.9927	0.9720	0.8960	0.8438	0.6480	0.5000
	2	0.9999	0.9990	0.9920	0.9844	0.9360	0.8750
4	0	0.8145	0.6561	0.4096	0.3164	0.1296	0.0625
	1	0.9860	0.9477	0.8192	0.7383	0.4752	0.3125
	2	0.9995	0.9963	0.9728	0.9493	0.8208	0.6875
	3	1.0000	0.9999	0.9984	0.9961	0.9744	0.9375
5	0	0.7738	0.5905	0.3277	0.2373	0.0778	0.0313
	1	0.9774	0.9185	0.7373	0.6328	0.3370	0.1875
	2	0.9988	0.9914	0.9421	0.8965	0.6826	0.5000
	3	1.0000	0.9995	0.9933	0.9844	0.9130	0.8125
	4	1.0000	1.0000	0.9997	0.9990	0.9898	0.9688
10	0	0.5987	0.3487	0.1074	0.0563	0.0060	0.0010
	1	0.9139	0.7361	0.3758	0.2440	0.0464	0.0107
	2	0.9885	0.9298	0.6778	0.5256	0.1673	0.0547
	3	0.9990	0.9872	0.8791	0.7759	0.3823	0.1719
	4	0.9999	0.9984	0.9672	0.9219	0.6331	0.3770
	5	1.0000	0.9999	0.9936	0.9803	0.8338	0.6230
	6		1.0000	0.9991	0.9965	0.9452	0.8281
	7			0.9999	0.9996	0.9877	0.9453
	8			1.0000	1.0000	0.9983	0.9893
	9					0.9999	0.9990
15	0	0.4633	0.2059	0.0352	0.0134	0.0005	0.0000
	1	0.8290	0.5490	0.1671	0.0802	0.0052	0.0005
	2	0.9638	0.8159	0.3980	0.2361	0.0271	0.0037
	3	0.9945	0.9444	0.6482	0.4613	0.0905	0.0176
	4	0.9994	0.9873	0.8358	0.6865	0.2173	0.0592
	5	0.9999	0.9977	0.9389	0.8516	0.4032	0.1509
	6	1.0000	0.9997	0.9819	0.9434	0.6098	0.3036
	7		1.0000	0.9958	0.9827	0.7869	0.5000
	8			0.9992	0.9958	0.9050	0.6964
	9			0.9999	0.9992	0.9662	0.8491
	10			1.0000	0.9999	0.9907	0.9408
	11				1.0000	0.9981	0.9824
	12					0.9997	0.9963
	13					1.0000	0.9995
	14						1.0000

Table 3.1. Continued.

n	x	0.05	0.10	0.20	0.25	0.40	0.50
20	0	0.3585	0.1216	0.0115	0.0032	0.0000	
	1	0.7358	0.3917	0.0692	0.0243	0.0005	0.0000
	2	0.9245	0.6769	0.2061	0.0913	0.0036	0.0002
	3	0.9841	0.8670	0.4114	0.2252	0.0160	0.0013
	4	0.9974	0.9568	0.6296	0.4148	0.0510	0.0059
	5	0.9997	0.9887	0.8042	0.6172	0.1256	0.0207
	6	1.0000	0.9976	0.9133	0.7858	0.2500	0.0577
	7		0.9996	0.9679	0.8982	0.4159	0.1316
	8		0.9999	0.9900	0.9591	0.5956	0.2517
	9		1.0000	0.9974	0.9861	0.7553	0.4119
	10			0.9994	0.9961	0.8725	0.5881
	11			0.9999	0.9991	0.9435	0.7483
	12			1.0000	0.9998	0.9790	0.8684
	13				1.0000	0.9935	0.9423
	14					0.9984	0.9793
	15					0.9997	0.9941
	16					1.0000	0.9987
	17						0.9998
	18						1.0000

$$\sum_{k=1}^{\infty} p_X(k) = p \sum_{k=1}^{\infty} q^{k-1} = p\{1 + q + q^2 + \cdots\} \overset{q<1}{=} \frac{p}{1-q} = 1.$$

Hence, p_X is a proper probability mass function (see Fig. 3.11).

ii) The distribution function of X is given, for any integer n, by

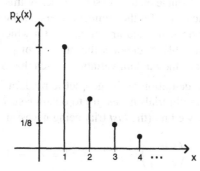

Figure 3.11. Probability mass function of a geometric random variable for which $p = 1/2$.

$$F_X(n) = \sum_{k=1}^{n} p_X(k) = p\{1 + q + \cdots + q^{n-1}\} = p\frac{1 - q^n}{1 - q} = 1 - q^n. \quad (3.25)$$

It follows that $P[X > n] = q^n$ for $n = 1, 2, \ldots$.

Proposition 3.4.3. *Let X be a geometric random variable. Then, it possesses the following* **memoryless** *property:*

$$P[X > k + j \mid X > j] = P[X > k] \quad \forall k, j \in \{1, 2, \ldots\}. \quad (3.26)$$

Proof. We have:

$$P[X > k + j \mid X > j] = \frac{P[\{X > k + j\} \cap \{X > j\}]}{P[X > j]}$$

$$= \frac{P[X > k + j]}{P[X > j]} = \frac{q^{k+j}}{q^j} = q^k = P[X > k]. \quad \square$$

Remark. We also say that the geometric distribution has the "no aging" or "lack of aging" property. In fact the geometric distribution is the only discrete distribution that possesses this property.

Remarks. i) The geometric distribution is sometimes defined as being the number X of Bernoulli trials *before* the first success. We then have: $S_X = \{0, 1, \ldots\}$ and the function p_X becomes

$$p_X(k) = q^k p \quad \text{for } k = 0, 1, \ldots.$$

If we denote this variable by Y, then we can write that $Y = X - 1$.

The choice of the random variable used when we propose a stochastic model for a given variable depends of course on the possible values of this variable. For example, if we look for a stochastic model for the number of persons *involved* in a car accident, then it is preferable to choose the random variable X for which $S_X = \{1, 2, \ldots\}$. On the other hand, if the variable X denotes the number of persons *injured* in a car accident, we should choose the random variable X such that $S_X = \{0, 1, \ldots\}$.

ii) We can generalize the definition of the geometric random variable as follows: let X be the number of Bernoulli trials necessary to obtain r successes. Then, we have: $S_X = \{r, r + 1, \ldots\}$ and we find (the *last* trial being necessarily a *success*) that

$$p_X(k) = \binom{k-1}{r-1} p^r q^{k-r} \quad \text{for } k = r, r + 1, \ldots. \quad (3.27)$$

We say that X has a **Pascal distribution** (see p. 2), or a **negative binomial distribution**, with parameters r and p. We write: $X \sim NB(r, p)$.

Poisson distribution[2]

Poisson's theorem. *Let $X \sim B(n, p)$. Suppose that $n \to \infty$ and $p \downarrow 0$ in such a way that $n \times p$ remains equal to α. Then, we find that*

$$p_X(k) \to \frac{e^{-\alpha}\alpha^k}{k!} \quad \text{for } k = 0, 1, \ldots . \tag{3.28}$$

Proof. Let $p_k := p_X(k)$. For $k = 0$, we have:

$$p_0 = q^n = (1 - p)^n = \left(1 - \frac{\alpha}{n}\right)^n \to e^{-\alpha} \quad \text{as } n \to \infty.$$

Next, we consider the ratio

$$\frac{p_{k+1}}{p_k} = \frac{\binom{n}{k+1}p^{k+1}q^{n-k-1}}{\binom{n}{k}p^k q^{n-k}} = \frac{(n-k)p}{(k+1)q} = \frac{\alpha}{k+1}\left[\left(\frac{n-k}{n}\right)\left(1 - \frac{\alpha}{n}\right)^{-1}\right].$$

Since the two factors inside the square brackets tend to 1 as n tends to infinity, we can write that

$$\lim_{n \to \infty} \frac{p_{k+1}}{p_k} = \frac{\alpha}{k+1}. \tag{3.29}$$

Using this result recurrently, we obtain Equation (3.28). Indeed, we have:

$$\lim_{n \to \infty} \frac{p_1}{p_0} = \frac{\alpha}{0+1} \Rightarrow p_X(1) \stackrel{n \to \infty}{\longrightarrow} \alpha e^{-\alpha},$$

so that

$$\lim_{n \to \infty} p_2 = \frac{\alpha}{1+1} \lim_{n \to \infty} p_1 = \frac{\alpha^2}{2}e^{-\alpha},$$

etc. $\qquad\qquad\qquad\qquad\qquad\qquad\qquad\qquad\qquad\qquad\qquad\qquad\qquad\square$

Definition 3.4.1. *If X is a discrete random variable with $S_X = \{0, 1, \ldots\}$ and for which*

$$p_X(k) = \frac{e^{-\alpha}\alpha^k}{k!} \quad \text{for } k = 0, 1, \ldots, \tag{3.30}$$

*then we say that X has a **Poisson distribution** with parameter $\alpha > 0$. We write: $X \sim Poi(\alpha)$.*

---

[2] Siméon Denis Poisson, 1781–1840, was born and died in France. He first studied medicine and, from 1798, mathematics at the École Polytechnique de Paris, where he taught from 1802 to 1808. His professors at the École Polytechnique were, among others, Laplace and Lagrange. In mathematics, his main results were his papers on definite integrals and Fourier series. The Poisson distribution appeared in his important book on probability theory published in 1837. He also published works on mechanics, electricity, magnetism and astronomy. His name is associated with various mathematical results.

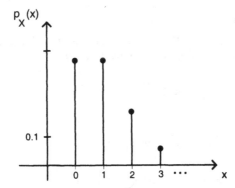

Figure 3.12. Probability mass function of a Poisson random variable for which $\alpha = 1$.

Properties. i) The function p_X above satisfies the two conditions required to be a proper probability mass function: it is non-negative and

$$\sum_{k=0}^{\infty} p_X(k) = e^{-\alpha} \sum_{k=0}^{\infty} \frac{\alpha^k}{k!} = e^{-\alpha} e^{\alpha} = 1$$

(see Fig. 3.12).

ii) As in the case of the binomial random variables, we can use a pocket calculator or a table to evaluate the distribution function or the probability mass function of X. Table 3.2 (p. 79) gives the value of the function $F_X(x)$ for a few values of the parameter α.

iii) We can show that $p_X(x)$ attains its maximum at $x = 0$ if $\alpha < 1$ and at $x = \text{int}(\alpha)$ if $\alpha > 1$. Finally, if α is an integer, then the maximum is attained at both $x = \alpha$ and $x = \alpha - 1$.

Poisson's approximation to the binomial distribution. We deduce from Poisson's theorem that if n is sufficiently *large* (> 20) and p *small* enough (< 0.05), then we can approximate the distribution of a binomial random variable with parameters n and p by that of a Poisson random variable with parameter $\alpha = np$. The approximation is generally better for the function F_X than for p_X (see Example 3.4.1).

Remark. Before proceeding with Poisson's approximation, we must check whether the parameter p of the binomial distribution is smaller than or equal to $1/2$. For example, suppose that $X \sim B(n = 20, p = 0.9)$. We have:

$$P[X = 15] \simeq 0.0319.$$

Now, if we replace X by a random variable Y that has a Poisson distribution with parameter $\alpha = 20 \times 0.9 = 18$, we obtain:

$$P[X = 15] \simeq P[Y = 15] \simeq 0.0786,$$

Table 3.2. Distribution function of the Poisson distribution.

x	α							
	0.5	1	1.5	2	5	10	15	20
0	0.6065	0.3679	0.2231	0.1353	0.0067	0.0000		
1	0.9098	0.7358	0.5578	0.4060	0.0404	0.0005		
2	0.9856	0.9197	0.8088	0.6767	0.1247	0.0028	0.0000	
3	0.9982	0.9810	0.9344	0.8571	0.2650	0.0103	0.0002	
4	0.9998	0.9963	0.9814	0.9473	0.4405	0.0293	0.0009	0.0000
5	1.0000	0.9994	0.9955	0.9834	0.6160	0.0671	0.0028	0.0001
6		0.9999	0.9991	0.9955	0.7622	0.1301	0.0076	0.0003
7		1.0000	0.9998	0.9989	0.8666	0.2202	0.0180	0.0008
8			1.0000	0.9998	0.9319	0.3328	0.0374	0.0021
9				1.0000	0.9682	0.4579	0.0699	0.0050
10					0.9863	0.5830	0.1185	0.0108
11					0.9945	0.6968	0.1848	0.0214
12					0.9980	0.7916	0.2676	0.0390
13					0.9993	0.8645	0.3632	0.0661
14					0.9998	0.9165	0.4657	0.1049
15					0.9999	0.9513	0.5681	0.1565
16					1.0000	0.9730	0.6641	0.2211
17						0.9857	0.7489	0.2970
18						0.9928	0.8195	0.3814
19						0.9965	0.8752	0.4703
20						0.9984	0.9170	0.5591
21						0.9993	0.9469	0.6437
22						0.9997	0.9673	0.7206
23						0.9999	0.9805	0.7875
24						1.0000	0.9888	0.8432
25							0.9938	0.8878
26							0.9967	0.9221
27							0.9983	0.9475
28							0.9991	0.9657
29							0.9996	0.9782
30							0.9998	0.9865
31							0.9999	0.9919
32							1.0000	0.9953

which is not a good approximation. On the other hand, let X^* be the number of *failures* among the 20 Bernoulli trials. In this case, we can approximate the distribution of X^* by that of $Z \sim \text{Poi}(\alpha = 20 \times 0.1 = 2)$ and write that

$$P[X = 15] = P[X^* = 5] \simeq P[Z = 5] \simeq 0.0361,$$

which is a much better approximation. Similarly, we have:

$$P[X \leq 15] = P[X^* \geq 5] \simeq 0.0432$$

and $P[Z \geq 5] \simeq 0.0527$, while $P[Y \leq 15] \simeq 0.2867$.

Example 3.4.1 Suppose that the number X of errors in a computer program submitted for the first time to a mainframe computer, by each student in a certain group, is a random variable that has a Poisson distribution with parameter $\alpha = 3$. Suppose also that all the computer programs submitted are independent of one another.

a) What is the probability that a given program contains no errors?

b) What is the probability that, among 20 programs submitted, there are at least three that contain no errors?

c) Calculate approximately the probability in b) with the help of a Poisson distribution.

d) What is the probability that the mainframe computer receives less than five programs, from a given time instant, before it receives a first program that contains no errors?

Solution. a) We seek $P[X = 0] = e^{-3} \simeq 0.05$.

Remark. The Poisson distribution as a model for X is actually an approximation, since the number of errors in a program cannot tend to infinity.

b) Let Y be the number of programs, among the 20 submitted, that contain no errors. Then Y has a binomial distribution with parameters $n = 20$ and $p = P[X = 0] \overset{a)}{\simeq}$ 0.05. We want

$$P[Y \geq 3] = 1 - P[Y \leq 2] \simeq 1 - \sum_{k=0}^{2} \binom{20}{k}(0.05)^k (0.95)^{20-k}$$
$$= 1 - [(0.95)^{20} + 20(0.05)(0.95)^{19} + 190(0.05)^2(0.95)^{18}]$$
$$\simeq 1 - (0.3585 + 0.3774 + 0.1887)$$
$$= 1 - 0.9246 = 0.0754.$$

c) We can write that

$$P[Y \geq 3] \simeq P[W \geq 3], \quad \text{where } W \sim \text{Poi}(\alpha = np \simeq 1)$$
$$= 1 - \sum_{k=0}^{2} \frac{e^{-1} 1^k}{k!} = 1 - e^{-1}\left(1 + 1 + \frac{1}{2}\right)$$
$$\simeq 1 - (0.3679 + 0.3679 + 0.1839)$$
$$= 1 - 0.9197 = 0.0803.$$

Remark. If we replace p by 0.01, then we find that $P[Y \geq 3] \simeq 0.0010$ and $P[W \geq 3] \simeq 0.0011$. On the other hand, here if $p = 0.1$, we obtain that $P[Y \geq 3] \simeq 0.3231$ and $P[W \geq 3] \simeq 0.3233$, which is even better (in error percentage). In general, for an n fixed, the approximation should improve when the parameter p decreases. In fact, if we calculate $P[Y = 0]$ and the approximations obtained with a Poisson distribution for $n = 20$ and $p = 1/100$, 1/20 and 1/10, we obtain the following pairs:

$$(0.8179, 0.8187), \quad (0.3585, 0.3679), \quad (0.1216, 0.1353).$$

We observe that the approximation deteriorates as p increases, the error percentages being equal to about 0.1%, 2.6% and 11.3%, respectively. But when we add many terms, the errors can cancel one another. For example, comparing the numerical values of the corresponding terms in b) and c), we notice that the probability $P[W = k]$ sometimes overestimates and sometimes underestimates the probability $P[Y = k]$, so that the errors partly cancel. That is why it is generally easier to approximate the distribution function F_X than the probability mass function p_X, as mentioned above.

d) Let U be the number of programs required to obtain a first program that contains no errors. We assume that there are at least five programs remaining to be received from the time instant considered. We seek

$$P[U < 6] = P[V < 6], \quad \text{where } V \sim \text{Geom}(p = P[X = 0] \simeq 0.05)$$
$$= P[V \leq 5] \simeq 1 - (0.95)^5 \simeq 0.2262.$$

Remark. The random variable U itself does not have a geometric distribution, since the group of students considered is not infinite. In fact, U is a *defective* random variable, because, if there are 20 students in the group (and if nobody has submitted his/her program yet), we can write that

$$\sum_{k=1}^{\infty} P[U = k] = \sum_{k=1}^{20} P[U = k] + 0 = \sum_{k=1}^{20} P[V = k]$$
$$\simeq 1 - (0.95)^{20} \simeq 0.6415 < 1.$$

Example 3.4.2 In a certain lottery, six balls are picked at random and without replacement among 49 balls numbered from 1 to 49. We win a prize if the combination that we have chosen contains at least three "good" numbers. A player decides to buy one ticket per drawing until he wins a prize. What is the probability that he must buy less than ten tickets?

Solution. Let M be the number of good numbers in the chosen combination, and let X be the number of tickets that the player will have to buy to win a prize for the first time. We have:

$$P[M \geq 3] = 1 - P[M \leq 2]$$
$$= 1 - \sum_{k=0}^{2} \frac{\binom{6}{k}\binom{43}{6-k}}{\binom{49}{6}}$$
$$= 1 - \frac{(1 \cdot 6, 096, 454) + (6 \cdot 962, 598) + (15 \cdot 123, 410)}{13, 983, 816}$$
$$= 1 - \frac{13, 723, 192}{13, 983, 816} \simeq 1 - 0.9814 = 0.0186.$$

Next, by independence of the drawings, we may write that X has a geometric distribution with $p \simeq 0.0186$. We want

$$P[X < 10] = P[X \leq 9] \simeq 1 - (1 - 0.0186)^9 \simeq 0.1555.$$

Remark. The random variable M has in fact a *hypergeometric* distribution with parameters $N = 49, n = 6$ and $m = 6$. If the *sampling fraction* n/N is small (< 0.1), then we can approximate the distribution of a hypergeometric random variable with parameters N, n and m by that of a binomial random variable with parameters n and $p = m/N$. In our case, we have: $n/N = 6/49 > 0.1$. The approximation to the probability $P[M \geq 3]$ would give

$$P[M \geq 3] \simeq P\left[B\left(6, \tfrac{6}{49}\right) \geq 3\right] \simeq 1 - 0.9724 = 0.0276.$$

3.5 Important Continuous Random Variables

Uniform distribution. Suppose that a number X is taken at random in the interval $[a, b]$. In this case, the probability that X is in an interval of length ϵ (small enough) around a point x, where $a < x < b$, must be the same for any x. It follows that the probability density function of X must be a constant. Using the condition

$$\int_a^b f_X(x)\, dx = 1,$$

we deduce that we must have:

$$f_X(x) = \frac{1}{b-a} \quad \text{for } a \leq x \leq b. \tag{3.31}$$

Definition 3.5.1. *Let X be a continuous random variable defined on the interval $[a, b]$ and whose probability density function is given by Equation* (3.31). *We say that X has a* **uniform distribution** *on the interval $[a, b]$ and we write that $X \sim U[a, b]$ (see Fig. 3.13, p. 83).*

Properties. i) The distribution function of X is given by (see Fig. 3.14, p. 83)

$$F_X(x) = \begin{cases} 0 & \text{if } x < a, \\ \displaystyle\int_a^x \frac{1}{b-a}\, dt = \frac{x-a}{b-a} & \text{if } a \leq x \leq b, \\ 1 & \text{if } x > b. \end{cases} \tag{3.32}$$

ii) Let $[c, d] \subset [a, b]$. Then, we have:

$$P[c < X \leq d] = \frac{d-c}{b-a}.$$

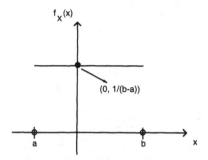

Figure 3.13. Probability density function of a random variable having a uniform distribution on the interval $[a, b]$.

Thus, the probability that X is in a given subinterval depends only on the length of this subinterval.

Example 3.5.1 Suppose that the number of power failures in the interval $[0, t]$, in a certain region, has a Poisson distribution with parameter λt, where $\lambda > 0$. Then, given that exactly one power failure occurred in the interval $[a, b] \subset [0, t]$, we can show that the moment at which this failure occurred is uniformly distributed on the interval $[a, b]$. We can also show that the waiting time between two power failures has an exponential distribution with parameter λ (see Definition 3.5.2).

Exponential distribution

Definition 3.5.2. *Let X be a continuous random variable defined on the interval $[0, \infty)$. If the density function of X is of the form*

$$f_X(x) = \lambda e^{-\lambda x} \quad for \; x \geq 0, \tag{3.33}$$

then we say that X has an **exponential distribution** *with parameter $\lambda > 0$. We write: $X \sim \text{Exp}(\lambda)$ (see Fig. 3.15, p. 84).*

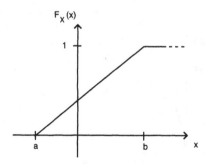

Figure 3.14. Distribution function of a random variable uniformly distributed on the interval $[a, b]$.

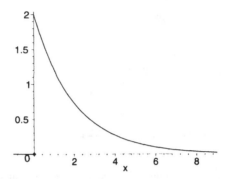

Figure 3.15. Density function of an exponential random variable with parameter $\lambda = 2$.

Properties. i) The function f_X above is a proper probability density function, since it is non-negative and

$$\int_0^\infty \lambda e^{-\lambda x}\, dx = -e^{-\lambda x}\big|_0^\infty = 1.$$

ii) The distribution function of X is given by

$$F_X(x) = \int_0^x \lambda e^{-\lambda t}\, dt = 1 - e^{-\lambda x} \quad \text{for } x \geq 0 \tag{3.34}$$

and $F_X(x) = 0$ for $x < 0$ (see Fig. 3.16). We then obtain that

$$P[X > x] = e^{-\lambda x} \quad \text{for } x \geq 0. \tag{3.35}$$

iii) Let $Y := \text{int}(X)+1$. Then, $S_Y = \{1, 2, \ldots\}$ and we can show (see Section 3.6) that $Y \sim \text{Geom}(p = 1 - e^{-\lambda})$.

iv) The exponential random variables are the only *continuous* random variables that possess the following *memoryless property*:

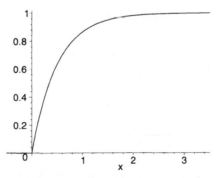

Figure 3.16. Distribution function of an exponential random variable with parameter $\lambda = 2$.

$$P[X > t + s \mid X > t] = P[X > s] \quad \forall s, t > 0. \tag{3.36}$$

The demonstration of this property of the exponential distribution is similar to that of Proposition 3.4.3, p. 76, for the geometric distribution. This property is used in reliability and in queueing theory.

v) The exponential distribution can be considered on the entire real line by defining

$$f_X(x) = \frac{\lambda}{2} e^{-\lambda |x|} \quad \text{for } x \in \mathbb{R},$$

where λ is a positive parameter. This random variable, called the **double exponential distribution**, is also known as the **Laplace distribution**.[3]

Example 3.5.2 Suppose that the lifetime X (in years) of a machine has an exponential distribution with parameter $\lambda = 1/3$. What is the probability that a three-year-old machine will still work at the end of three additional years?

Solution. We want

$$P[X > 6 \mid X > 3] = P[X > 6 - 3] = e^{-1} \simeq 0.3679.$$

Note that the answer would be the same if the machine were brand-new, or six years old, etc. If we apply this property to the lifetime of a car, for instance, it is clear that the assumption that we make is not entirely realistic. Indeed, the probability that a new car lasts at least three years is surely larger than the probability that a ten-year-old car lasts at least three additional years. However, the assumption of an exponential lifetime can be acceptable for a shorter period of time. For example, the probability that a one-year-old car lasts at least two additional years should be almost equal to the probability that a brand-new car lasts at least two years.

Gamma distribution

Definition 3.5.3. *The* **gamma function***, denoted by* $\Gamma(\cdot)$*, is defined (for* $\alpha > 0$*) by*

$$\Gamma(\alpha) = \int_0^\infty x^{\alpha-1} e^{-x} \, dx. \tag{3.37}$$

Doing the above integral by parts, we find that

$$\Gamma(\alpha) = (\alpha - 1)\Gamma(\alpha - 1) \quad \text{if } \alpha > 1. \tag{3.38}$$

---

[3] Pierre Simon (Marquis de) Laplace, 1749–1827, was born and died in France. In addition to being a mathematician and an astronomer, he was also a minister and a senator. He was made count by Napoléon and marquis by Louis XVIII. He participated in the organization of the École Polytechnique de Paris. His main works were on astronomy and on the calculus of probabilities: the *Traité de mécanique céleste*, published in five volumes, from 1799, and the *Théorie analytique des probabilités*, the first edition of which appeared in 1812. Many mathematical formulas bear his name.

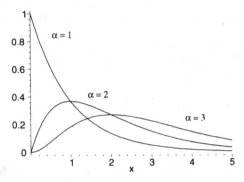

Figure 3.17. Probability density functions of various random variables having a gamma distribution with $\lambda = 1$.

It follows that if $\alpha = n \in \{2, 3, \dots\}$, then

$$\Gamma(\alpha) = \Gamma(n) = (n-1)\Gamma(n-1) = (n-1)(n-2)\Gamma(n-2)$$
$$= \cdots = (n-1)(n-2)\cdots 1 \cdot \Gamma(1) = (n-1)! \cdot 1 = (n-1)!.$$

In addition, we can show that $\Gamma(1/2) = \sqrt{\pi}$.

Definition 3.5.4. *Let X be a continuous non-negative random variable. If its probability density function is of the form*

$$f_X(x) = \frac{(\lambda x)^{\alpha-1}\lambda e^{-\lambda x}}{\Gamma(\alpha)} \quad for\ x \geq 0, \tag{3.39}$$

then we say that X has a **gamma distribution** *with parameters $\alpha > 0$ and $\lambda > 0$. We write: $X \sim G(\alpha, \lambda)$.*

Remarks. i) The parameter α is a *shape* parameter, while λ is a *scale* parameter (see Fig. 3.17). Because the shape of the density function f_X changes a lot when the parameter α takes on various values, the gamma distribution is a very useful model for the applications.

ii) The gamma distribution generalizes the exponential distribution since if we take $\alpha = 1$ in the function f_X, we obtain: $f_X(x) = \lambda e^{-\lambda x}$ for $x \geq 0$. The gamma distribution is also used in reliability and it appears in the *stochastic process* known as the Poisson process.

iii) If $\alpha = n/2$ and $\lambda = 1/2$, the gamma distribution is also called the **chi-square distribution** with n **degrees of freedom**, where n is a positive integer. This distribution is used in statistics, particularly in Pearson's *goodness-of-fit test*, which will be seen in Chapter 6.

iv) Finally, if α is a positive integer, the gamma distribution is sometimes called the **Erlang distribution**,[4] which is very important in the theory of queues.

---

[4] Agner Krarup Erlang, 1878–1929, was born and died in Denmark. He was first educated by his father, who was a schoolmaster. He studied mathematics and natural sciences at the

Properties. i) Since $\Gamma(\alpha)$ is positive (for $\alpha > 0$), the function f_X is non-negative. Moreover, we have:

$$\int_0^\infty f_X(x)\,dx = \frac{\lambda^\alpha}{\Gamma(\alpha)} \int_0^\infty x^{\alpha-1} e^{-\lambda x}\,dx$$

$$\overset{y=\lambda x}{=} \frac{1}{\Gamma(\alpha)} \int_0^\infty y^{\alpha-1} e^{-y}\,dy = \frac{\Gamma(\alpha)}{\Gamma(\alpha)} = 1.$$

Thus, f_X is a proper probability density function.

ii) There is no simple formula that gives the distribution function of X in the general case. However, if α is an integer, then we can show, integrating by parts repeatedly, that

$$F_X(x) = 1 - \sum_{k=0}^{\alpha-1} e^{-\lambda x} \frac{(\lambda x)^k}{k!} \quad \text{for } x \geq 0. \tag{3.40}$$

That is, we have:

$$P[X \leq x] = 1 - P[Y \leq \alpha - 1] = P[Y \geq \alpha], \quad \text{where } Y \sim \text{Poi}(\lambda x). \tag{3.41}$$

Remark. We can also write that $P[X < x] = P[Y \geq \alpha]$, because X is a *continuous* random variable. However, since Y is a *discrete* random variable, we have: $P[X \leq x] \neq P[Y > \alpha]$. Note, additionally, that Equation (3.41) is *exact*, and not only an approximation. We will see in Chapter 5 that the time required for n events of a Poisson process to occur has a gamma distribution, whereas the number of events that occur in a given interval has a Poisson distribution, whence the above relationship between the two distributions.

Example 3.5.3 Suppose that the daily consumption of electricity (in millions of kilowatt-hours) in a certain region is a random variable X having a gamma distribution with parameters $\alpha = 3$ and $\lambda = 1/2$. Suppose also that the production capacity is equal to 12 millions of kilowatt-hours. Calculate the probability that the demand exceeds the capacity on a given day.

Solution. We can write that

$$P[X > 12] = P[Y < 3], \quad \text{where } Y \sim \text{Poi}(12/2 = 6)$$

$$= e^{-6}\left(1 + 6 + \frac{6^2}{2!}\right) \simeq 0.0620.$$

If we make the assumption (which is probably not correct in practice) that the probability that the demand exceeds the capacity on a given day does not depend

University of Copenhagen and taught in schools for several years. After meeting the chief engineer for the Copenhagen telephone company, he joined this company in 1908. He then started to apply his knowledge of probability theory to the resolution of problems related to telephone calls. He was also interested in mathematical tables.

on what happened previously, that is, the assumption that the days are independent, then the probability that the demand exceeds the capacity on two consecutive days is given by approximately $(0.0620)^2 \simeq 0.0038$. Finally, let N be the number of days in a given week on which the demand exceeds the production capacity. Under the previous assumptions, we can write that N has a binomial distribution with parameters $n = 7$ and $p \simeq 0.0620$.

Remark. Another continuous random variable that generalizes the exponential distribution and that is useful in reliability is the *Weibull distribution*[5] with parameters $\beta > 0, \gamma \in \mathbb{R}$ and $\delta > 0$, whose probability density function is

$$f_X(x) = \frac{\beta}{\delta}\left(\frac{x - \gamma}{\delta}\right)^{\beta-1} \exp\left[-\left(\frac{x - \gamma}{\delta}\right)^{\beta}\right] \quad \text{for } x \geq \gamma. \tag{3.42}$$

Note that the exponential distribution is the particular case when $\beta = 1, \gamma = 0$ and $\delta = 1/\lambda$. The parameter β is a *shape* parameter, γ is a *position* parameter and δ is a *scale* parameter.

Gaussian distribution

Definition 3.5.5. *Let X be a random variable whose set of possible values is $S_X = (-\infty, \infty)$. If the probability density function of X is of the form*

$$f_X(x) = \frac{1}{\sqrt{2\pi}\sigma} \exp\left\{-\frac{(x - \mu)^2}{2\sigma^2}\right\} \quad \text{for } -\infty < x < \infty, \tag{3.43}$$

then we say that X has a **normal** *or* **Gaussian distribution** *with parameters μ and σ^2, where $\sigma > 0$. We write: $X \sim N(\mu, \sigma^2)$.*

Remark. This probability distribution is also sometimes called the *Laplace–Gauss distribution*,[6] or the *Laplace distribution* (name used for the double exponential distribution as well; see p. 85).

Properties. i) Since σ is positive, it is clear that the function f_X is non-negative. To show that its integral on the entire real line is equal to 1, we consider I^2, where

---

[5] E. H. Wallodi Weibull, 1887–1979, was born in Sweden and died in France. His family originated from Germany. He started his career in the Swedish Coast Guard. After completing his doctorate, he worked as an inventor and consulting engineer for many companies in Sweden and in Germany. He published many papers on the strength of materials, fatigue and reliability, in addition to his papers, from 1939, on the distribution that bears his name. The Weibull distribution is used in many applications.

[6] Carl Friedrich Gauss, 1777–1855, was born and died in Germany. Considered as one of the greatest geniuses of all time, he carried out many works on astronomy and physics, in addition to his important mathematical discoveries. In the field of mathematics, he was interested, in particular, in algebra and geometry. He also discovered the *method of least squares*, which he used to make astronomical predictions. It is in fact for this method that he introduced the *law of errors*, that now bears his name, as a model for the errors in astronomical observations.

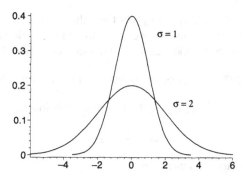

Figure 3.18. Probability density functions of Gaussian random variables with $\mu = 0$.

$$I := \frac{1}{\sqrt{2\pi}\sigma} \int_{-\infty}^{\infty} \exp\left\{ -\frac{(x-\mu)^2}{2\sigma^2} \right\} dx = \frac{1}{\sqrt{2\pi}} \int_{-\infty}^{\infty} \exp(-t^2/2)\, dt,$$

by making the change of variable $t = (x - \mu)/\sigma$. Then, using polar coordinates, we can write that

$$I^2 = \frac{1}{2\pi} \int_{-\infty}^{\infty}\int_{-\infty}^{\infty} \exp[-(t^2+s^2)/2]\, dt\, ds = \frac{1}{2\pi} \int_{0}^{\infty}\int_{0}^{2\pi} r\exp(-r^2/2)\, d\theta\, dr.$$

We easily find that $I^2 = 1$. Then, knowing that $I \geq 0$ (because $f_X(x) \geq 0$), we deduce that $I = 1$.

ii) The parameter μ is a *position* parameter, while σ is a *scale* parameter. Furthermore, all Gaussian distributions have the same general shape, namely that of a *bell* (see Fig. 3.18).

iii) The function f_X is symmetrical with respect to μ; that is,

$$f_X(x + \mu) = f_X(-x + \mu).$$

Moreover, $f_X(x)$ attains its maximum at $x = \mu$ and has two inflection points: at $x = \mu \pm \sigma$.

iv) If $\mu = 0$ and $\sigma = 1$, then we say that X has a *standard* Gaussian distribution. Its probability density function is given by

$$\phi(z) := \frac{1}{\sqrt{2\pi}} \exp(-z^2/2) \quad \text{for } -\infty < z < \infty \tag{3.44}$$

and its distribution function is denoted by Φ.

Now, if $X \sim N(\mu, \sigma^2)$, we find that its distribution function can be expressed in terms of Φ as follows:

$$F_X(x) = \Phi\left(\frac{x-\mu}{\sigma}\right). \tag{3.45}$$

Thus, any probability involving a Gaussian distribution can be obtained by using the $N(0, 1)$ distribution. To evaluate the function Φ, we can use a statistical software package or a table. Table 3.3 (p. 91) gives the value of the function $\Phi(z)$ for $z = 0.00\ldots(0.01)\ldots3.99$ (by symmetry, we have: $\Phi(-z) = 1 - \Phi(z)$). From $z = 3.90$, we can write that $\Phi(z) \simeq 1.0000$. We find, with more precision, that

$$\Phi(4) \simeq 0.99997, \quad \Phi(5) \simeq 0.9999997 \quad \text{and} \quad \Phi(6) \simeq 0.999999999.$$

Alternatively, we can make use of the following relationship:

$$\Phi(z) = 1 - Q(z), \tag{3.46}$$

where $Q(z)$ is given, approximately, by

$$Q(z) \simeq \left[\frac{\pi}{(\pi - 1)z + (z^2 + 2\pi)^{1/2}} \right] \frac{1}{\sqrt{2\pi}} \exp(-z^2/2) \quad \text{for } z > 0. \tag{3.47}$$

Moreover, we have: $Q(0) = 1/2$ and $Q(-z) = 1 - Q(z)$. Finally, we also find tables of the function $Q^{-1}(p)$ (the *inverse* function of Q) for $0 < p < 1$ (see Table 3.4, p. 92). For instance, $Q^{-1}(0.05) \simeq 1.645$.

Remark. The Gaussian distribution is often used as a model in situations in which we know, for example, that the random variable X that we consider cannot be negative. This model can nevertheless be a good approximation to the true (generally unknown) model, as long as μ and σ are such that $F_X(x) \simeq 0$ if $x < 0$. In fact, the Gaussian distribution is useful mainly because of the *central limit theorem*, which will be seen in Chapter 4.

Example 3.5.4 Suppose that the lifetime X of a microprocessor has a Gaussian distribution with parameters μ and σ^2. Calculate the probability $P[\mu - k\sigma \leq X \leq \mu + k\sigma]$, where $k \in \{1, 2, \ldots\}$.

Solution. We have:

$$P[\mu - k\sigma \leq X \leq \mu + k\sigma]$$
$$\stackrel{\text{cont.}}{=} F_X(\mu + k\sigma) - F_X(\mu - k\sigma) = \Phi(k) - \Phi(-k) = 2\Phi(k) - 1,$$

which is independent of μ and σ. For example, if $k = 1$, we find that $\Phi(1) \simeq 0.8413$, so that the probability that we seek is approximately equal to 0.683.

Remark. If X is a continuous random variable such that $S_X = (0, \infty)$, and if $Y := \ln X$ has a normal distribution with parameters μ and σ^2, then we say that X has a **lognormal distribution** (with parameters μ and σ^2). We can show (see Theorem 3.6.1, p. 94) that the probability density function of X is given by

$$f_X(x) = \frac{1}{\sqrt{2\pi}\sigma x} \exp\left\{ -\frac{(\ln x - \mu)^2}{2\sigma^2} \right\} \quad \text{for } x > 0. \tag{3.48}$$

This random variable is important in many fields, notably in reliability.

Table 3.3. Values of the function $\Phi(z)$.

z	+0.00	+0.01	+0.02	+0.03	+0.04	+0.05	+0.06	+0.07	+0.08	+0.09
0.0	0.5000	0.5040	0.5080	0.5120	0.5160	0.5199	0.5239	0.5279	0.5319	0.5359
0.1	0.5398	0.5438	0.5478	0.5517	0.5557	0.5596	0.5636	0.5675	0.5714	0.5753
0.2	0.5793	0.5832	0.5871	0.5910	0.5948	0.5987	0.6026	0.6064	0.6103	0.6141
0.3	0.6179	0.6217	0.6255	0.6293	0.6331	0.6368	0.6406	0.6443	0.6480	0.6517
0.4	0.6554	0.6591	0.6628	0.6664	0.6700	0.6736	0.6772	0.6808	0.6844	0.6879
0.5	0.6915	0.6950	0.6985	0.7019	0.7054	0.7088	0.7123	0.7157	0.7190	0.7224
0.6	0.7257	0.7291	0.7324	0.7357	0.7389	0.7422	0.7454	0.7486	0.7517	0.7549
0.7	0.7580	0.7611	0.7642	0.7673	0.7704	0.7734	0.7764	0.7794	0.7823	0.7852
0.8	0.7881	0.7910	0.7939	0.7967	0.7995	0.8023	0.8051	0.8078	0.8106	0.8133
0.9	0.8159	0.8186	0.8212	0.8238	0.8264	0.8289	0.8315	0.8340	0.8365	0.8389
1.0	0.8413	0.8438	0.8461	0.8485	0.8508	0.8531	0.8554	0.8577	0.8599	0.8621
1.1	0.8643	0.8665	0.8686	0.8708	0.8729	0.8749	0.8770	0.8790	0.8810	0.8830
1.2	0.8849	0.8869	0.8888	0.8907	0.8925	0.8944	0.8962	0.8980	0.8997	0.9015
1.3	0.9032	0.9049	0.9066	0.9082	0.9099	0.9115	0.9131	0.9147	0.9162	0.9177
1.4	0.9192	0.9207	0.9222	0.9236	0.9251	0.9265	0.9279	0.9292	0.9306	0.9319
1.5	0.9332	0.9345	0.9357	0.9370	0.9382	0.9394	0.9406	0.9418	0.9429	0.9441
1.6	0.9452	0.9463	0.9474	0.9484	0.9495	0.9505	0.9515	0.9525	0.9535	0.9545
1.7	09554	0.9564	0.9573	0.9582	0.9591	0.9599	0.9608	0.9616	0.9625	0.9633
1.8	0.9641	0.9649	0.9656	0.9664	0.9671	0.9678	0.9686	0.9693	0.9699	0.9706
1.9	0.9713	0.9719	0.9726	0.9732	0.9738	0.9744	0.9750	0.9756	0.9761	0.9767
2.0	0.9772	0.9778	0.9783	0.9788	0.9793	0.9798	0.9803	0.9808	0.9812	0.9817
2.1	0.9821	0.9826	0.9830	0.9834	0.9838	0.9842	0.9846	0.9850	0.9854	0.9857
2.2	0.9861	0.9864	0.9868	0.9871	0.9875	0.9878	0.9881	0.9884	0.9887	0.9890
2.3	0.9893	0.9896	0.9898	0.9901	0.9904	0.9906	0.9909	0.9911	0.9913	0.9916
2.4	0.9918	0.9920	0.9922	0.9925	0.9927	0.9929	0.9931	0.9932	0.9934	0.9936
2.5	0.9938	0.9940	0.9941	0.9943	0.9945	0.9946	0.9948	0.9949	0.9951	0.9952
2.6	0.9953	0.9955	0.9956	0.9957	0.9959	0.9960	0.9961	0.9962	0.9963	0.9964
2.7	0.9965	0.9966	0.9967	0.9968	0.9969	0.9970	0.9971	0.9972	0.9973	0.9974
2.8	0.9974	0.9975	0.9976	0.9977	0.9977	0.9978	0.9979	0.9979	0.9980	0.9981
2.9	0.9981	0.9982	0.9982	0.9983	0.9984	0.9984	0.9985	0.9985	0.9986	0.9986
3.0	0.9987	0.9987	0.9987	0.9988	0.9988	0.9989	0.9989	0.9989	0.9990	0.9990
3.1	0.9990	0.9991	0.9991	0.9991	0.9992	0.9992	0.9992	0.9992	0.9993	0.9993
3.2	0.9993	0.9993	0.9994	0.9994	0.9994	0.9994	0.9994	0.9995	0.9995	0.9995
3.3	0.9995	0.9995	0.9995	0.9996	0.9996	0.9996	0.9996	0.9996	0.9996	0.9997
3.4	0.9997	0.9997	0.9997	0.9997	0.9997	0.9997	0.9997	0.9997	0.9997	0.9998
3.5	0.9998	0.9998	0.9998	0.9998	0.9998	0.9998	0.9998	0.9998	0.9998	0.9998
3.6	0.9998	0.9998	0.9999	0.9999	0.9999	0.9999	0.9999	0.9999	0.9999	0.9999
3.7	0.9999	0.9999	0.9999	0.9999	0.9999	0.9999	0.9999	0.9999	0.9999	0.9999
3.8	0.9999	0.9999	0.9999	0.9999	0.9999	0.9999	0.9999	0.9999	0.9999	0.9999
3.9	1.0000	1.0000	1.0000	1.0000	1.0000	1.0000	1.0000	1.0000	1.0000	1.0000

Table 3.4. Values of the function $Q^{-1}(p)$ for some values of p.

p	0.10	0.05	0.01	0.005	0.001	0.0001	0.00001
$Q^{-1}(p)$	1.282	1.645	2.326	2.576	3.090	3.719	4.265

3.6 Transformations

Let X be a random variable and let $Y := g(X)$, where g is a real-valued function defined on the real line. Then Y is also a random variable, because the *composition* of two functions is again a function. We want to obtain the functions F_Y and f_Y (or p_Y). For example, the distribution of the random variable V representing the speed of a particle may be known, but we might be interested in determining the distribution of its energy E, which is given by $E = mV^2/2$, where m is the mass of the particle.

I) Discrete case. a) If X is a discrete random variable, then Y is a discrete random variable as well. Let $x_{k(1)}, x_{k(2)}, \ldots$ be the values of X for which

$$g[x_{k(j)}] = y_k \quad \text{for } j = 1, 2, \ldots.$$

Then, we can write that

$$p_Y(y_k) = \sum_{j=1}^{\infty} p_X(x_{k(j)}) \quad \text{for } k = 1, 2, \ldots. \tag{3.49}$$

For example, let $S_X = \{-2, 0, 2\}$ and

x	-2	0	2
$p_X(x)$	1/4	1/4	1/2

If $Y := X^2$, then the values -2 and 2 of X correspond to the same value, 4, of Y, and we find that

y	0	4	Σ
$p_Y(y)$	1/4	3/4	1

In the particular case when the function g is bijective, we simply have:

$$p_Y(y_k) = p_X(x_k), \quad \text{where } y_k = g(x_k) \ \forall k. \tag{3.50}$$

For instance, if $Z := -X^3$ in the above example, then to the values $-2, 0$ and 2 of X correspond respectively the values $8, 0$ and -8 of Z, and we have:

z	-8	0	8	Σ
$p_Z(z)$	1/2	1/4	1/4	1

Example 3.6.1 Let $X \sim \text{Geom}(1/2)$ and

$$Y := \begin{cases} 1 \text{ if } X \text{ is even,} \\ 0 \text{ if } X \text{ is odd.} \end{cases}$$

We have:

$$p_Y(1) = \sum_{k=1}^{\infty} p_X(2k) = \sum_{k=1}^{\infty} (1/2)^{2k-1}(1/2) = 1/4 + (1/4)^2 + \cdots = 1/3.$$

It follows that $p_Y(0) = 1 - (1/3) = 2/3$. Therefore, we have:

y	0	1	Σ
$p_Y(y)$	2/3	1/3	1

and

y	0	1
$F_Y(y)$	2/3	1

Now, let $Z := X - 1$. Then, we have: $S_Z = \{0, 1, \dots\}$ and

$$p_Z(k) = p_X(k+1) = (1/2)^{(k+1)-1}(1/2) = (1/2)^{k+1} \quad \text{for } k = 0, 1, \dots .$$

b) If X is a continuous random variable, but g is a staircase function, that is, a function that takes at most a denumerably infinite number of values, then Y is a discrete random variable. We have:

$$p_Y(y_k) := P[Y = y_k] = P[X \in A], \tag{3.51}$$

where A is the event equivalent to $\{Y = y_k\}$ in S_X.

Example 3.6.2 Let $X \sim \text{Exp}(\lambda)$ and $Y := \text{int}(X)+1$. We already mentioned in Section 3.5 that $Y \sim \text{Geom}(p = 1 - e^{-\lambda})$. Indeed, in this case, the event equivalent to $\{Y = k\}$ is $\{k - 1 \leq X < k\}$ for $k \in \{1, 2, \dots\}$, so that

$$p_Y(k) := P[Y = k] = P[\text{int}(X) = k - 1] = P[k - 1 \leq X < k]$$

$$= \int_{k-1}^{k} \lambda e^{-\lambda x} \, dx = -e^{-\lambda x}\Big|_{k-1}^{k} = e^{-\lambda(k-1)}(1 - e^{-\lambda}) = q^{k-1}p,$$

where $p := 1 - e^{-\lambda}$.

II) Continuous case. If X is a continuous random variable and g is a continuous function, then Y is a continuous random variable as well. To obtain the distribution function of Y, we must find the probability of the event $A \subset S_X$ which is equivalent to $\{Y \leq y\}$. Next, we differentiate $F_Y(y)$ to obtain $f_Y(y)$. We can also make use of the following theorem.

Theorem 3.6.1. *Suppose that the equation $y = g(x)$ has n (real) solutions: $x_1, \ldots,$ x_n. Then, the probability density function of Y is given by*

$$f_Y(y) = \sum_{k=1}^{n} f_X(x_k)|dx_k/dy|. \tag{3.52}$$

Remarks. i) If $n = 1$, so that the function g is *bijective*, then we have:

$$f_Y(y) = f_X(g^{-1}(y))|dg^{-1}(y)/dy|, \tag{3.52a}$$

where g^{-1} is the *inverse* function of g. For example, if $y = g(x) = e^x$, then $x = g^{-1}(y) = \ln y$.

ii) Moreover, if $g(x) = x^2$, then we can write, in general, that

$$f_Y(y) = \frac{1}{2\sqrt{y}}[f_X(-\sqrt{y}) + f_X(\sqrt{y})]. \tag{3.52b}$$

However, if X is always positive, for instance, then $f_X(-\sqrt{y}) \equiv 0$.

Proof (of Theorem 3.6.1). Consider the case when the function g is bijective and increasing. Then, we have:

$$P[y < Y \le y + dy] = P[x < X \le x + dx], \quad \text{where } g(x) = y$$

$$\Rightarrow \quad f_Y(y)|dy| = f_X(x)|dx| \Rightarrow f_Y(y) = f_X(x)|dx/dy|.$$

The proof in the general case is similar. □

Example 3.6.3 Let $X \sim N(0, 1)$ and $Y := X^2$. Then, we have:

$$F_Y(y) = P[X^2 \le y] \stackrel{y \ge 0}{=} P[-y^{1/2} \le X \le y^{1/2}] = \Phi(y^{1/2}) - \Phi(-y^{1/2})$$

$$\Rightarrow \quad f_Y(y) = \frac{1}{2}y^{-1/2}[\phi(y^{1/2}) + \phi(-y^{1/2})] = \frac{1}{\sqrt{2\pi}}y^{-1/2}e^{-y/2}$$

for $y \ge 0$. In this example, the event $\{-y^{1/2} \le X \le y^{1/2}\}$ is equivalent to $\{Y \le y\}$, and the equation $y = x^2$ has two solutions: $x_1 = -y^{1/2}$ and $x_2 = y^{1/2}$. Moreover, the random variable Y has a $G(\frac{1}{2}, \frac{1}{2})$ distribution or a *chi-square* distribution with 1 *degree of freedom*.

Example 3.6.4 Let $X \sim N(\mu, \sigma^2)$ and $Y := (X - \mu)/\sigma$. In this case, the transformation is *bijective*. We have:

$$f_Y(y) = f_X(\mu + \sigma y)|d(\mu + \sigma y)/dy| = \frac{1}{\sqrt{2\pi}\sigma} \exp(-y^2/2)|\sigma| = \phi(y).$$

That is, Y has a standard Gaussian distribution. In general, if $Z := aX + b$, then Z has a Gaussian distribution with parameters $a\mu + b$ and $a^2\sigma^2$.

Example 3.6.5 Let $X \sim U(0, 1]$ and $Y := -2 \ln X$. Then, we find that Y has an exponential distribution with parameter $\lambda = 1/2$. This kind of result is used in *simulation*.

Remark. We must always give the set of possible values of the new random variable Y, that is, S_Y. In the case when the function g is strictly *increasing* (respectively *decreasing*), we can show that if $S_X = [a, b]$, then

$$S_Y = [g(a), g(b)] \quad (\text{resp. } [g(b), g(a)]).$$

3.7 Mathematical Expectation and Variance

Definition 3.7.1. *The* **mathematical expectation**, *or the* **mean**, *of a random variable X is denoted by $E[X]$ and is defined by*

$$E[X] \equiv \mu_X = \begin{cases} \displaystyle\sum_{k=1}^{\infty} x_k p_X(x_k) & \text{if } X \text{ is discrete,} \\ \displaystyle\int_{-\infty}^{\infty} x f_X(x) \, dx & \text{if } X \text{ is continuous.} \end{cases} \tag{3.53}$$

Remarks. i) In general, we assume that the integral converges absolutely. That is, we assume that the integral of the absolute value of $x f_X(x)$ converges; similarly in the discrete case. Otherwise, the expected value of X does not exist. However, we must distinguish between the case when $E[X] = \infty$ or $-\infty$, and that when $E[X] = \infty - \infty$. In the first case, we can simply say that the mathematical expectation of X is infinite (in absolute value), whereas in the second case it does not exist.

ii) The mathematical expectation of the random variable X is a *weighted* average of its possible values, the weight associated with a given value being equal to its probability (in the discrete case) or to the value of the probability density function of X at this point (in the continuous case).

iii) The expected value of X corresponds to the *center of gravity* of the distribution.

iv) In the case when X is a random variable of mixed type, its expected value can be obtained by combining the two formulas in the definition of $E[X]$. For example, let $S_X = \{-1\} \cup [0, 1]$. If $P[X = -1] = 1/4$ and

$$f_X(x) = 3/4 \quad \text{for } 0 \leq x \leq 1,$$

then we can write that

$$E[X] = (-1) \cdot \frac{1}{4} + \int_0^1 x \frac{3}{4} \, dx = -\frac{1}{4} + \frac{3}{8} = \frac{1}{8}.$$

We can also make use of the Dirac delta function in the density function of X and employ the formula for continuous X as follows:

$$f_X(x) = \begin{cases} \frac{1}{4}\delta(0) & \text{if } x = -1, \\ \frac{3}{4} & \text{if } x \in [0, 1], \\ 0 & \text{elsewhere} \end{cases}$$

$$\Rightarrow \quad E[X] = \int_{-\infty}^{\infty} x f_X(x)\, dx = \frac{1}{4}(-1) + \int_0^1 x \frac{3}{4}\, dx = \frac{1}{8}.$$

Finally, another way of expressing the density function of X is with the help of the indicator function I_A, defined by

$$I_A(x) = \begin{cases} 1 & \text{if } x \in A, \\ 0 & \text{otherwise.} \end{cases}$$

Then, we have:

$$f_X(x) = \frac{1}{4}\delta(x + 1) + \frac{3}{4}I_{[0,1]}(x) \quad \forall x \in \mathbb{R}.$$

v) The notion of mathematical expectation can be generalized by defining $E[X \mid A]$, the *conditional (mathematical) expectation* of X, given an event A. We only have to replace $f_X(x)$ by $f_X(x \mid A)$ or $p_X(x)$ by $p_X(x \mid A)$ in the definition of $E[X]$.

Proposition 3.7.1. *The mathematical operator E is linear. That is, if c_k is a constant and g_k is a real-valued function for $k = 1, 2, \ldots, n$, then*

$$E\left[\sum_{k=1}^{n} c_k g_k(X)\right] = \sum_{k=1}^{n} c_k E[g_k(X)]. \tag{3.54}$$

Proof. This result follows directly from the fact that the integral is a linear operator, and so is the summation. □

We deduce from the definition of $E[X]$ and the preceding proposition that $E[c] = c$ and $E[aX + b] = aE[X] + b$ for any constants a, b and c.

Remark. A (real) constant c can be considered as a degenerate random variable C whose probability mass function is

$$p_C(x) = \begin{cases} 1 & \text{if } x = c, \\ 0 & \text{otherwise,} \end{cases}$$

or whose probability density function is

$$f_C(x) = \delta(x - c) \quad \text{for all } x \in \mathbb{R}.$$

Proposition 3.7.2. *a) Let X be a non-negative continuous random variable. Then, we have:*

$$E[X] = \int_0^{\infty} P[X > x]\, dx = \int_0^{\infty} [1 - F_X(x)]\, dx. \tag{3.55}$$

b) If X is a discrete random variable whose possible values are included in the set $\{0, 1, 2, \ldots\}$, then

$$E[X] = \sum_{k=0}^{\infty} P[X > k]. \tag{3.56}$$

Proof. a) We have:

$$E[X] = \int_0^\infty x f_X(x)\,dx = \int_0^\infty \int_0^x f_X(x)\,dt\,dx$$

$$= \int_0^\infty \int_t^\infty f_X(x)\,dx\,dt = \int_0^\infty P[X > t]\,dt.$$

b) The proof is similar to the previous one. ◻

Remark. We can also prove the result in a) as follows. By integrating by parts, we find that

$$\int_0^\infty [1 - F_X(x)]\,dx = x[1 - F_X(x)]|_0^\infty + \int_0^\infty x f_X(x)\,dx$$

$$= \lim_{x \to \infty} x[1 - F_X(x)] + E[X].$$

Now, by l'Hospital's rule (H.R.), we have (assuming that $E[X^2] < \infty$):

$$\lim_{x \to \infty} x[1 - F_X(x)] = \lim_{x \to \infty} \frac{1 - F_X(x)}{1/x} \overset{\text{H.R.}}{=} \lim_{x \to \infty} x^2 f_X(x) = 0,$$

since $x^2 f_X(x)$ must converge (toward 0) when x tends to infinity (otherwise the integral of $x^2 f_X(x)$ from 0 to infinity would diverge).

Example 3.7.1 Let $X \sim \text{Poi}(\alpha)$. We have:

$$E[X] = \sum_{k=0}^\infty k e^{-\alpha} \frac{\alpha^k}{k!} = e^{-\alpha} \sum_{k=1}^\infty \frac{\alpha^k}{(k-1)!} = e^{-\alpha}\alpha \sum_{k=1}^\infty \frac{\alpha^{k-1}}{(k-1)!} = e^{-\alpha}\alpha e^\alpha = \alpha.$$

Example 3.7.2 Let $X \sim \text{Exp}(\lambda)$. Making use of the preceding proposition, we obtain (see Equation (3.35)):

$$E[X] = \int_0^\infty P[X > x]\,dx = \int_0^\infty e^{-\lambda x}\,dx = -\frac{e^{-\lambda x}}{\lambda}\bigg|_0^\infty = \frac{1}{\lambda}.$$

We can demonstrate the following proposition or use it as a definition.

Proposition 3.7.3. *Let X be a random variable and let $Y := g(X)$. Then, we have:*

$$E[Y] = \begin{cases} \displaystyle\sum_{k=1}^\infty g(x_k) p_X(x_k) & \text{if } X \text{ is discrete,} \\ \displaystyle\int_{-\infty}^\infty g(x) f_X(x)\,dx & \text{if } X \text{ is continuous.} \end{cases}$$

Remarks. i) Let X be a continuous random variable and g be the indicator function of the event $\{X \in A\}$; that is,

$$g(X) = \begin{cases} 1 \text{ if } X \in A, \\ 0 \text{ otherwise.} \end{cases}$$

We can write that

$$E[g(X)] = 1 \cdot P[g(X) = 1] = P[X \in A] = \int_A f_X(x)\,dx = \int_{-\infty}^{\infty} g(x) f_X(x)\,dx,$$

which corresponds to the above formula. Note that here $g(X)$ has a Bernoulli distribution with parameter $p = P[X \in A]$.

ii) Let $B_1, \ldots, B_n$ be a partition of a sample space S (for which $P[B_i] > 0$ for $i = 1, \ldots, n$). We can write that

$$E[g(X)] = \sum_{i=1}^{n} E[g(X) \mid B_i] P[B_i].$$

For example, if X is a random variable of discrete type, then

$$p_X(x_k) := P[X = x_k] = \sum_{i=1}^{n} P[X = x_k \mid B_i] P[B_i].$$

It follows that

$$E[g(X)] = \sum_{k=1}^{\infty} g(x_k) p_X(x_k) = \sum_{k=1}^{\infty} \sum_{i=1}^{n} g(x_k) P[X = x_k \mid B_i] P[B_i]$$

$$= \sum_{i=1}^{n} \left\{ \sum_{k=1}^{\infty} g(x_k) P[X = x_k \mid B_i] \right\} P[B_i] = \sum_{i=1}^{n} E[g(X) \mid B_i] P[B_i].$$

Example 3.7.3 Let $X \sim U[0, 2]$. Then the probability density function of $Y := X^2$ is given by

$$f_Y(y) = f_X(x) \left| \frac{dx}{dy} \right| = f_X(\sqrt{y}) \left| \frac{d\sqrt{y}}{dy} \right|$$

$$= \frac{1}{2} \left| \frac{1}{2\sqrt{y}} \right| = \frac{1}{4\sqrt{y}} \quad \text{if } 0 \le y \le 4,$$

because the transformation $y = x^2$ is bijective here, since $S_X = [0, 2]$. It follows that

$$E[Y] = \int_0^4 y \frac{1}{4\sqrt{y}}\,dy = \frac{1}{4} \frac{y^{3/2}}{3/2} \Big|_0^4 = \frac{4}{3}.$$

Now, we calculate (directly)

$$E[X^2] = \int_0^2 x^2 \frac{1}{2}\,dx = \frac{x^3}{6} \Big|_0^2 = \frac{4}{3}.$$

Thus, Proposition 3.7.3 enables us to obtain the mean of the random variable Y much more easily.

The Variance of a Random Variable

Definition 3.7.2. *The* **variance** *of a random variable X is defined by*

$$\text{VAR}[X] \equiv \sigma_X^2 = E[(X - E[X])^2] \quad (\geq 0). \tag{3.57}$$

Remarks. i) The variance is a measure of *variability* of the random variable about its mean. It is the *moment of inertia* of the distribution. Moreover, the variance of a random variable cannot be negative. In fact, the variance $\text{VAR}[X]$ is strictly positive, unless X is a constant. If $X \equiv c$, then $E[X] = E[c] = c$ and $\text{VAR}[X] = \text{VAR}[c] = E[(c - c)^2] = 0$.

ii) We also define the *standard deviation* of X by

$$\text{STD}[X] \equiv \sigma_X = (\text{VAR}[X])^{1/2}. \tag{3.58}$$

The units of measure of X and of its standard deviation are the same.

iii) We generalize the notion of variance by considering the *conditional variance* of X, given an event A, defined by

$$\text{VAR}[X \mid A] = E[(X - E[X \mid A])^2 \mid A]. \tag{3.59}$$

We can show (see Proposition 3.7.5, p. 100) that

$$\text{VAR}[X \mid A] = E[X^2 \mid A] - (E[X \mid A])^2.$$

In fact, to calculate the conditional variance of a random variable X, given an event A, it is often easier to first find the probability density (or mass) function of $Y := X \mid A$, and then to use the definition of $\text{VAR}[Y]$. For example, if $X \sim U[-1, 1]$ and if $A = \{X \leq 0\}$, we find that $Y := X \mid A \sim U[-1, 0]$ and it is then easy to calculate $\text{VAR}[Y]$ (see Example 3.7.5, p. 100).

iv) We can calculate the variance of X as follows:

$$\text{VAR}[X] = \sum_{i=1}^{n} E[(X - E[X])^2 \mid B_i] P[B_i],$$

where the events $B_1, \ldots, B_n$ form a partition of a sample space S and $P[B_i] > 0$ $\forall i$. However, in general

$$\text{VAR}[X] \neq \sum_{i=1}^{n} \text{VAR}[X \mid B_i] P[B_i],$$

because $E[(X - E[X])^2 \mid B_i] \neq \text{VAR}[X \mid B_i]$.

Proposition 3.7.4. *We have:* $\text{VAR}[aX + b] = a^2 \text{VAR}[X]$.

Proof. By definition,

$$\text{VAR}[aX + b] = E[((aX + b) - (aE[X] + b))^2]$$
$$= a^2 E[(X - E[X])^2] = a^2 \text{VAR}[X]. \qquad \square$$

Proposition 3.7.5. *We have:* $\text{VAR}[X] = E[X^2] - (E[X])^2$.

Proof. We deduce from the definition of $\text{VAR}[X]$ that

$$\text{VAR}[X] = E[X^2] - 2E[XE[X]] + E[(E[X])^2]$$
$$= E[X^2] - 2E[X]E[X] + (E[X])^2 = E[X^2] - (E[X])^2$$

(since the fact that $E[X]$ is a constant implies that $E[g(E[X])] = g(E[X])$ for any function g). $\qquad \square$

Example 3.7.4 Let X be a random variable having a Bernoulli distribution with parameter p. We have:

$$E[X^n] = 1^n p + 0^n q = p \quad \text{for any } n \geq 1.$$

It follows that

$$\text{VAR}[X] = E[X^2] - (E[X])^2 = p - p^2 = p(1 - p) = pq. \qquad (3.60)$$

Example 3.7.5 Let $X \sim U[a, b]$. Then, by symmetry, we can write that the mean of X is given by $(a + b)/2$. In addition, we have:

$$E[X^2] = \int_a^b x^2 \frac{1}{b - a} \, dx = \frac{b^3 - a^3}{3(b - a)} = \frac{b^2 + ab + a^2}{3}.$$

Making use of Proposition 3.7.5, we find that

$$\text{VAR}[X] = \frac{(b - a)^2}{12}. \qquad (3.61)$$

Table 3.5 gives the mean and the variance of the various probability distributions defined in Sections 3.4 and 3.5.

Definition 3.7.3. a) *The **kth moment** (or **order k moment**) of X about the origin is defined by* $\mu'_k = E[X^k]$ *for* $k = 0, 1, 2, \dots$.

b) *The **kth central moment** or **kth moment about the mean** of X is given by* $\mu_k = E[(X - E[X])^k]$ *for* $k = 0, 1, \dots$.

Particular cases. The mean is the first moment of X about the origin, while the variance is the second central moment of X. Moreover, $\mu'_0 = \mu_0 = 1$ and $\mu_1 = 0$.

Two other quantities, in addition to the mean and the variance, that characterize the distribution of a random variable are the *skewness* and the *kurtosis* coefficients.

Table 3.5. Means and variances of the main probability distributions.

Distribution	Parameters	Mean	Variance
Bernoulli	p	p	pq
Binomial	n and p	np	npq
Hypergeometric	N, n and m	$n \cdot \frac{m}{N}$	$n \cdot \frac{m}{N} \cdot \left(1 - \frac{m}{N}\right) \cdot \left(\frac{N-n}{N-1}\right)$
Geometric	p	$1/p$	q/p^2
Pascal	r and p	r/p	rq/p^2
Poisson	α	α	α
Uniform	$[a, b]$	$(a+b)/2$	$(b-a)^2/12$
Exponential	λ	$1/\lambda$	$1/\lambda^2$
Laplace	λ	0	$2/\lambda^2$
Gamma	α and λ	α/λ	α/λ^2
Weibull	γ, δ and β	$\gamma + \delta\Gamma(1 + \beta^{-1})$	$\delta^2\left[\Gamma(1 + 2\beta^{-1}) - \Gamma^2(1 + \beta^{-1})\right]$
Gaussian	μ and σ^2	μ	σ^2
Lognormal	μ and σ^2	$e^{\mu + \frac{1}{2}\sigma^2}$	$e^{2\mu + \sigma^2}(e^{\sigma^2} - 1)$

Definition 3.7.4. a) *The* **skewness**, β_1, *of a random variable X is the quantity (without units of measure)* μ_3/σ^3.

b) *The* **kurtosis** *of X is given by* $\beta_2 = \mu_4/\sigma^4$.

Remark. In the case of the Gaussian distribution, we find that $\beta_1 = 0$ (since it is symmetrical) and $\beta_2 = 3$.

To complete this section, we give two results that enable us to calculate upper bounds, in terms of the mean and the variance of the random variable X, for the probabilities of events defined with respect to S_X. We can of course use these results to obtain lower bounds for the probabilities of the complements of the events in question.

Theorem 3.7.1. (Markov's inequality)[7] *Let X be a random variable that can take only non-negative values. Then, we have:*

$$P[X \geq a] \leq \frac{E[X]}{a} \quad \forall a > 0. \tag{3.62}$$

---

[7] Andrei Andreyevich Markov, 1856–1922, was born and died in Russia. He was a professor at St. Petersburg University. His first works were on number theory and mathematical analysis. He proved the *central limit theorem* under quite general conditions. His study of what is now called *Markov chains* initiated the theory of stochastic processes. He was also interested in poetry.

Proof. Suppose that X is a discrete random variable. Then, we can write that

$$E[X] = \sum_{x_k < a} x_k p_X(x_k) + \sum_{x_k \geq a} x_k p_X(x_k) \geq \sum_{x_k \geq a} x_k p_X(x_k)$$

$$\geq \sum_{x_k \geq a} a p_X(x_k) = a P[X \geq a].$$

The proof is similar in the case when X is a random variable of continuous type, or of mixed type. □

The following inequality can in fact be considered as a corollary to Markov's inequality.

Theorem 3.7.2. (Bienaymé–Chebyshev's inequality)[8,9] *Let X be a random variable whose mean $E[X] = \mu$ and whose variance $\mathrm{VAR}[X] = \sigma^2$ exist. Then, for any $a > 0$, we have:*

$$P[|X - \mu| \geq a] \leq \frac{\sigma^2}{a^2}. \tag{3.63}$$

Proof. Since $(X - \mu)^2$ is a non-negative random variable, we deduce from Markov's inequality that

$$P[(X - \mu)^2 \geq a^2] \leq \frac{E[(X - \mu)^2]}{a^2} \Rightarrow P[|X - \mu| \geq a] \leq \frac{\mathrm{VAR}[X]}{a^2}. \quad \square$$

Example 3.7.6 Let $X \sim \mathrm{Exp}(\lambda)$. We have already found that $E[X] = 1/\lambda$. We can also show that $\mathrm{VAR}[X] = 1/\lambda^2$. It follows that

$$P[|X - \lambda^{-1}| \geq k] \leq \frac{\lambda^{-2}}{k^2} = \frac{1}{k^2 \lambda^2} \quad \text{for } k = 1, 2, \ldots.$$

Suppose that $\lambda = 1$ and $k = 2$. Then, we have: $P[|X - 1| \geq 2] \leq 1/4$.

Now, since we assumed that the distribution of X is known, we can calculate directly

---

[8] Irénée-Jules Bienaymé, 1796–1878, was born and died in France. He studied at the École Polytechnique de Paris and taught in a military academy for two years. Next, he became Inspector General of Finances. In 1848, he was named professor of probability at the Sorbonne. A friend of Chebyshev, he translated his works from Russian into French. The work of Laplace, in particular the *Théorie analytique des probabilités*, had a great influence on his scientific thinking.

[9] Pafnuty Lvovich Chebyshev, 1821–1894, was born and died in Russia. Son of a Russian nobleman, he was a professor at St. Petersburg University. He is considered the founder of the great Russian school of mathematics of the twentieth century. In addition to his contributions to number theory and to the theory of approximation of functions, he was also very interested in mechanics. By using the inequality that bears his name, he gave a simple proof of the *law of large numbers*. Moreover, he worked intensively on the *central limit theorem*.

$$P[|X - 1| \geq 2] \overset{\text{inc.}}{=} P[X \geq 3] + \underbrace{P[X \leq -1]}_{0} = P[X \geq 3] = e^{-3} \simeq 0.0498.$$

We notice that the value of the upper bound obtained with Bienaymé–Chebyshev's inequality is about five times greater than the exact value. However, we must remember that the bound in question is valid for any random variable whose mean and variance are both equal to 1.

3.8 Transforms

Definition 3.8.1. *The **characteristic function** of the random variable X is defined by*

$$\phi_X(\omega) = E[e^{j\omega X}], \tag{3.64}$$

where $j := \sqrt{-1}$ *and* $\omega \in \mathbb{R}$.

Continuous case. In this case, we can write that

$$\phi_X(\omega) = \int_{-\infty}^{\infty} e^{j\omega x} f_X(x)\, dx. \tag{3.65}$$

Thus, $\phi_X(\omega)$ is the *Fourier transform*[10] of the density function $f_X(x)$. It follows that

$$f_X(x) = \frac{1}{2\pi} \int_{-\infty}^{\infty} e^{-j\omega x} \phi_X(\omega)\, d\omega. \tag{3.66}$$

Hence, we can state that the characteristic function of a random variable *characterizes* entirely the distribution of this variable.

Discrete case. By definition, we have:

$$\phi_X(\omega) = \sum_{k=1}^{\infty} e^{j\omega x_k} p_X(x_k). \tag{3.67}$$

In the particular case when the x_k's are integers, we can write that

$$\phi_X(\omega) = \sum_{n=-\infty}^{\infty} e^{j\omega n} p_X(n). \tag{3.68}$$

---

[10] Jean Baptiste Joseph (Baron) Fourier, 1768–1830, was born and died in France. He first studied for the priesthood, but he did not take his vows. Later, he taught at the Collège de France and then at the École Polytechnique. He took part in the French Revolution and became friends with Napoléon. In 1822, he published his main work, the *Théorie analytique de la chaleur*, in which he made wide use of the series that now bear his name. However, he is not the inventor of these series.

Thus, the right-hand member of the preceding equation can be considered as the *complex Fourier series* of the function $\phi_X(\omega)$ (which is periodic with period 2π because $e^{2\pi nj} = 1 \ \forall n \in \mathbf{Z}$). Then, we have:

$$p_X(n) = \frac{1}{2\pi} \int_0^{2\pi} e^{-j\omega n} \phi_X(\omega) \, d\omega \quad \text{for } n = 0, \pm 1, \pm 2, \ldots . \tag{3.69}$$

Proposition 3.8.1. *Let X be a random variable for which all the moments about the origin exist and are finite. Then the order n moment of X, about the origin, is given by*

$$E[X^n] = (-j)^n \frac{d^n}{d\omega^n} \phi_X(\omega) \Big|_{\omega=0} . \tag{3.70}$$

Proof. We simply have to use the series expansion of $e^{j\omega X}$:

$$\phi_X(\omega) \equiv E[e^{j\omega X}] = E[1 + j\omega X + (j\omega X)^2/2! + \cdots].$$

Then, if we assume that we can interchange the summation and the mathematical expectation, we obtain:

$$\phi_X(\omega) = 1 + j\omega E[X] + \frac{1}{2!}(j\omega)^2 E[X^2] + \cdots .$$

Differentiating n times the two members of the above equation, and evaluating at $\omega = 0$, we obtain Equation (3.70). For instance, we have:

$$\phi_X(0) = 1 \Rightarrow E[X^0] = \phi_X(\omega) \Big|_{\omega=0} .$$

Thus, the formula is valid for $n = 0$. Note that $\phi_X(0) = 1$ for any random variable X, which is also a consequence of the definition $\phi_X(\omega) = E[e^{j\omega X}]$.

Similarly, for $n = 1$, we have:

$$\phi_X'(\omega) \Big|_{\omega=0} = jE[X] + \frac{j^2}{2!}(2\omega)E[X^2] + \frac{j^3}{3!}(3\omega^2)E[X^3] + \cdots \Big|_{\omega=0} = jE[X]$$

$$\Rightarrow \quad E[X] = -j\frac{d}{d\omega}\phi_X(\omega) \Big|_{\omega=0} ,$$

etc.   □

Example 3.8.1 Let $X \sim \text{Poi}(\alpha)$. We have:

$$\phi_X(\omega) = \sum_{k=0}^{\infty} e^{j\omega k} e^{-\alpha} \frac{\alpha^k}{k!} = e^{-\alpha} \sum_{k=0}^{\infty} \frac{(e^{j\omega}\alpha)^k}{k!} = e^{-\alpha} \exp(e^{j\omega}\alpha).$$

It follows that

$$E[X] = -j\frac{d}{d\omega}e^{-\alpha}\exp(e^{j\omega}\alpha)\big|_{\omega=0} = -je^{-\alpha}\exp(e^{j\omega}\alpha)e^{j\omega}\alpha j\big|_{\omega=0} = \alpha,$$

which corresponds to the result obtained in Example 3.7.1. Likewise, we find that

$$E[X^2] = \alpha^2 + \alpha.$$

Then, making use of Proposition 3.7.5, we obtain:

$$\text{VAR}[X] = E[X^2] - (E[X])^2 = \alpha^2 + \alpha - (\alpha)^2 = \alpha.$$

So, in the case of the Poisson distribution, the parameter α is both the mean and the variance of the distribution. A similar result is obtained with the exponential distribution, for which the mean and the standard deviation are both equal to $1/\lambda$.

Example 3.8.2 Let $X \sim N(\mu, \sigma^2)$. Then,

$$\begin{aligned}
\phi_X(\omega) &= \frac{1}{\sqrt{2\pi}\sigma}\int_{-\infty}^{\infty} e^{j\omega x}\exp\left\{-\frac{(x-\mu)^2}{2\sigma^2}\right\}dx \\
&= \frac{1}{\sqrt{2\pi}\sigma}\int_{-\infty}^{\infty}\exp\left\{-\frac{[x-(j\omega\sigma^2+\mu)]^2 - 2j\omega\sigma^2\mu + (\omega\sigma^2)^2}{2\sigma^2}\right\}dx \\
&= \exp\left(j\omega\mu - \frac{1}{2}\omega^2\sigma^2\right) \times P[-\infty < N(\mu + j\omega\sigma^2, \sigma^2) < \infty] \\
&= \exp\left(j\omega\mu - \frac{1}{2}\omega^2\sigma^2\right).
\end{aligned}$$

Hence, we can write that

$$E[X^n] = (-j)^n\frac{d^n}{d\omega^n}\exp\left(j\omega\mu - \frac{1}{2}\omega^2\sigma^2\right)\bigg|_{\omega=0}.$$

We find, in particular, that

$$E[X] = -j\exp\left(j\omega\mu - \frac{1}{2}\omega^2\sigma^2\right)(j\mu - \omega\sigma^2)\big|_{\omega=0} = \mu$$

and

$$. \; E[X^2] = -\exp\left(j\omega\mu - \frac{1}{2}\omega^2\sigma^2\right)[(j\mu - \omega\sigma^2)^2 - \sigma^2]\big|_{\omega=0} = \mu^2 + \sigma^2,$$

so that $\text{VAR}[X] = \sigma^2$.

Note that since we have used the symbols μ_X and σ_X^2 to denote the mean and the variance of the random variable X, in general, then, to be coherent, in the case of a Gaussian distribution with parameters μ and σ^2 the constants μ and σ^2 had to represent the mean and the variance of this Gaussian distribution, respectively.

Remark. We have written above that the integral in the computation of the characteristic function of $X \sim N(\mu, \sigma^2)$ can be expressed in terms of the probability that a random variable $Y \sim N(\mu + j\omega\sigma^2, \sigma^2)$ takes a value in the interval $(-\infty, \infty)$. Of course, a Gaussian distribution cannot have a complex mean, since we only consider *real* variables. However, from a purely mathematical point of view, the fact that the constant $\mu + j\omega\sigma^2$ is complex does not change anything in the computation of the integral.

Now, when X is a discrete random variable whose possible values are positive integers or zero, we work with its *probability generating function*.

Definition 3.8.2. *Let X be a random variable taking its values in $\{0, 1, \ldots\}$. The* **(probability) generating function** *of X is defined by*

$$G_X(z) = E[z^X] = \sum_{k=0}^{\infty} z^k p_X(k). \tag{3.71}$$

Remarks. i) The generating function of X is in fact the *z-transform* (with a different sign in the exponent) of the sequence $p_0, p_1, \ldots$, where $p_k := p_X(k)$ $\forall k$.

ii) Since $p_0 + p_1 + p_2 + \cdots = 1$, the radius of convergence of the power series is greater than or equal to 1.

iii) We can write that $\phi_X(\omega) = G_X(e^{j\omega})$. It follows that

$$E[X] = -j\frac{d}{d\omega}G_X(e^{j\omega})\Big|_{\omega=0} = -jG'_X(e^{j\omega})e^{j\omega}j\Big|_{\omega=0} = G'_X(1). \tag{3.72}$$

Likewise, we find that $E[X^2] = G''_X(1) + G'_X(1)$. Hence, we have:

$$VAR[X] = G''_X(1) + G'_X(1) - [G'_X(1)]^2. \tag{3.73}$$

For example, let X be a discrete random variable whose probability mass function is given by the following table:

x	0	1	3	Σ
$p_X(x)$	1/2	1/3	1/6	1

We have:

$$G_X(z) = z^0\frac{1}{2} + z^1\frac{1}{3} + z^3\frac{1}{6} = \frac{z^3}{6} + \frac{z}{3} + \frac{1}{2}.$$

Then,

$$G'_X(1) = \frac{z^2}{2} + \frac{1}{3}\Big|_{z=1} = \frac{5}{6},$$

which corresponds to the result obtained by using the definition of $E[X]$:

$$E[X] = 0 + 1 \cdot \frac{1}{3} + 3 \cdot \frac{1}{6} = \frac{5}{6}.$$

iv) Proceeding as in Proposition 3.8.1, we show that

$$p_X(k) = \frac{1}{k!}\frac{d^k}{dz^k}G_X(z)\Big|_{z=0}. \tag{3.74}$$

Indeed,

$$G_X(z) := E[z^X] = p_X(0) + zp_X(1) + z^2 p_X(2) + \cdots,$$

which implies that

$$G_X(0) = p_X(0),$$
$$G'_X(0) = p_X(1) + 2zp_X(2) + \cdots|_{z=0} = p_X(1),$$
$$G''_X(0) = 2p_X(2) + 6zp_X(3) + \cdots|_{z=0} = 2p_X(2),$$

etc. We can also express $p_X(k)$ in terms of a complex integral as follows:

$$p_X(k) = \frac{1}{2\pi j}\oint_{|z|=r}\frac{G_X(z)}{z^{k+1}}\,dz, \quad \text{where } r < 1.$$

Example 3.8.3 Suppose that

$$G_X(z) = \frac{pz}{1-qz}, \quad \text{where } |z| < 1/q.$$

We have: $p_X(0) = G_X(0) = 0$. Moreover, since we can write that

$$G_X(z) = pz(1 + qz + (qz)^2 + \cdots),$$

we find that

$$p_X(k) = \frac{1}{k!}p(k!q^{k-1}) = q^{k-1}p \quad \text{for } k = 1, 2, \ldots.$$

That is, the function $G_X(z)$ above is the generating function of the geometric distribution with parameter p. We also deduce from the function G_X that

$$E[X] = \frac{1}{p} \quad \text{and} \quad \text{VAR}[X] = \frac{1-p}{p^2}.$$

Finally, another function that enables us to obtain the moments of a random variable is the *moment generating function*.

Definition 3.8.3. *Let X be a random variable. The* **moment generating function** *of X is defined by*

$$M_X(s) = E[e^{-sX}], \tag{3.75}$$

if the mathematical expectation exists.

Remarks. i) If X is a non-negative continuous random variable, then $M_X(s)$ is the *Laplace transform* of the function $f_X(x)$.

ii) We have: $\phi_X(\omega) = M_X(-j\omega)$. It follows that

$$\frac{d^n}{d\omega^n}\phi_X(\omega)\Big|_{\omega=0} = \frac{d^n}{d\omega^n}M_X(-j\omega)\Big|_{\omega=0} = (-j)^n\frac{d^n}{d\omega^n}M_X(\omega)\Big|_{\omega=0}.$$

Then, we deduce from Proposition 3.8.1 that

$$E[X^n] = (-1)^n\frac{d^n}{ds^n}M_X(s)\Big|_{s=0}. \tag{3.76}$$

In fact, if we define the moment generating function of X by $E[e^{sX}]$ (as many authors do), then the term $(-1)^n$ disappears in the previous formula.

Example 3.8.4 If $X \sim \text{Exp}(\lambda)$, then we find that

$$M_X(s) = \int_0^\infty e^{-sx}\lambda e^{-\lambda x}\,dx = -\frac{\lambda}{\lambda+s}e^{-(s+\lambda)x}\Big|_0^\infty$$

$$= \frac{\lambda}{\lambda+s} \quad \text{if } s > -\lambda.$$

This formula enables us to check that $E[X] = 1/\lambda$ and $\text{VAR}[X] = 1/\lambda^2$. Note that the function $M_X(s)$ does not exist (that is, the integral diverges) if $s \le -\lambda$, whereas the characteristic function of X, $\phi_X(\omega)$, exists for any real value ω.

3.9 Reliability

Definition 3.9.1. *Let T be a non-negative continuous random variable that represents the lifetime of a system (or a machine, etc.). The* **reliability function** *$R(t)$ is defined by*

$$R(t) = P[T > t] \quad (= 1 - F_T(t)). \tag{3.77}$$

The **mean lifetime** of the system, $E[T]$, can be obtained as follows (see Proposition 3.7.2, p. 96):

$$E[T] = \int_0^\infty R(t)\,dt. \tag{3.78}$$

Remark. The symbol $MTTF$ (*mean time to failure*) is sometimes used for the quantity $E[T]$. We can also define the symbols $MTBF$ and $MTTR$, that is, the *mean time between failures* and the *mean time to repair*, respectively. We have: $MTBF = MTTF + MTTR$.

Definition 3.9.2. *The* **failure rate function** *of the system is denoted by $r(t)$ and is given by*

$$r(t) = f_T(t \mid T > t). \tag{3.79}$$

Remark. The function $r(t)$, multiplied by dt, is approximately equal to the probability that a machine, that is still working at time t, breaks down in the interval $(t, t + dt]$, because

$$f_T(t \mid T > t)dt \simeq P[t < T \leq t + dt \mid T > t].$$

Now, the probability that the lifetime of the system is smaller than or equal to s, given that the system is still functioning at time t, is

$$P[T \leq s \mid T > t] := F_T(s \mid T > t) = \begin{cases} 0 & \text{if } s \leq t, \\ \dfrac{F_T(s) - F_T(t)}{1 - F_T(t)} & \text{if } s > t. \end{cases}$$

It follows that

$$f_T(s \mid T > t) = \frac{d}{ds} F_T(s \mid T > t) = \frac{f_T(s)}{1 - F_T(t)} \quad \text{if } s > t. \tag{3.80}$$

Thus, since $f_T(t) = -R'(t)$, we can also write (taking the limit as s decreases to t) that

$$r(t) = \frac{f_T(t)}{1 - F_T(t)} = -\frac{R'(t)}{R(t)}. \tag{3.81}$$

Example 3.9.1 If $X \sim \text{Exp}(\lambda)$, then the failure rate function is *constant*:

$$r(t) = \frac{\lambda e^{-\lambda t}}{e^{-\lambda t}} = \lambda.$$

Proposition 3.9.1. *There is a bijective relationship between the functions $R(t)$ and $r(t)$:*

$$R(t) = \exp\left[-\int_0^t r(s)\, ds \right]. \tag{3.82}$$

Proof. We have:

$$\int_0^t r(s)\, ds = -\int_0^t \frac{R'(s)}{R(s)}\, ds = -\ln R(t) + \ln R(0).$$

Now, $R(0) = 1$, because T is a non-negative continuous random variable. Hence, Equation (3.82) follows at once. ☐

Remark. We then deduce from Example 3.9.1 that the exponential distribution is the only one for which the failure rate function is constant.

Reliability of Systems in Series and in Parallel

a) Systems in series. If a system is made up of n subsystems placed *in series* and operating independently from one another, then the reliability function of the system is given by

$$R(t) = \prod_{k=1}^{n} R_k(t) = \exp\left[-\int_0^t [r_1(s) + \cdots + r_n(s)]\,ds \right], \qquad (3.83)$$

where $R_k(t)$ is the reliability function of the subsystem k, for $k = 1, \ldots, n$. This result follows from the fact that $T > t$ if and only if $T_k > t$ for all values of k, where T_k is the lifetime of the subsystem k.

Example 3.9.2 If T_k has an $\mathrm{Exp}(\lambda_k)$ distribution $\forall k$, then

$$R(t) = \exp[-(\lambda_1 + \cdots + \lambda_n)t].$$

Thus, it is as if there were a single component, whose lifetime has an exponential distribution with parameter $\lambda := \lambda_1 + \cdots + \lambda_n$.

b) Systems in parallel. In the case when the system is made up of n subsystems placed *in parallel* and operating independently from one another, we have: $T \leq t \Leftrightarrow T_k \leq t \; \forall k$. It follows that the reliability function of the system is given by

$$R(t) = 1 - \prod_{k=1}^{n}[1 - R_k(t)]. \qquad (3.84)$$

Remarks. i) We assume above that all the components begin to function at time $t = 0$. This is called *active redundancy*. Suppose now that there are n components placed in parallel, but only component no. 1 functions at first. When it fails, component no. 2 takes over, and so on. This type of redundancy is called *standby (or passive) redundancy*. If the components operate independently from one another, we can show (see Chapter 4) that

$$E[T] = \sum_{k=1}^{n} E[T_k]$$

and

$$\mathrm{VAR}[T] \stackrel{\mathrm{ind.}}{=} \sum_{k=1}^{n} \mathrm{VAR}[T_k].$$

In the particular case when T_k has an $\mathrm{Exp}(\lambda)$ distribution $\forall k$, we can also show (see Chapter 4, again) that T has a gamma distribution with parameters n and λ. It follows that

$$R(t) := P[T > t] = P[\mathrm{Poi}(\lambda t) \leq n - 1] = \sum_{k=0}^{n-1} e^{-\lambda t}(\lambda t)^k / k! \quad \text{for } t \geq 0.$$

ii) Suppose next that a system is made up of n components, but that k (working) components are sufficient for the system to function, where $0 < k \leq n$. Then, if all the components have the same reliability function $R_1(t)$ and are all independent, the reliability function of the system is given by

$$R(t) = \sum_{i=k}^{n} \binom{n}{i} [R_1(t)]^i [1 - R_1(t)]^{n-i} = 1 - \sum_{i=0}^{k-1} \binom{n}{i} [R_1(t)]^i [1 - R_1(t)]^{n-i}.$$

iii) Finally, we can of course consider systems made up of a number of subsystems placed in series and others placed in parallel. For instance, if there are two subsystems in parallel followed by another subsystem placed in series, then we can write that

$$R(t) = \{1 - [1 - R_1(t)][1 - R_2(t)]\} R_3(t).$$

3.10 Exercises, Problems, and Multiple Choice Questions

Solved Exercises

Exercise no. 1 (3.3)

Let X be the execution time (in minutes) of a program submitted to a central computer. When a program contains at least one "fatal error," it is not executed (execution time is equal to zero). Moreover, there is a one-minute time limit imposed on the execution time (the execution will stop at the end of one minute, even if it is not finished). Finally, we suppose that the distribution function of X, which is a random variable of mixed type, is

$$F_X(x) = \begin{cases} 0 & \text{if } x < 0, \\ 0.2 & \text{if } x = 0, \\ (3x + c)/5 & \text{if } 0 < x < 1, \\ 1 & \text{if } x \geq 1. \end{cases}$$

Calculate
a) the constant c;
b) the probability $P[\{X = 0\} \cup \{X = 1\}]$;
c) the function $f_X(x \mid 0 < X < 1)$.

Solution

a) Since F_X is right-continuous, we can write that

$$P[X \leq 0] = 0.2 = \frac{3 \cdot 0 + c}{5} \Rightarrow c = 1.$$

b) $P[\{X = 0\} \cup \{X = 1\}] = P[X = 0] + P[X = 1]$
$$\stackrel{a)}{=} 0.2 + \left[1 - \frac{3 \cdot 1 + 1}{5}\right] = 0.2 + 0.2 = 0.4.$$

c) $$f_X(x \mid 0 < X < 1) = \frac{d}{dx} F_X(x \mid 0 < X < 1)$$

$$= \frac{\frac{d}{dx}[F_X(x) - F_X(0)]}{P[0 < X < 1]} \quad \text{if } 0 < x < 1$$

$$\overset{a)}{=} \frac{3/5}{0.8 - 0.2} = 1.$$

It follows that

$$f_X(x \mid 0 < X < 1) = \begin{cases} 1 \text{ if } 0 < x < 1, \\ 0 \text{ elsewhere}. \end{cases}$$

Exercise no. 2 (3.3)

Let

$$F_X(x) = \begin{cases} 0 & \text{if } x < 0, \\ 1/4 & \text{if } 0 \le x < 1, \\ 1/2 & \text{if } 1 \le x < 4, \\ c & \text{if } x \ge 4. \end{cases}$$

a) Find the constant c. Justify your answer.

b) Calculate $P[1 \le X < 5]$.

c) Find $p_X(1) + p_X(2)$.

d) Obtain $f_X(x \mid X \le 1)$.

Solution

a) Since $\lim_{x \to \infty} F_X(x) = 1$, we deduce that $c = 1$.

b) $P[1 \le X < 5] = F_X(5^-) - F_X(1^-) = c - \frac{1}{4} \overset{a)}{=} 1 - \frac{1}{4} = \frac{3}{4}$.

c) We calculate

$$p_X(1) = F_X(1) - F_X(1^-) = \frac{1}{2} - \frac{1}{4} = \frac{1}{4}$$

and

$$p_X(2) = F_X(2) - F_X(2^-) = \frac{1}{2} - \frac{1}{2} = 0.$$

Thus, $p_X(1) + p_X(2) = \frac{1}{4}$.

d) We have:

x	0	1	4	Σ
$p_X(x)$	$\frac{1}{4}$	$\frac{1}{4}$	$\frac{1}{2}$	1

$\Rightarrow$

x	0	1	Σ
$p_X(x \mid X \le 1)$	$\frac{1}{2}$	$\frac{1}{2}$	1

$$\Rightarrow \quad f_X(x \mid X \le 1) = \frac{1}{2}\delta(x) + \frac{1}{2}\delta(x - 1) \quad \forall x \in \mathbb{R}.$$

Exercise no. 3 (3.6) A well-balanced (or non-biased) die is tossed until a "6" is obtained.

a) Given that the first three tosses produced a "1," what is the probability that the die will have to be tossed more than five times in all?

b) Let X be the number of tosses required to obtain a first "6." We define $Y = \min\{X, 5\}$. That is, Y is the minimum between X and the number 5. Calculate $p_Y(y)$.

Solution

a) We seek $P[X > 5 \mid X > 3]$, where $X \sim \text{Geom}(p = 1/6)$. We have:

$$P[X > 5 \mid X > 3] = P[X > 2] = \left(\frac{5}{6}\right)^2 = \frac{25}{36}.$$

b) We can write that

$$P[Y = y] = P[X = y] \quad \text{if } y = 1, 2, 3, 4$$

and

$$P[Y = 5] = P[X \geq 5] = P[X > 4] = \left(\frac{5}{6}\right)^4.$$

Thus, we have:

y	1	2	3	4	5
$p_Y(y)$	$\frac{1}{6}$	$\left(\frac{5}{6}\right)\left(\frac{1}{6}\right)$	$\left(\frac{5}{6}\right)^2\left(\frac{1}{6}\right)$	$\left(\frac{5}{6}\right)^3\left(\frac{1}{6}\right)$	$\left(\frac{5}{6}\right)^4$

Exercise no. 4 (3.7)

Let

$$f_X(x) = \begin{cases} 2xe^{-x^2} & \text{if } x > 0, \\ 0 & \text{elsewhere.} \end{cases}$$

a) We define $Y = 2 \ln X$. Calculate $f_Y(y)$.
b) Let $Z := 1/X$. Calculate $E[Z]$.

Solution

a)

$$f_Y(y) = f_X(x)\left|\frac{de^{y/2}}{dy}\right| = 2\left(e^{y/2}\right)e^{-\left(e^{y/2}\right)^2}\left|\frac{1}{2}e^{y/2}\right| = e^y e^{-e^y}$$

$$= \exp\{y - e^y\} \quad \text{if } y \in \mathbb{R}, \text{ since } 2 \ln 0 = -\infty \text{ and } 2 \ln \infty = \infty.$$

b)

$$E[Z] = E\left[\frac{1}{X}\right] = \int_0^\infty \frac{1}{x} 2xe^{-x^2} dx = 2\sqrt{2\pi} \int_0^\infty \frac{1}{\sqrt{2\pi}} e^{-x^2} dx$$

$$= 2\sqrt{2\pi} \, (1/2)^{1/2} \int_0^\infty \frac{1}{\sqrt{2\pi} \, (1/2)^{1/2}} e^{-\frac{x^2}{2(1/2)}} dx$$

$$= \frac{2\sqrt{2\pi}}{\sqrt{2}} \underbrace{P\left[N\left(0, \frac{1}{2}\right) > 0\right]}_{1/2} = \sqrt{\pi} \simeq 1.77.$$

Exercise no. 5 (3.7)
Let X be the time (in hours) a computer terminal is used during one workday.
a) Suppose that X has (approximately) a Gaussian $N(4, 2)$ distribution. Find the
number x_0 such that, for 90% of the days, X is greater than x_0.
Hint. We have: $Q^{-1} (0.10) \simeq 1.28$.
b) Suppose that $X \sim U[0, 8]$. Calculate i) the probability density function of $Y := e^X$
and ii) the mathematical expectation of Y^2.

Solution

a) We want

$$P[X > x_0] = 0.90 \Rightarrow P\left[N(0, 1) > \frac{x_0 - 4}{\sqrt{2}}\right] \simeq 0.90 \Rightarrow Q\left(\frac{x_0 - 4}{\sqrt{2}}\right) \simeq 0.90$$

$$\Rightarrow Q\left(\frac{-x_0 + 4}{\sqrt{2}}\right) \simeq 0.10 \Rightarrow \frac{-x_0 + 4}{\sqrt{2}} \simeq 1.28 \Rightarrow x_0 \simeq 2.19.$$

b) i) We calculate

$$f_Y(y) = f_X(x) \left|\frac{d \ln y}{dy}\right| = \frac{1}{8}\left|\frac{1}{y}\right| = \frac{1}{8y} \quad \text{if } e^0 \le y \le e^8$$

$$\Leftrightarrow \quad f_Y(y) = \begin{cases} (8y)^{-1} & \text{if } 1 \le y \le e^8, \\ 0 & \text{elsewhere.} \end{cases}$$

ii) We have:

$$E[Y^2] = E[e^{2X}] = \int_0^8 e^{2x} \frac{1}{8} dx = \frac{e^{2x}}{16}\Big|_0^8 = \frac{e^{16} - 1}{16}.$$

Or:

$$E[Y^2] = \int_1^{e^8} y^2 \frac{1}{8y} dy = \frac{y^2}{16}\Big|_1^{e^8} = \frac{e^{16} - 1}{16}.$$

Exercise no. 6 (3.7)

Four calls are received at random, according to a uniform distribution, at a central dispatch location during a one-hour period. What is the mathematical expectation of the number of calls received during the first half hour of the hour considered, knowing that at least one of the four calls was received during these first 30 minutes?

Solution

Let X be the number of calls received during the first half hour. Then, we can write that $X \sim B(n = 4, p = 1/2)$. We seek $E[X \mid X \geq 1]$. We have:

$$P[X \geq 1] = 1 - P[X = 0] = 1 - \left(\frac{1}{2}\right)^4 = \frac{15}{16}.$$

It follows that

$E[X \mid X \geq 1]$

$$= \sum_{k=1}^{4} k\, P[X = k \mid X \geq 1] = \frac{16}{15} \sum_{k=1}^{4} k\, p_X(k)$$

$$= \frac{16}{15} \sum_{k=1}^{4} k\binom{4}{k}\left(\frac{1}{2}\right)^4 = \frac{1}{15}\{1 \times 4 + 2 \times 6 + 3 \times 4 + 4 \times 1\} = \frac{32}{15} \simeq 2.13.$$

Exercise no. 7 (3.7)

Let X be the time (in minutes) required by a technician to assemble a certain part.

a) We define $Y = e^X$. Calculate the probability density function of Y if $X \sim U(0, 1)$.

b) Suppose that $X \sim \text{Exp}(1)$. Calculate $P[X^2 > 4 \mid X^2 > 1]$.

c) Suppose that $X \sim N(1, \sigma_X^2)$.

   i) Calculate $P[|X - 1| < 1]$ if $\sigma_X^2 = 0.25$.

   ii) Find σ_X^2 if $P[X < 2] = 0.99865$.

Hint. $Q(1) \simeq 1.59 \times 10^{-1}$, $Q(2) \simeq 2.28 \times 10^{-2}$, $Q(3) \simeq 1.35 \times 10^{-3}$ and $Q(4) \simeq 3.17 \times 10^{-5}$.

Solution

a) $f_Y(y) = f_X(x)\left|\dfrac{dx}{dy}\right| = 1 \left|\dfrac{d \ln y}{dy}\right| = \dfrac{1}{y}$ if $1 < y < e$ ($= 0$ elsewhere).

b) $P\left[X^2 > 4 \mid X^2 > 1\right] = P[X > 2 \mid X > 1]$,   because $X \geq 0$

$$= P[X > 2 - 1] = e^{-1(1)} = e^{-1} \simeq 0.3679.$$

c) i) $P[|X - 1| < 1]$

$$= P[-1 < X - 1 < 1] = P\left[-\frac{1}{0.5} < Z_0 < \frac{1}{0.5}\right], \quad \text{where } Z_0 \sim N(0, 1)$$

$$= \Phi\,(2) - \Phi\,(-2) = 2\Phi\,(2) - 1 = 2\,[1 - Q\,(2)] - 1 \simeq 2(0.9772) - 1$$
$$= 0.9544.$$

ii) $P\,[X < 2] = \Phi\left(\dfrac{2 - 1}{\sigma_X}\right) = 0.99865 \Rightarrow Q\left(\dfrac{1}{\sigma_X}\right) = 1.35 \times 10^{-3}$

$$\Rightarrow \dfrac{1}{\sigma_X} \simeq 3 \Rightarrow \sigma_X \simeq \dfrac{1}{3} \Rightarrow \sigma_X^2 \simeq \dfrac{1}{9}.$$

Exercise no. 8 (3.7)

Let

$$f_X(x) = \dfrac{6x}{(1 + x)^4} \quad \text{for } x > 0 \quad (= 0 \text{ elsewhere}).$$

a) Calculate the probability density function of $Y := 1/X$.
b) Use part a) to compute $E\,[X]$.
c) Calculate $f_X\,(x \mid X < 1)$.
d) Calculate the median of X.

Solution

a) $f_Y\,(y) = f_X\,(1/y)\left|\dfrac{d\,(1/y)}{dy}\right| = \dfrac{6y^{-1}}{\left(1 + y^{-1}\right)^4}\left|-\dfrac{1}{y^2}\right| = \dfrac{6y}{(1 + y)^4}$ if $y > 0$.

b) We have:

$$E\,[X] \overset{a)}{=} E\,[Y] = E\left[\dfrac{1}{X}\right] = \int_0^\infty \dfrac{1}{x}\dfrac{6x}{(1 + x)^4}dx = 6\left.\dfrac{(1 + x)^{-3}}{-3}\right|_0^\infty = 2.$$

c) $P\,[X < 1] = \displaystyle\int_0^1 \dfrac{6x}{(1 + x)^4}dx = 6x\left.\dfrac{(1 + x)^{-3}}{-3}\right|_0^1 + \int_0^1 2\,(1 + x)^{-3}\,dx$

$$= -\dfrac{2}{2^3} + 2\left.\dfrac{(1 + x)^{-2}}{-2}\right|_0^1 = -\dfrac{1}{4} - \dfrac{1}{2^2} + 1 = \dfrac{1}{2}$$

$$\Rightarrow \quad f_X\,(x \mid X < 1) = \begin{cases} 0 & \text{if } x \leq 0, \\ 12x\,(1 + x)^{-4} & \text{if } 0 < x < 1, \\ 0 & \text{if } x \geq 1. \end{cases}$$

d) We deduce from part c) that $F_X\,(1) = 1/2$. It follows that $x_{0.5} \equiv x_m = 1$.

Exercise no. 9 (3.7)

Let X be the service time (in minutes) at a counter. We suppose that $E\,[X] = 3$.
a) Give a lower bound (> 0) for $P\,[X < 6]$. Justify your answer.
b) Suppose that $X \sim G(\alpha = 3, \lambda = 1)$. Calculate $P\,[X > 5 \mid X \geq 3]$.
c) Suppose that $X \sim N(3, 1)$. Calculate
   i) the 90th percentile of $Y := 2X - 4$;
   ii) $P\left[W = 2 \mid W^2 = 4\right]$, where $W := X - 3$.

Hint. We have: $Q(1.2815) \simeq 0.1$ and $Q(2.3263) \simeq 0.01$.

Solution

a) We have, by Markov's inequality:

$$P[X \geq 6] \leq \frac{E[X]}{6} = \frac{1}{2} \Rightarrow P[X < 6] \geq 1 - \frac{1}{2} = \frac{1}{2}.$$

b)

$$P[X > 5 \mid X \geq 3] = \frac{P[X > 5]}{P[X \geq 3]}.$$

We have:

$$P[X \leq 5] = P[\text{Poi}(1 \cdot 5) \geq 3] = 1 - e^{-5}\left\{1 + 5 + \frac{5^2}{2!}\right\}$$

and

$$P[X \leq 3] = P[\text{Poi}(3) \geq 3] = 1 - e^{-3}\left\{1 + 3 + \frac{3^2}{2!}\right\}.$$

It follows that

$$P[X > 5 \mid X \geq 3] \overset{\text{cont.}}{=} \frac{18.5e^{-5}}{8.5e^{-3}} = e^{-2}\frac{18.5}{8.5} \simeq 0.2946.$$

c) i) $Y \sim N(2 \times 3 - 4, 4 \times 1) \equiv N(2, 4)$

$$\Rightarrow P[Y \leq y_{0.9}] = 0.9 \Leftrightarrow \Phi\left(\frac{y_{0.9} - 2}{2}\right) = 0.9 \Leftrightarrow Q\left(\frac{y_{0.9} - 2}{2}\right) = 0.1$$

$$\Leftrightarrow \frac{y_{0.9} - 2}{2} \simeq 1.2815 \Leftrightarrow y_{0.9} \simeq 4.563.$$

ii) $W \sim N(0, 1)$

$$\Rightarrow P[W = 2 \mid W^2 = 4] = P[W = 2 \mid \{W = 2\} \cup \{W = -2\}] = 1/2,$$

by symmetry.

Exercise no. 10 (3.7)

We suppose that the length T_j (in minutes) of the telephone calls received during the day at a certain emergency number has a $U(0, 4]$ distribution, whereas the length T_N (in minutes) of those received during the night at the same emergency number has (approximately) an $\text{Exp}(1)$ distribution. Moreover, 80% of the calls are received during the day (and 20% during the night).

a) Calculate the probability that a given call lasts more than one minute.

b) What is the average length of the calls received during the night and lasting more than one minute?

c) Calculate $P[T_j = 1 \mid \{T_j = 1\} \cup \{T_j = 2\}]$.

Solution

a) Let T be the length of a given call. We can write that

$$P[T > 1] = P[T_J > 1](0.8) + P[T_N > 1](0.2)$$
$$= (3/4)(0.8) + e^{-1(1)}(0.2) = 0.6 + (0.2)e^{-1} \simeq 0.6736.$$

b) We seek

$$E[T_N \mid T_N > 1] = 1 + E[T_N], \quad \text{by the memoryless property of } T_N$$
$$= 1 + 1 = 2 \text{ minutes.}$$

c) By symmetry (since $f_{T_J}(1) = f_{T_J}(2) = 1/4$), we can write that

$$P[T_J = 1 \mid \{T_J = 1\} \cup \{T_J = 2\}] = \frac{1}{2}.$$

Exercise no. 11 (3.7)

The input to a communication channel is a certain voltage v and the output is a voltage $X = v + B$, where B has a standard Gaussian distribution. We use this communication channel to send binary information as follows: to send a 0, we input a voltage $v = -2$, while to send a 1, we input a voltage $v = +2$. The receiver decides that a 0 has been sent if the voltage X he receives is negative, and that a 1 has been sent if X is positive or zero.

a) Calculate $P[X = v]$.

b) Let $Y := e^X$. Obtain the probability density function of Y.

c) Calculate i) VAR$[X]$ and ii) VAR$[X \mid B = 0]$.

d) Find the probability that the receiver makes a detection error, if the probability of sending a 0 is equal to p.

Solution

a) $P[X = v] = P[v + B = v] = P[B = 0] = 0$, because B is a continuous random variable.

b) $Y := e^X$, where $X = v + B \sim N(v, 1) \Rightarrow y > 0$ and

$$f_Y(y) = f_X(\ln y) \left| \frac{d \ln y}{dy} \right| = \frac{1}{\sqrt{2\pi}} \exp\left\{ -\frac{1}{2}(\ln y - v)^2 \right\} \cdot \frac{1}{y} \quad \text{for } y \in \mathbb{R}^+.$$

c) i) VAR$[X]$ = VAR$[v + B]$ = VAR$[B]$ = 1.
    ii) VAR$[X \mid B = 0]$ = VAR$[v + 0 \mid B = 0]$ = VAR$[v]$ = 0.

d) Let F = the receiver makes a detection error, and E_k = a k is sent, for $k = 0, 1$. We have:

$$P[F] = P[F \mid E_0]P[E_0] + P[F \mid E_1]P[E_1]$$

$$= P[-2 + B \geq 0]p + P[2 + B < 0](1 - p)$$
$$= P[N(0, 1) \geq 2]p + P[N(0, 1) < -2](1 - p)$$
$$= Q(2)p + \Phi(-2)(1 - p) = 1 - \Phi(2) \stackrel{\text{Tab. 3.3}}{\simeq} 1 - 0.9772 = 0.0228.$$

Remark. For this last question, we could consider the constant v as a random variable V that takes on the value -2 (respectively $+2$) with probability p (resp. $1 - p$) and write that

$$P[F] = P[X \geq 0 \mid V = -2]p + P[X < 0 \mid V = +2](1 - p).$$

However, this formula will only be given in Chapter 4 (on random vectors).

Exercise no. 12 (3.8) Let X be a discrete random variable whose set of possible values is $S_X = \{0, 1, 2, \ldots\}$ and for which $M_X(s) = \exp\{\alpha(e^{-s} - 1)\}$. Calculate a) $p_X(0)$; b) the standard deviation of $2X$.

Solution

a) We have:

$$G_X(z) = M_X(-\ln z) = \exp\left\{\alpha\left(e^{\ln z} - 1\right)\right\} = \exp\{\alpha(z - 1)\}.$$

We find (see Equation (3.74) and also Example 3.8.1) that $X \sim \text{Poi}(\alpha)$, so that $p_X(0) = e^{-\alpha}$.

b) Since $X \sim \text{Poi}(\alpha)$, we may write that

$$\text{STD}[2X] = (\text{VAR}[2X])^{1/2} = (4\alpha)^{1/2} = 2\sqrt{\alpha}.$$

Exercise no. 13 (3.8)

Let

x	0	1	3
$p_X(x)$	1/8	3/8	1/2

Use the generating function of X to calculate its standard deviation.

Solution

We have:

$$G_X(z) := E[z^X] = z^0 \frac{1}{8} + z^1 \frac{3}{8} + z^3 \frac{1}{2} = \frac{1}{8} + \frac{3}{8}z + \frac{1}{2}z^3.$$

Then,

$$\text{VAR}[X] = G_X''(1) + G_X'(1) - \left(G_X'(1)\right)^2$$
$$= (3z)|_{z=1} + \left(\frac{3}{8} + \frac{3}{2}z^2\right)\Big|_{z=1} - \left(\frac{3}{8} + \frac{3}{2}\right)^2$$

$$= 3 + \frac{15}{8} - \left(\frac{15}{8}\right)^2 \simeq 1.359.$$

It follows that STD$[X] \simeq 1.166$.

Exercise no. 14 (3.8)

The probability mass function of the discrete random variable X is given by the following table:

x	-1	0	$1/2$	1
$p_X(x)$	$1/8$	$1/8$	$1/4$	$1/2$

a) Calculate $P[X = \frac{3}{4}] + F_X(\frac{3}{4})$.

b) Let $Y := X^2$. Calculate i) $p_Y(y)$ and ii) the standard deviation of Y.

c) Calculate the characteristic function of X at $\omega = \pi$. Simplify your answer as much as possible.

d) Can we calculate the generating function of X? Justify your answer.

Solution

a) We have: $P[X = 3/4] = 0$, because $3/4 \notin S_X$. Moreover,

$$F_X(3/4) = P[X \leq 3/4] = P[X \leq 1/2] = 1/2.$$

Thus,

$$P[X = 3/4] + F_X(3/4) = 1/2.$$

b) i) We have:

y	0	$\frac{1}{4}$	1	Σ
$p_Y(y)$	$\frac{1}{8}$	$\frac{1}{4}$	$\frac{5}{8}$	1

ii) We calculate

$$E[Y] = 0 + \frac{1}{4} \times \frac{1}{4} + 1 \times \frac{5}{8} = \frac{11}{16} \quad \text{and} \quad E\left[Y^2\right] = 0 + \frac{1}{16} \times \frac{1}{4} + 1 \times \frac{5}{8} = \frac{41}{64}.$$

Then,

$$\text{VAR}[Y] = \frac{41}{64} - \left(\frac{11}{16}\right)^2 = \frac{43}{256} \quad \text{and} \quad \text{STD}[Y] = \frac{\sqrt{43}}{16} \simeq 0.4098.$$

c) $\qquad \phi_X(\omega) := E\left[e^{j\omega X}\right] = e^{-j\omega} \cdot \frac{1}{8} + 1 \cdot \frac{1}{8} + e^{j\omega/2} \cdot \frac{1}{4} + e^{j\omega} \cdot \frac{1}{2}$

$$\Rightarrow \phi_X(\pi) = \frac{1}{8}\left\{e^{-j\pi} + 1 + 2e^{j\pi/2} + 4e^{j\pi}\right\} = \frac{1}{8}\{-1 + 1 + 2j - 4\}$$

$$= -\frac{1}{2} + \frac{j}{4}.$$

d) We cannot calculate $G_X(z)$, because this function is only defined when the elements of S_X are all positive integers or zero, which is not the case here.

Remark. We can calculate $E[z^X]$. However, we cannot call this mathematical expectation the generating function of the random variable X.

Exercise no. 15 (3.8)

Let $X \sim N(0, 1)$. We define $Y = e^{X^2}$ and $Z = 2X - 1$.

a) Obtain $f_Y(y)$.

b) Calculate $E[1/Y]$.

c) i) Obtain the characteristic function of Z. ii) Use $\phi_Z(\omega)$ to calculate $E[Z^2]$.

Hint. We have: $\phi_X(\omega) = e^{-\omega^2/2}$.

Solution

a) We have: $y = e^{x^2} \Leftrightarrow x = \pm\sqrt{\ln y}$, where $y \geq 1$. It follows that

$$f_Y(y) = f_X\left(-\sqrt{\ln y}\right)\left|-\frac{d}{dy}\sqrt{\ln y}\right| + f_X\left(\sqrt{\ln y}\right)\left|\frac{d}{dy}\sqrt{\ln y}\right|$$

$$= 2 \cdot \frac{1}{2\sqrt{\ln y}\, y}\left[\frac{1}{\sqrt{2\pi}}\exp\left\{-\frac{(\sqrt{\ln y})^2}{2}\right\}\right]$$

$$= \frac{1}{\sqrt{2\pi}}\frac{1}{y\sqrt{\ln y}}\exp\left\{-\frac{\ln y}{2}\right\} = \frac{1}{\sqrt{2\pi}}y^{-3/2}(\ln y)^{-1/2} \quad \text{for } y \geq 1.$$

b) $E[1/Y] = E[e^{-X^2}] = \int_{-\infty}^{\infty} e^{-x^2}\frac{1}{\sqrt{2\pi}}e^{-x^2/2}dx = \int_{-\infty}^{\infty}\frac{1}{\sqrt{2\pi}}e^{-3x^2/2}dx$

$$= \sqrt{\frac{1}{3}}\int_{-\infty}^{\infty}\frac{1}{\sqrt{2\pi}\sqrt{1/3}}\exp\left\{-\frac{x^2}{2(1/3)}\right\}dx = \sqrt{\frac{1}{3}}P\left[N\left(0, \tfrac{1}{3}\right) \in \mathbb{R}\right]$$

$$= 3^{-1/2} \simeq 0.5774.$$

c) i) $\quad \phi_Z(\omega) := E[e^{j\omega Z}] = E[e^{j\omega(2X-1)}] = e^{-j\omega}E[e^{j2\omega X}] = e^{-j\omega}\phi_X(2\omega)$

$$= e^{-j\omega}e^{-(2\omega)^2/2} = e^{-j\omega-2\omega^2}.$$

ii) We can write that

$$E[Z^2] = -\phi_Z''(\omega)\big|_{\omega=0} = -\frac{d}{d\omega}\left\{e^{-j\omega-2\omega^2}(-j-4\omega)\right\}\Big|_{\omega=0}$$

$$= -e^{-j\omega-2\omega^2}[(-j-4\omega)^2 - 4]\big|_{\omega=0} = -e^0[(-j)^2 - 4] = 5.$$

Unsolved Problems

Problem no. 1

Let

$$F_X(x) = \begin{cases} 0 & \text{if } x < -1, \\ 1/3 & \text{if } -1 \le x < 0, \\ 3/4 & \text{if } 0 \le x < 2, \\ 1 & \text{if } 2 \le x. \end{cases}$$

Calculate a) $p_X(0) + p_X(1)$; b) $F_X(0 \mid X > -1)$.

Problem no. 2

Let

$$f_X(x) = \begin{cases} 1/2 & \text{if } 0 < x < 1, \\ 1/(2x) & \text{if } 1 \le x < e. \end{cases}$$

Calculate a) $P[X < 1 \mid X < 2]$; b) $f_X(x \mid 1 < X < e)$.

Problem no. 3

Let X be a random variable having a $B(n = 2, p = 1/4)$ distribution. Calculate
a) $P[X \ge 1 \mid X \le 1]$; b) $E[X \mid X > 0]$.

Problem no. 4

Let $X \sim \text{Exp}(\lambda = 2)$. a) Calculate $P[X^2 < 9]$. b) Use Bienaymé–Chebyshev's
inequality to compute $P\left[\left|X - \frac{1}{2}\right| \ge 2\right]$. Compare with the exact value.

Problem no. 5

Let $X \sim G(\alpha = 3, \lambda = 1/2)$. Calculate a) $P[X < 5]$; b) the standard deviation
of $Y := 2X - 4$.

Problem no. 6

A machine produces parts whose diameter X (in centimeters) has approximately
a Gaussian $N(\mu, \sigma^2 = (0.01)^2)$ distribution.
a) What should the value of μ be so that no more than 1% of the parts have a diameter
greater than 3 cm?
b) Suppose that $\mu = 3$. Calculate the probability density function of $Y := |X - 3|$.

Problem no. 7

Let X be a discrete random variable such that $p_X(0) = e^{-\lambda}$, where $\lambda > 0$, and

$$p_X(x) = \frac{e^{-\lambda}\lambda^{|x|}}{2(|x|!)} \quad \text{for } x = \dots, -2, -1, 1, 2, \dots.$$

Calculate a) $E[|X|]$; b) $E[X \mid X > 0]$ if $\lambda = 1$.

Problem no. 8

Let

$$f_X(x) = \begin{cases} \frac{1}{4}e^{x/2} & \text{if } x < 0, \\ \frac{1}{4}e^{-x/2} & \text{if } x \ge 0. \end{cases}$$

Calculate a) $E\left[X^2\right]$; b) the characteristic function of $|X|$.

Problem no. 9
Let $X \sim U[-1, 1]$. Calculate a) the kurtosis of the random variable X;
b) $\mathrm{VAR}\,[X \mid X > 0]$.

Problem no. 10
Let

$$F_X\,(x) = \begin{cases} 0 & \text{if } x < -1, \\ 1/8 & \text{if } -1 \le x < 1, \\ 3/8 & \text{if } 1 \le x < 2, \\ 1 & \text{if } 2 \ge x. \end{cases}$$

a) Calculate the probability density function of X.
b) Let $A := \{X > 1\}$. Calculate $F_X\,(x \mid A)$.

Problem no. 11
Let

$$F_X\,(x) = \begin{cases} 0 & \text{if } x < -1, \\ 1/2 & \text{if } x = -1, \\ (x + 4)/6 & \text{if } -1 < x < 2, \\ 1 & \text{if } 2 \ge x. \end{cases}$$

a) Calculate $P\,[-1 < X \le 1.5]$.
b) Let $A := \{X \ge 1\}$. Calculate $f_X\,(x \mid A)$.

Problem no. 12
The number of power failures that occur in a certain region during a one-year period has a Poisson distribution with parameter $\alpha = 3$. The length (in hours) of a power failure has an exponential distribution with parameter $\lambda = 1/2$.
a) Calculate the probability that a power failure that has already lasted two hours will end within the next 30 minutes.

b) Assuming independence of the power failures from year to year, calculate the probability that, during the next ten years, there will be at least one year during which exactly one power failure will occur.

Problem no. 13
Let $X \sim N\,(0, 1)$. We have: $\phi_X\,(\omega) = e^{-\omega^2/2}$. Calculate $\mathrm{VAR}\left[X^2\right]$.

Problem no. 14
The time T (in months) that elapses between two car accidents at a certain intersection has an exponential distribution with parameter $\lambda = 1/4$. The number of persons involved in an accident has a geometric distribution with parameter $p = 1/3$.
a) Calculate the probability that a given accident involves at least four persons, knowing that more than two persons are involved.

b) Let A_k = the time elapsed between the $(k - 1)$st and kth accidents at this intersection is greater than one year, for $k = 2, 3, \ldots$. We assume that the events A_k are (globally) independent. What is the probability that, among the events $A_2, \ldots, A_{21}$, exactly two will occur?

Problem no. 15

Let X be the lifetime (in months) of an electronic component.

a) Suppose that X has (approximately) a Gaussian N(20, 16) distribution. Calculate $P[|X| < 24]$.

Hint. We have: $Q(1) \simeq 0.159$.

b) Suppose that $X \sim G(\alpha = 2, \lambda)$; that is,

$$f_X(x) = \lambda^2 x e^{-\lambda x} \quad \text{if } x > 0 \quad (= 0 \text{ elsewhere}).$$

Calculate i) the probability density function of $Y := 1/X$ and ii) the mathematical expectation of Y.

Problem no. 16

A box contains three transistors, denoted by A, B and C. Transistor A was made by a machine that produces 3% defectives, transistor B by a machine that produces 5% defectives and transistor C by a machine that produces 7% defectives. Ten transistors are taken, at random and with replacement. Let N be the number of defective transistors obtained.

a) Calculate $P[N = 0]$.

b) Does N have a binomial distribution? If so, give its parameters. Otherwise, justify.

c) Calculate the mathematical expectation of N, given that N is less than or equal to 1.

Problem no. 17

Let

$$f_X(x) = \begin{cases} kx(1 - 2x) & \text{if } 0 < x \leq 1/2, \\ 0 & \text{elsewhere,} \end{cases}$$

where $k > 0$.

a) Find the constant k.

b) Calculate the distribution function of X.

c) Calculate the variance of X.

d) Let $Y := 1/X$. Calculate i) the probability density function of Y and ii) its mathematical expectation.

Problem no. 18

a) Let

$$f_X(x) = K e^{-x^2} \quad \text{if } x \in \mathbb{R},$$

where $K > 0$.

i) Find the constant K.

ii) Calculate $f_X(x \mid X > 0)$.

b) Let

$$f_Y(y) = \begin{cases} 2e^{-2(y-1)} & \text{if } y \geq 1, \\ 0 & \text{elsewhere.} \end{cases}$$

i) Calculate the characteristic function of Y.

ii) Use the function calculated in i) to obtain the mean of Y.

Problem no. 19

A crate holds two boxes that contain ten items each. All the items were made by the same machine. Suppose that the probability that an item made by this machine is defective is equal to 0.05, independently from one item to another.

a) What is the probability that there are less than three defective items among the 20 items considered, knowing that there is at least one defective among these 20 items?

b) Use a Poisson distribution to compute (approximately) the probability in a).

Problem no. 20

Let X be the waiting time (in minutes) before being served at a certain counter. We suppose that the distribution function of X is given by

$$F_X(x) = \begin{cases} 0 & \text{if } x < 0, \\ \frac{1}{4} & \text{if } x = 0, \\ 1 - \frac{3}{x+4} & \text{if } x > 0. \end{cases}$$

a) Calculate $P[X = 0 \mid X < 1]$.

b) Calculate the probability density function of X.

c) Let $Y := X \mid \{X > 0\}$. Calculate i) $f_Y(y)$ and ii) $E[Y]$.

Problem no. 21

Let X be a discrete random variable whose probability mass function is given by the following table:

x	1	4	9
$p_X(x)$	$\frac{1}{8}$	$\frac{5}{8}$	$\frac{1}{4}$

a) Calculate the variance of $\sqrt{X}$.

b) Let $Z := X \mid \{X > 1\}$. Calculate the probability density function of Z.

c) i) Calculate the moment generating function of X.

ii) Use the function computed in i) to obtain $E[X^2]$.

Problem no. 22

A discrete random variable X has the following probability mass function:

x	-2	-1	0	1
$p_X(x)$	$\frac{1}{8}$	$\frac{1}{4}$	$\frac{1}{8}$	$\frac{1}{2}$

a) Calculate $F_X(0) + F_X(2)$.

b) Let $Y := -2X$. Obtain the standard deviation of Y.

c) Let $Z := X^2$. Find $f_Z(z)$.

d) Calculate $E\left[\frac{1}{X+3}\right]$.

e) Calculate $E[X \mid X < 0]$.

Problem no. 23

A small company has five telephone lines. We suppose that the probability that a given line is free at 11:00 a.m. is equal to 0.5 for each line, independently from one another. Moreover, this probability is the same for each workday, that is, from Monday through Friday.

a) We consider a period of four weeks with five workdays each. Calculate the probability that over this time period there are at least two days during which not a single telephone line is free at 11:00 a.m.

b) Use a Poisson distribution to compute (approximately) the probability in a).

c) If a man calls every day, exactly at 11:00 a.m., what is the average number of telephone calls that he will have to make to reach the company, knowing that on his first trial all the lines were busy?

d) During a given five-workday week, the lines were all busy at 11:00 a.m. on exactly two days. What is the probability that these two days were the Monday and Tuesday of that week?

Problem no. 24

Let X be the length (in minutes) of a telephone call received by a computer consultant.

a) Suppose first that $X \sim U[1, 5]$. Calculate $P\left[X^2 < 9\right]$.

b) Suppose next that $X \sim G(\alpha = 3, \lambda = 1)$. Calculate $P[X \geq 4]$.

c) Suppose finally that $X \sim N(3, 1)$ (approximately). If ten independent calls are received by the consultant during a certain period of time, what is the probability that exactly one of these calls lasts more than three minutes?

d) If the characteristic function of X is given by

$$\phi_X(\omega) = \frac{k}{1 - 4j\omega},$$

what is the value of the constant k?

Problem no. 25

Let

$$p_X(x) = [x(x+1)]^{-1} \quad \text{if } x = 1, 2, \ldots,$$
$$= 0 \quad \text{otherwise.}$$

a) Calculate $P[2 \leq X \leq 4]$.

b) Let
$$Y := \begin{cases} 1 \text{ if } X < 4, \\ 0 \text{ if } X \geq 4. \end{cases}$$

Find $F_Y(y)$.
c) Calculate $E[X \mid X < 4]$.
d) Calculate $E[X^2]$.

Problem no. 26

A point X is taken in the interval $[-1, 1]$ in such a way that it is twice as likely that X is positive. We suppose that

$$f_X(x) = \begin{cases} 1/3 \text{ if } -1 \leq x \leq 0, \\ 2/3 \text{ if } 0 < x \leq 1, \\ 0 \text{ elsewhere.} \end{cases}$$

Calculate
a) the probability density function $f_Y(y)$, where $Y := X^2$;
b) the variance $\mathrm{VAR}[-2X + 4]$;
c) the characteristic function $\phi_X(\omega)$;
d) the conditional probability density function $f_X(x \mid X \leq 0)$.

Problem no. 27

A computer generates random numbers according to an exponential distribution with parameter $\lambda = 1/3$. The numbers generated are independent of one another.
a) What is the probability that each of the first three numbers generated is greater than 9?

b) Let N be the number of random numbers greater than 9 among the first 20 numbers generated. Calculate the variance of N.

c) Use a Poisson distribution to calculate (approximately) the probability that exactly one of the first ten numbers generated is greater than 9.

d) Knowing that the first number generated is greater than 9, what is the probability that it is greater than 12?

Problem no. 28

A machine is made up of two types of parts: A and B. The number X of parts of type A that fail during a given period has a Poisson distribution with parameter $\alpha = 4$. In the case of the type B parts, the number Y of failures has the following probability mass function:

$$p_Y(k) = \left(\frac{3}{4}\right)^k \left(\frac{1}{4}\right) \quad \text{for } k = 0, 1, 2, \ldots .$$

The failures of the type A and type B parts are independent.
a) Calculate the probability that there is one failure (in all) during the period considered.

b) Calculate $E[X \mid X < 3]$.

c) We consider ten independent periods, each having the same length as the period above. What is the probability that during at least two of these periods there are no failures of the type B parts?

Problem no. 29

The time T that a technician spends performing the periodical maintenance of a certain machine is divided into two parts: a fixed period of 5 minutes made up of customary checks, followed by a random period X (in minutes) which depends on the repair work that the technician has to do.

a) Suppose that $X \sim U[0, 20]$. Calculate the probability density function of T.

b) Let $X \sim G(\alpha = 5, \lambda = 1/2)$. Calculate $E[T^2]$.

c) Suppose that $X \sim N(10, 4)$.

   i) Calculate the probability $P[|T - 15| \geq 6]$.

   ii) Compare the answer in i) with the bound given by Bienaymé–Chebyshev's inequality.

Hint. We have: $Q(1.5) \simeq 0.0668$ and $Q(3) \simeq 0.00135$.

Problem no. 30

Let

$$f_X(x) = \begin{cases} x/k & \text{if } 0 \leq x \leq k, \\ 0 & \text{elsewhere}, \end{cases}$$

where $k > 0$ is a constant.

a) Find the constant k.

b) Calculate the distribution function of X.

c) Calculate the moment generating function of X.

d) Let $Y := 1/X$. Calculate $\text{VAR}[\sqrt{Y}]$.

Problem no. 31

An airline company has realized that, on average, 4% of the people making reservations on a certain flight do not show up for the flight. Consequently, the policy of this company is to accept 75 reservations for a flight that can only hold 73 passengers. What is the probability that there is a seat available for each passenger who arrives for the flight? (Make the necessary assumptions.)

Problem no. 32

An exam grade, out of 100 points, can be considered as having a Gaussian $N(60, 100)$ distribution.

a) What is the average grade in this class?

b) What is the proportion of students whose grades are between 55% and 75%?

c) Where should the threshold for an A be placed, if we want 10% of the students to get an A?

Problem no. 33

In information theory, Hartley's formula expresses the quantity I of information of a message as a function of the signal-to-noise ratio S/N as follows:

$$2^I = 1 + \frac{S}{N}.$$

Let us write that $X = 2^I$.

a) Suppose that $S = 1$ and that N has a uniform distribution on the interval $(0, 1)$. Calculate i) the probability density function of X; ii) the expected value of X.

b) Suppose that $N = 1$ and that S has a uniform distribution on the interval $[1, 2]$. Calculate the moment generating function of X.

Problem no. 34

Let X be the number of different versions of a program that a computer science student has to write to obtain a program that works. We suppose that X has a geometric distribution with parameter $p = 1/4$. We consider a class with 20 students, who must each write a program individually.

a) Calculate $P[4 \leq X < 6 \mid X > 2]$.

b) Let N be the number of students, among the 20, who have to write exactly two versions of their programs to obtain a version that works. Calculate $P[N \leq 1]$.

c) Use Poisson's approximation to calculate the probability in b).

Problem no. 35

In information theory, the differential entropy of a continuous random variable X is defined by $H_X = -E[\ln f_X(X)]$. Moreover, the relative entropy of the continuous random variables X and Y is given by

$$H(f_X; f_Y) := \int_{-\infty}^{\infty} f_X(x) \ln(f_X(x)/f_Y(x)) \, dx.$$

a) Calculate H_X if $X \sim \text{Exp}(\lambda)$.

b) Let $Y := e^X$, where $X \sim U[0, 1]$. Find i) the probability density function of the random variable Y; ii) H_Y.

c) Calculate $H(f_X; f_Y)$ if $X \sim U[0, 1]$ and $Y \sim \text{Exp}(1)$.

Problem no. 36

a) Let $X \sim N(0, 1)$. We define $Y = |X|$.

   i) Calculate $P[Y < 2]$.

Hint. We have: $Q(1) \simeq 0.159$ and $Q(2) \simeq 0.0228$.

   ii) Calculate $\text{VAR}[Y]$.

b) Let $Z \sim G(\alpha = 2, \lambda = 1)$. Obtain $E[Z^3]$ by making use of the characteristic function of Z.

Hint. We have: $\phi_Z(\omega) = (1 - j\omega)^{-2}$.

Problem no. 37

Let X be the number of unsuccessful interviews that a graduating student must have before getting a first job. According to the data collected, we accept that $E[X] = 2.5$.

a) Calculate the maximum value of $P[X > 5]$.

b) If we assume that $X = Y - 1$, where Y has a geometric distribution, what value of p must we choose? Justify your answer.

c) If we have found that $VAR[X] \simeq 2.5$, is it better to use a geometric distribution as in b) or a Poisson distribution as a model for X? Justify your answer.

d) Suppose that X has a Poisson distribution. Calculate $P[X = 0 \mid X \leq 1]$.

Problem no. 38

   Let X be the lifetime (in years) of a certain machine.

a) Calculate $P[X^2 > 4]$ if $X \sim U[0, 10]$.

b) If the value of x that maximizes the probability density function of X is $x = 4$ years, is it better to use an exponential distribution or a gamma distribution as a model for X? Justify your answer.

c) Suppose that $X \sim N(4, 1)$ (approximately).
   i) Calculate $P[3 < X < 5]$.
   ii) Find the number x_0 such that $P[2X < x_0] = 0.9$.

Hint. We have: $Q^{-1}(0.1) \simeq 1.2815$ and $Q^{-1}(0.9) \simeq -1.2815$.

Problem no. 39

   We have a computer program that enables us to test whether a software program contains bugs or not. We suppose that, each time the program is used, the probability that it detects the presence of a given bug is equal to 0.95, independently from time to time. Furthermore, we suppose that the number N of bugs that a given software program contains has a Poisson distribution with parameter $\alpha = 0.5$, and that the bugs are independent.

a) Calculate $E[N \mid N \leq 2]$.

b) What is the probability that we must use the program less than three times to detect the presence of a given bug?

c) Calculate the probability that the program does not detect the presence of any bugs
   i) in a software program that has in fact k bugs, where $k \in \{0, 1, 2, \ldots\}$;
   ii) in any software program.

Problem no. 40

   Let
$$f_X(x) = \begin{cases} x^2/9 & \text{if } 0 < x < 3, \\ 0 & \text{elsewhere.} \end{cases}$$

We define $Y = int(X)$, where *int* denotes the integer part. Calculate

a) the probability density function of Y;

b) the generating function of Y;

c) $VAR[1/X]$;

d) $F_X(x \mid X < 1)$.

Problem no. 41

a) The random variable X has a Gaussian $N(0, 1)$ distribution. We define $Y = X^2$.

   i) Calculate the characteristic function of Y.

   ii) Give the distribution of Y, as well as its parameters.

b) Let Z be a random variable that has a $G(\alpha, \lambda)$ distribution. We define $W = aZ+b$, where $a \neq 0$ and b are constants.

   i) Calculate the characteristic function of W.

   ii) For what values of a and b does the random variable W have a gamma distribution?

Hint. We have: $\phi_Z(\omega) = \left(1 - \frac{j\omega}{\lambda}\right)^{-\alpha}$.

Problem no. 42

   Let

$$f_X(x) = \begin{cases} 0 & \text{if } x < 0, \\ c & \text{if } 0 \le x < 1, \\ 2c(2-x) & \text{if } 1 \le x \le 2, \\ 0 & \text{if } x > 2, \end{cases}$$

where c is a positive constant. Calculate

a) the constant c;

b) the distribution function of X;

c) the probability density function of X, given that $X < 1$;

d) the median of X;

e) the 95th percentile of X.

Problem no. 43

   A computer system user works at home and connects to the network by modem. We suppose that the probability that the user manages to connect to the network is equal to 0.95, independently from one attempt to another.

a) Let N be the number of successful connection attempts among 50 attempts. Calculate $P[N = 50 \mid N \ge 48]$.

b) Use a Poisson distribution to calculate (approximately) the probability $P[N = 49]$, where N is defined in a).

c) Let M be the number of attempts needed to manage to connect to the network. Calculate the variance of M, knowing that the first attempt was unsuccessful.

Problem no. 44

a) Let $X \sim U(0, 1]$. We define $Y = \ln X$. Calculate

   i) the distribution function of Y;

   ii) the moment generating function of Y.

b) Let $Z \sim G(\alpha = 2, \lambda = 1/2)$. Use the characteristic function of Z to obtain VAR$[Z^2]$.

Hint. We have: $\phi_Z(\omega) = (1 - 2j\omega)^{-2}$.

Problem no. 45

The users of a certain computer network can work at home with the help of a modem. Let T be the duration (in minutes) of a given user's work session. We suppose that

$$F_T(t) = \begin{cases} 0 & \text{if } t < 0, \\ t/100 & \text{if } 0 \le t < 90, \\ 1 & \text{if } t \ge 90. \end{cases}$$

a) What is the probability that a given user reaches the maximum duration of a work session? Justify your answer.

b) Let $S := T \mid \{T < 60\}$. That is, S is the duration of a work session, given that it is less than 60 minutes. Calculate i) the probability density function $f_S(s)$; ii) $E[S^2]$.

c) Let

$$U := \begin{cases} 0 & \text{if } T < 30, \\ 1 & \text{if } 30 \le T < 60, \\ 2 & \text{if } T \ge 60. \end{cases}$$

Calculate i) $p_U(u)$ and ii) the generating function of U^2.

Problem no. 46

Let X be a discrete random variable whose set of possible values is $S_X = \{-2, 0, 2\}$, and let $A = \{X \le 0\}$. We suppose that

$$F_X(x \mid A) = \begin{cases} 0 & \text{if } x < -2, \\ 3/5 & \text{if } -2 \le x < 0, \\ 1 & \text{if } x \ge 0. \end{cases}$$

Calculate a) $P[X^2 \le 1 \mid A]$; b) $E[X \mid A]$; c) $f_X(x \mid A)$; d) $F_X(x \mid A^c)$; e) $p_X(2)$ if $P[X = 0] = 2P[X = 2]$.

Problem no. 47

Let p be the probability that an individual in a population has a certain virus. To detect the presence of the virus, we analyze blood samples taken from each of the N members of this population. We assume that the N members of the population are independent.

a) i) Calculate the mean and the variance of the number of persons who have the virus, given that the first two blood tests performed were positive.

ii) Use Poisson's approximation to calculate the probability that at least three persons have the virus, if $N = 1000$ and $p = 0.01$.

b) Suppose now that instead of proceeding as above, we rather decide to combine the samples taken from k persons before analyzing them. If the test is negative, then the k persons do not have the virus. On the other hand, if the test is positive, then the k persons must be tested again individually. Calculate the average number of tests required with this method, if we suppose that $N = 500$, $p = 0.01$ and $k = 10$.

Multiple Choice Questions

Question no. 1

We suppose that the probability that a telephone call lasts more than five minutes is equal to 0.1, independently from one call to another.

A) Calculate the probability that, among 20 calls taken at random, there are more than 18 calls that do not last more than five minutes.

a) 0   b) 0.3917   c) 0.6083   d) 0.6769   e) 1

B) Calculate approximately the probability in A) using a Poisson distribution.

a) 0   b) 0.3233   c) 0.4060   d) 0.6767   e) 1

C) Calculate the probability that it takes less than five calls to obtain a first call that lasts more than five minutes.

a) 0.0001   b) 0.3439   c) 0.4095   d) 0.6561   e) 0.9999

D) What is the probability that, among five calls taken at random, the longest lasts less than five minutes?

a) 0.00001   b) 0.1   c) 0.4095   d) 0.5905   e) 0.9

Question no. 2

Let X be the time (in days) required to repair a machine. We suppose that the average repair time is equal to four days and the standard deviation to two days.

A) What is, at most (and with as much accuracy as possible), the probability that the repair time is smaller than one day or greater than seven days?

a) 4/9   b) 2/3   c) 1   d) we cannot calculate it   e) none of these answers

B) Suppose that X has a uniform distribution on the interval $[a, b]$. Find the constant a.

a) 0   b) 2   c) 4   d) we cannot calculate it   e) none of these answers

C) Suppose that X has a gamma distribution with parameters $\alpha = 4$ and $\lambda = 1$. Calculate $P[X < 4]$.

a) 0.2381   b) 0.3528   c) 0.5665   d) 0.5768   e) 0.6288

D) Suppose that X has a Gaussian distribution. Find the number x_0 such that $P[|X - 4| < x_0] = 0.99$.

Hint. We have: $Q^{-1}(0.01) \simeq 2.326$ and $Q^{-1}(0.005) \simeq 2.576$.

a) 2.326   b) 2.576   c) 4.652   d) 5.152   e) 10.304

Question no. 3

Let

$$F_X(x) = \begin{cases} \frac{1}{2}e^{\lambda x} & \text{if } x \le 0, \\ 1 - \frac{1}{2}e^{-\lambda x} & \text{if } x > 0, \end{cases}$$

where $\lambda > 0$. Calculate $F_X(1 \mid X > 0)$.

a) $\frac{1}{2}e^{-\lambda}$ b) $e^{-\lambda}$ c) $1 - e^{-\lambda}$ d) $1 - \frac{1}{2}e^{-\lambda}$ e) $2 - e^{-\lambda}$

Question no. 4
  Let

x	-1	0	1
$p_X(x)$	1/8	3/8	1/2

Calculate $F_X(0) + F_X(1/2)$.

a) 3/8   b) 1/2   c) 3/4   d) 7/8   e) 1

Question no. 5
  A computer generates independent observations of a random variable that has a Poisson distribution with parameter $\alpha = 1$. Let N be the number of observations, among the first 100 observations, that are greater than 1. Calculate VAR[N].

a) $200(1 - 2e^{-1})e^{-1}$ b) $100(1 - e^{-1})e^{-1}$ c) $100(1 - 2e^{-1})e^{-1}$
d) $200(1 - e^{-1})e^{-1}$ e) none of these answers

Question no. 6
  Let X be a random variable that has an exponential distribution with parameter λ. What is the value of λ if the 90th percentile of X is equal to 1?

a) 0.1   b) 0.9   c) 1   d) ln 10   e) ln 90

Question no. 7
  Suppose that X is a continuous non-negative random variable such that $E[X] = 1$. We define $p = P[X > 1]$. Let p_1 be the value of p if X has a uniform distribution, and p_2 be the value of p if $X = Y^2$, where $Y \sim N(0, 1)$. Calculate $p_1 + p_2$.

a) 0.318   b) 0.5   c) 0.659   d) 0.818   e) 1

Question no. 8
  Let

x	-1	0	1
$p_X(x)$	1/2	1/4	1/4

Calculate STD[X^2].

a) 0.1875   b) 0.4330   c) 0.6875   d) 0.8292   e) 1.1456

Question no. 9
  We define $Y = |X|$, where $X \sim N(0, 1)$. What is, according to Markov's inequality, the largest value that the probability $P[Y \geq 2/\sqrt{\pi}]$ can take?

a) 1/2   b) 0.5642   c) 0.7071   d) 0.8862   e) 1

Question no. 10
  Let $\phi_X(\omega) = e^{-\omega^2}$. We define $Y = 2X - 1$. Calculate $\phi_Y(\omega)$.

a) $e^{-\omega^2}$ b) $2e^{-\omega^2} - 1$ c) $e^{-4\omega^2}$ d) $e^{-j\omega - 2\omega^2}$ e) $e^{-j\omega - 4\omega^2}$

Question no. 11
  Let $X \sim N(0, 1)$. We define

$$Y = \begin{cases} 0 \text{ if } X < 0, \\ 1 \text{ if } 0 \le X \le 1, \\ 2 \text{ if } X > 1. \end{cases}$$

Calculate the generating function of Y.

a) $0.341z + 0.159z^2$ b) $1 + 0.341z + 0.159z^2$ c) $\frac{1}{2} + 0.341z + 0.159z^2$
d) $\frac{1}{2} + 0.841z + z^2$ e) $1 + 0.841z + z^2$

Question no. 12
A certain school bought 20 computers so that its students can connect to the Internet. The school distributed access codes to the 200 students registered in the computer science course. We estimate that each student having an access code has a 0.2 probability of wishing to connect to the Internet at noon on a given day, independently from one student to another and from day to day.

A) Use Poisson's approximation to calculate the probability p that all the computers are occupied, at noon, on a given day.

a) 0.9990   b) 0.9992   c) 0.9994   d) 0.9996   e) 0.9998

Hint. If $M \sim \text{Poi}(40)$, then $p_M(20) \simeq 0.00019$ and $F_M(20) \simeq 0.00037$.

Answer the following questions by assuming that the answer in A) is $p = 0.9$.

B) What is the probability that all the computers are occupied, at noon, on at least two days among the five days of a given school week?

a) 0.00046   b) 0.00856   c) 0.99144   d) 0.99954   e) 0.99990

C) What is the probability that the first day, from the beginning of the term, on which all the computers are occupied, at noon, is the second or third day of this term?

a) 0.009   b) 0.0099   c) 0.099   d) 0.1539   e) 0.171

Question no. 13
Let X be the utilization time of a computer during an eight-hour workday. We propose four models for the distribution of X: $X_1 \sim \text{U}[0, 8]$, $X_2 \sim \text{Exp}(1/4)$, $X_3 \sim G\left(2, \frac{1}{2}\right)$ and $X_4 = 8e^{-X_2}$.

A) Among the above models, which are the only ones that can be the true model?

a) X_1 b) X_1, X_4 c) X_2, X_3 d) X_1, X_2, X_3 e) all

B) Suppose that $X \sim \text{Exp}(1/4)$. Calculate $m := E[X \mid X > 1]$ and $v := \text{VAR}[X \mid X > 1]$.

a) $m = 4$ and $v = 4$ b) $m = 4$ and $v = 5$ c) $m = 4$ and $v = 16$
d) $m = 5$ and $v = 16$ e) $m = 5$ and $v = 17$

C) Suppose that $X \sim G\left(2, \frac{1}{2}\right)$. Calculate $P[X \ge 3.5]$.

a) 0.3041   b) 0.4779   c) 0.5578   d) 0.7440   e) 0.8088

Question no. 14
An information source generates letters taken at random among the letters a, b, c, d and e. We suppose that $P[\{a\}] = 1/2$, $P[\{b\}] = 1/4$, $P[\{c\}] = 1/8$ and

$P[\{d\}] = P[\{e\}] = 1/16$. Moreover, a data compression system transforms the letters into binary strings as follows: $a = 1, b = 01, c = 001, d = 0001$ and $e = 0000$. Let X be the length of a binary string emitted by the data compression system. Calculate

A) $F_X(2^-) + F_X(5^-)$;

a) 1/2   b) 3/4   c) 1   d) 3/2   e) 7/4

B) $E[X|X > 1]$;

a) 11/8   b) 15/8   c) 9/4   d) 5/2   e) 11/4

C) $\text{VAR}[X^2]$.

a) 71/64   b) 5041/4096   c) 11/4   d) 1615/64   e) 42

Question no. 15

Let

$$F_X(x) = \begin{cases} 0 & \text{if } x < 1, \\ 1/4 & \text{if } x = 1, \\ x^2/4 & \text{if } 1 < x \le 2, \\ 1 & \text{if } x > 2. \end{cases}$$

Calculate $P[X = 1] + P[X < 2]$.

a) 1/4   b) 1/2   c) 3/4   d) 1   e) 5/4

Question no. 16

Let X be a continuous random variable for which

$$f_X(x \mid X \le 1) = 2x \quad \text{if } 0 \le x \le 1.$$

Which of the following functions can be the function $f_X(x)$?

1) $2x$ if $0 \le x \le 1$ 2) x if $0 \le x \le \sqrt{2}$
3) $x/4$ if $0 \le x \le 2\sqrt{2}$ 4) $4x$ if $0 \le x \le 1/\sqrt{2}$
5) x if $0 \le x \le 1$ and $1/2$ if $1 < x \le 2$

a) 1 and 2 only   b) 1, 2 and 3 only
c) all, except 4   d) all
e) none of these answers

Question no. 17

We suppose that the number X of particles emitted by a radioactive source during a one-hour period has a Poisson distribution with parameter $\alpha = 1/2$, independently from hour to hour. Let N be the number of hours during which no particles are emitted, among the 24 hours of a given day. What distribution does N have?

a) $\text{Poi}(e^{-1/2})$ b) $\text{Poi}(12)$ c) $B(24, \frac{1}{2})$ d) $B(24, \frac{1}{2}e^{-1/2})$

e) $B(24, e^{-1/2})$

Question no. 18

Boxes contain 20 objects each. We examine the contents of the boxes until we find one that contains no defective objects. Let N be the number of boxes that we must examine to end the random experiment. What distribution does N have if the probability that an object is defective is equal to $1/10$, independently from one object to another?

a) Geom$(1/10)$ b) Geom$\left((9/10)^{20}\right)$ c) B$(20, 1/10)$ d) B$\left(20, (9/10)^{20}\right)$
e) none of these answers

Question no. 19

Calculate $P[X^2 < 9]$, where X is a random variable that has a gamma distribution with parameters $\alpha = 2$ and $\lambda = 1$.

a) e^{-3} b) $4e^{-3}$ c) $1 - e^{-3}$ d) $1 - 4e^{-3}$ e) $1 - \frac{17}{2}e^{-3}$

Question no. 20

The lifetime (in months) of brand A components has (approximately) a Gaussian N(50, 100) distribution and that of brand B components a Gaussian N(60, 100) distribution. A component is taken at random from a box containing 10 brand A and 20 brand B components. Calculate the probability that the component selected lasts at least 50 months.

Hint. We have: $Q(1) \simeq 0.159$ and $Q(2) \simeq 0.0228$.

a) 0.6705   b) 0.7273   c) 0.7386   d) 0.8181   e) none of these answers

Question no. 21

We define $Y = 1/X^2$, where X is a random variable having a uniform distribution on the interval $[1, 3]$. Calculate $f_Y(y)$.

a) $\frac{1}{4}y^{-3/2}$ if $1/9 \le y \le 1$ b) $\frac{1}{2}y^{-3/2}$ if $1/9 \le y \le 1$ c) $\frac{1}{2}$ if $1 \le y \le 3$

d) $\frac{9}{8}$ if $1/9 \le y \le 1$ e) 2 if $-1 \le y \le 1$

Question no. 22

Let
$$f_X(x) = xe^{-x} \quad \text{if } x > 0.$$

Calculate $E[X^{-2}]$.

a) 1/6   b) 1/4   c) 6   d) ∞ e) $\infty - \infty$

Question no. 23

Let X be a random variable having a gamma distribution with parameters $\alpha = 30$ and $\lambda = 20$. According to Markov's inequality, what is the minimum value of $P[X \le 2]$?

a) 1/6   b) 1/4   c) 1/2   d) 3/4   e) 5/6

Question no. 24

A discrete random variable N has the following generating function:

$$G_N(z) = \frac{1}{6} + \frac{1}{3}z^2 + \frac{1}{2}z^4.$$

Calculate VAR$[N]$.

a) 2   b) 8/3   c) 20/3   d) 28/3   e) none of these answers

Question no. 25
    Let
$$\phi_X(\omega) = k\frac{e^{j\omega} - 1}{\omega},$$
where k is a constant. Calculate $E[2^X]$.
a) $1/\ln 2$ b) $2/\ln 2$ c) $\ln 2$ d) $2\ln 2$ e) none of these answers

Question no. 26
    Let
$$F_X(x) = \begin{cases} 0 & \text{if } x < 1, \\ (x-1)/2 & \text{if } 1 \le x < 2, \\ x^2/8 & \text{if } 2 \le x < 2\sqrt{2}, \\ 1 & \text{if } x \ge 2\sqrt{2}. \end{cases}$$

A) Calculate $F_X\left(x \,\Big|\, \frac{3}{2} \le X \le 2\right)$ for $x \in \left[\frac{3}{2}, 2\right]$.

a) $\dfrac{x-1}{2}$ b) $x - \frac{3}{2}$ c) $x - 1$ d) $2(x - 1)$ e) $2x - 3$

B) Calculate $E[X]$.
a) 0.97   b) 1.47   c) 1.97   d) 2.00   e) 2.47

C) Find the 90th percentile of X.
a) 0.9   b) 1.8   c) 2.0   d) 2.8   e) none of these answers

D) Calculate $\text{VAR}\left[X \,\Big|\, X < \frac{3}{2}\right]$.

a) 1/96   b) 1/48   c) 1/12   d) 1/4   e) 1/2

E) Let $Y := 1/X$. Obtain $F_Y(y)$ for $\frac{1}{2} < y \le 1$.

a) $\frac{3}{2} - \frac{1}{2y}$ b) $\frac{1}{2}(\frac{1}{y} - 1)$ c) $\frac{3}{2} - \frac{y}{2}$ d) $\frac{1}{2}(y - 1)$ e) none of these answers

F) Let $W := \text{int}(X)$. That is, W is the integer part of X. Calculate the generating function $G_W(z)$ of W.

a) $\frac{1}{2}(1 + z)$ b) $\frac{1}{3}(1 + z + z^2)$ c) $\frac{1}{2}(z + z^2)$ d) $\frac{1}{3}(z + z^2 + z^3)$
e) $\frac{1}{4}(1 + z + z^2 + z^3)$

Question no. 27
    A certain city is supplied with electricity through a single high tension line. The number X of failures, during a one-year period, has a Poisson distribution with parameter 3. Moreover, the duration Y (in hours) of a given failure has an exponential distribution with mean 2. Finally, we assume that all the failures are independent.

A) Let N be the number of years, over a 25-year period, during which there has been exactly one failure. Calculate the variance of N.
a) 1.2   b) 2.0   c) 3.0   d) 3.2   e) 4.0

B) Let M be the number of failures required, from now, to obtain a second failure that lasts more than two hours. Calculate $E[M]$.

a) $1/e$ b) $2/e$ c) e d) $2e$ e) none of these answers

C) Calculate $E[X|X \geq 1]$.

a) 3   b) 3.16   c) 3.32   d) 3.48   e) 3.64

D) Let $Z := Y \mid \{Y < 1\}$. Calculate the moment generating function $M_Z(s)$ of Z at $s = -1/2$.

a) $1/[2(1 - e^{-1/2})]$ b) $1/(1 - e^{-1/2})$ c) $1 - e^{-1/2}$ d) $2(1 - e^{-1/2})$
e) none of these answers

E) Suppose that there have been exactly three failures during a given year. We can show that the total duration T (in hours) of these failures has a $G(\alpha = 3, \lambda = 1/2)$ distribution. Calculate $P[T < 6.5]$.

a) 0.37   b) 0.41   c) 0.59   d) 0.63   e) 0.84

F) We consider 30 failures. We can show that the total duration D (in hours) of these failures has approximately a Gaussian $N(60, 120)$ distribution. Find the number d such that $P[D < d] = 0.1$.

Hint. We have: $Q(1) \simeq 0.159$ and $Q^{-1}(0.1) \simeq 1.2815$.

a) 46   b) 74   c) 129   d) 154   e) 214

Question no. 28
A) Let T be the time (in minutes) that a customer waits before being served at a counter when there is at least one customer in the queue when he/she arrives. We suppose that

$$P[T \leq t] = t/10 \quad \text{for } 0 < t \leq 10. \quad (*)$$

i) Let $X = 1$ if the customer waits more than two minutes and $X = 0$ otherwise. Calculate the function p_X if the queue is not empty when the customer arrives.

a) $p_X(x) = \begin{cases} 1/10 \text{ if } x = 0, \\ 9/10 \text{ if } x = 1. \end{cases}$ b) $p_X(x) = \begin{cases} 2/10 \text{ if } x = 0, \\ 8/10 \text{ if } x = 1. \end{cases}$

c) $p_X(x) = \begin{cases} 3/10 \text{ if } x = 0, \\ 7/10 \text{ if } x = 1. \end{cases}$ d) $p_X(x) = \begin{cases} 1/2 \text{ if } x = 0, \\ 1/2 \text{ if } x = 1. \end{cases}$

e) $p_X(x) = \begin{cases} 7/10 \text{ if } x = 0, \\ 3/10 \text{ if } x = 1. \end{cases}$

ii) Calculate $f_T(t \mid T > 2)$ for $2 < t \leq 10$.

a) $t/8$ b) $t/7$ c) $1/10$ d) $1/8$ e) $1/7$

B) Suppose now that the probability that the queue is empty when the customer arrives is equal to $1/5$. Let Y be the waiting time of the customer. Calculate, making use of the assumption $(*)$ in A), the distribution function of Y for $0 \leq y \leq 10$.

a) $0.08\,y$ b) $0.1\,y$ c) $0.2 + 0.08\,y$ d) $0.1 + 0.09\,y$ e) 1

Question no. 29

A box contains ten transistors, of which two are defective. Two transistors are taken at random and with replacement. Let X be the number of defective transistors among the two picked. We repeat this random experiment until the two transistors picked are two defective transistors. Let N be the number of repetitions required to obtain a first repetition for which the two transistors picked are defective.

A) Give the distribution of X, as well as its parameter(s). Likewise for N.

a) $X \sim B\left(2, \frac{1}{5}\right)$; $N \sim$ Geom(0.04)   b) $X \sim B\left(10, \frac{1}{5}\right)$; $N \sim B\left(2, \frac{1}{5}\right)$

c) $X \sim B\left(2, \frac{1}{5}\right)$; $N \sim B\left(10, \frac{1}{5}\right)$ d) $X \sim B\left(10, \frac{1}{5}\right)$; $N \sim$ Poi(2)

e) $X \sim B\left(2, \frac{1}{5}\right)$; $N \sim$ Geom(0.2)

B) Calculate $P[N \geq 4 \mid N > 1]$.

a) 0.0784   b) 0.1153   c) 0.8847   d) 0.9216   e) 0.9600

C) Give the distribution of N, as well as its parameter(s), if the two transistors are taken without replacement (but the transistors are put back into the box after the two draws if they are not both defective).

a) Geom$\left(\frac{1}{45}\right)$ b) Geom$\left(\frac{1}{25}\right)$ c) $B\left(10, \frac{1}{10}\right)$ d) $B\left(10, \frac{1}{5}\right)$ e) Poi(2)

Question no. 30

Let X be a random variable having a $G(\alpha, \lambda)$ distribution.

A) Calculate $P[X \geq 1]$ if $\alpha = 2$ and $\lambda = 3$.

a) e^{-3} b) $4e^{-3}$ c) $8.5e^{-3}$ d) $1 - 4e^{-3}$ e) $1 - e^{-3}$

B) Suppose that $\alpha = 1$ and $\lambda = 2$. Calculate $E\left[X^2 \mid X^2 > 1\right]$.

a) 1   b) 3/2   c) 2   d) 9/4   e) 5/2

C) Let $Z := X^2$. Find $f_Z(z)$ for $z > 0$, if $\alpha = 2$ and $\lambda = 1$.

a) $\frac{1}{2}e^{-\sqrt{z}}$ b) $\frac{\sqrt{z}}{2}e^{-\sqrt{z}}$ c) $\sqrt{z}e^{-\sqrt{z}}$ d) $z^2 e^{-2z}$ e) $z^2 e^{-z^2}$

Question no. 31

Let X be a standard Gaussian random variable.

A) Find the number x_0 for which $P\left[X^2 > x_0\right] = 0.05$.

Hint. We have: $Q^{-1}(0.05) \simeq 1.645$ and $Q^{-1}(0.025) \simeq 1.960$.

a) 1.645   b) 1.960   c) 3.290   d) 3.842   e) 3.920

B) Calculate $E\left[X^2(X+1)\right]$.

a) 0   b) 1   c) 2   d) 3   e) 5

C) Find the number(s) a for which the characteristic function of $Y := aX + 1$ is given by $\exp(j\omega - 2\omega^2)$.

a) -1 b) 1   c) ± 1 d) 2   e) ± 2

Hint. We have: $\phi_X(\omega) = e^{-\omega^2/2}$.

D) Calculate the generating function of W, that is, $G_W(z) := E[z^W]$, where

$$W := \begin{cases} 1 \text{ if } X > 0, \\ 0 \text{ if } X \leq 0. \end{cases}$$

a) $\frac{1}{2}$ b) $\frac{1}{2}z$ c) $\frac{1}{2}(1+z)$ d) $1+z$ e) $1+z+z^2$

Question no. 32

Let X be the utilization time (in months) of a laser printer until its first break-down.

A) Suppose that $X \sim \text{Exp}(1/3)$. Calculate $E[|X - 3|]$.

a) $3e^{-1}$ b) $3(1 - e^{-1})$ c) $6e^{-1}$ d) 3 e) $6(1 - e^{-1})$

B) Suppose that $X \sim G(\alpha = 2, \lambda = 1)$. We define $Y = 1/(X+1)$. Find $f_Y(1/2)$.

a) $\frac{1}{2}e^{-1}$ b) e^{-1} c) $2e^{-1}$ d) $4e^{-1}$ e) $8e^{-1}$

C) Suppose that the time Z (in months) required for the 25th breakdown to occur has approximately a Gaussian $N(75, 225)$ distribution. Calculate $P[Z < 90 \mid Z \geq 60]$.

Hint. We have: $Q(1) \simeq 0.159$.

a) 0.1891 b) 0.3180 c) 0.6820 d) 0.8109 e) 0.8410

Question no. 33

A) Let N be a random variable having a geometric distribution with parameter p. Calculate $F_N(n \mid N > 1)$ for $n = 2, 3, \ldots$.

a) $1 - q^{n-1}$ b) $1 - q^n$ c) $1 - q^{n+1}$ d) $1 - p^{n-1}$ e) $1 - p^n$

B) Let $X \sim U[0, 1]$.

i) Obtain the characteristic function of $Y := \ln(X + 1)$.

Hint. We have: $\phi_X(\omega) = \frac{e^{j\omega}-1}{j\omega}$.

a) $\dfrac{2^{j\omega} - 1}{j\omega}$ b) $\dfrac{2^{j\omega+1} - 1}{j\omega + 1}$ c) $\dfrac{2^{j\omega}}{j\omega + 1}$ d) $\dfrac{2^{j\omega} - 1}{2j\omega}$ e) $\dfrac{2^{j\omega} - 1}{j\omega \ln 2}$

ii) Let

$$W := \begin{cases} 1 \text{ if } X \leq 0.3, \\ 2 \text{ if } X > 0.3. \end{cases}$$

Calculate the generating function of $V := 2W$.

a) $z(0.3 + 0.7z)$ b) $2z(0.3 + 0.7z)$ c) $2z(0.3 + 1.4z)$ d) $[z(0.3 + 0.7z)]^2$
e) $z^2(0.3 + 0.7z^2)$

Question no. 34

A number N is taken at random in the set $\{0, 1, 2, \ldots, 9\}$. Calculate $p_N(5) + F_N(5.5)$.

a) 0.6 b) 0.65 c) 0.7 d) 0.75 e) 0.8

Question no. 35

Let

$$f_X(x) = 2x \quad \text{if } 0 < x < 1.$$

Find $f_X(1/4 \mid X < 1/2)$.

a) 0.5 b) 1 c) 1.5 d) 2 e) 2.5

Question no. 36

Let

$$f_X(x) = \begin{cases} 1/2 & \text{if } 0 \le x < 1, \\ kx & \text{if } 1 \le x \le 2, \end{cases}$$

where k is a positive constant. Calculate the 60th percentile of X.

a) 1.16   b) 1.26   c) 1.36   d) 1.46   e) 1.56

Question no. 37

Suppose that $X \sim B(n = 5, p = 0.2)$. Calculate $P[X = 1 | X \le 1]$.

a) 1/2   b) 5/9   c) 2/3   d) 7/9   e) 8/9

Question no. 38

We take independent observations of a random variable X having a Poisson distribution with parameter $\alpha = 2$ until we obtain "0". Let N be the number of observations required to end the random experiment. Calculate $\text{VAR}[N - 1]$.

a) 43.21   b) 44.21   c) 45.21   d) 46.21   e) 47.21

Question no. 39

We consider a random variable X having an exponential distribution with parameter $\lambda = 1$. Calculate $E[X \mid 1 < X < 2]$.

a) 1.27   b) 1.32   c) 1.37   d) 1.42   e) 1.47

Question no. 40

Suppose that $X \sim G(\alpha = 3, \lambda = 1)$. We define $Y = \sqrt{X}$. Find $f_Y(2)$.

a) $32e^{-4}$ b) $64e^{-4}$ c) $32e^{-2}$ d) $64e^{-2}$ e) $32e^{-1}$

Question no. 41

Find the number x_0 for which

$$P[X < -x_0] + P[X > x_0] = 1/2,$$

where X is a standard Gaussian random variable.

a) $Q^{-1}(1/4)$ b) $1 - Q^{-1}(1/4)$ c) $\frac{1}{2}Q^{-1}(1/4)$ d) $2Q^{-1}(1/4)$
e) $1 - 2Q^{-1}(1/4)$

Question no. 42

Let X be a random variable having a Bernoulli distribution with parameter $p = 3/4$. Calculate $\phi_X(1)$.

a) $\frac{1}{4}e^j$ b) $\frac{3}{4}e^j$ c) $\frac{1}{4}(1 + e^j)$ d) $\frac{1}{4}(3 + e^j)$ e) $\frac{1}{4}(1 + 3e^j)$

Question no. 43

Suppose that the lifetime of a system is a random variable T whose probability density function is given by

$$f_T(t) = 1/2 \quad \text{if } 1 < t < 3.$$

Calculate the failure rate $r(t)$ at time $t = 2$.

a) 1/4   b) 1/2   c) 1   d) 3/2   e) 2

Question no. 44

We consider a system made up of two components placed in parallel and a third component placed in series. What is the reliability of the system at time $t = 3$ if the three components function independently and all have the reliability function $R(t) = e^{-t}$ for $t \geq 0$?

a) 0.0038   b) 0.0048   c) 0.0058   d) 0.0068   e) 0.0078

Question no. 45

Let

$$F_X(x) = \begin{cases} 0 & \text{if } x < -1, \\ (x+1)/4 & \text{if } -1 \leq x \leq 1, \\ 1/2 & \text{if } 1 < x < 2, \\ 3/4 & \text{if } 2 \leq x < 3, \\ 1 & \text{if } x \geq 3. \end{cases}$$

A) Calculate $F_X(0 \mid X < 2)$.

a) 1/8   b) 1/4   c) 1/3   d) 1/2   e) 3/4

B) Calculate $E[X]$.

a) 0   b) 3/4   c) 5/4   d) 7/4   e) 9/4

Question no. 46

An exam is made up of five multiple choice questions. For each question, five answers are proposed. Every correct answer gives two points, while 1/2 point is deducted for each wrong answer, up to a maximum of -1. Finally, the exam is over 7 (that is, the maximum grade is 7) and the minimum grade is 0. A student answers every question at random. Let X be the grade of this student.

A) Calculate $E[X]$.

a) 1.327   b) 1.427   c) 1.527   d) 1.627   e) 1.727

B) What is the mathematical expectation of X, given that the first three answers are correct?

a) 5.67   b) 5.72   c) 5.77   d) 5.82   e) 5.87

Question no. 47

We have a random numbers generator for a random variable X having a uniform distribution on the interval $(0, 1)$.

A) We want to use a transformation $Y = g(X)$ to obtain observations of a random variable having the following probability density function:

$$f_Y(y) = 1/y^2 \quad \text{if } y > 1.$$

What transformation $g(X)$ can we use?

a) $1/X^2$ b) $e^{1/X} - e + 1$ c) $1 - \ln X$ d) $1/X$ e) $1 - 2\ln X$

B) We generate independent observations of X until we have obtained two observations greater than 0.9. What is the probability that we have to generate exactly three observations?

a) 0.009   b) 0.018   c) 0.027   d) 0.036   e) 0.045

Question no. 48

Let
$$f_X(x) = \sqrt{2/\pi}\, x^2 e^{-x^2/2} \quad \text{for } x > 0.$$

A) Use a linear interpolation, based on the fact that $Q(1) \simeq 0.159$ and $Q(2) \simeq 0.02275$, to calculate approximately the median of X.

a) 1.48   b) 1.52   c) 1.56   d) 1.60   e) 1.64

B) We find that $E[1/X] = \sqrt{2/\pi}$. Calculate VAR$[1/X]$.

a) 0.3234   b) 0.3334   c) 0.3434   d) 0.3534   e) 0.3634

Question no. 49

A) Let $X \sim N(1, 1)$. Calculate $P[X^2 - 2X > 0]$.

a) $Q(1)$ b) $2Q(1)$ c) $1 - Q(1)$ d) $1 - 2Q(1)$ e) $2(1 - Q(1))$

B) With the help of Bienaymé–Chebyshev's inequality, calculate an upper bound for $Q(\sqrt{2})$.

a) 1/4   b) 1/3   c) 1/2   d) 2/3   e) 3/4

Question no. 50

A) Let X be a random variable having a Poisson distribution with parameter $\alpha = 1/3$. Calculate $E[X!]$.

a) 1/3   b) 2/3   c) $e^{-1/3}$ d) $\frac{2}{3}e^{-1/3}$ e) $\frac{3}{2}e^{-1/3}$

B) The characteristic function of a random variable X having an exponential distribution with parameter λ is given by $\phi_X(\omega) = \lambda/(\lambda - j\omega)$. Use this formula to calculate $E[-Y(Y + 1)^2/3]$, if Y has a gamma distribution with parameters $\alpha = 1$ and $\lambda = 1$.

a) $-11/3$ b) $-7/9$ c) $-3/4$ d) 1/3   e) 3

Question no. 51

We suppose that the lifetime X of every component of the system in Figure 3.19, has a probability density function $f_X(x) = xe^{-x^2/2}$ for $x \geq 0$. Moreover, we assume that the components operate independently from one another.

A) What is the failure rate of a component at time $x = 2$?

a) e^{-2} b) e^2 c) $2e^2$ d) 1/2   e) 2

B) What is the value of the reliability function of the system at time $x = 2$?

a) $e^{-2} - 2e^{-6}$ b) $2e^{-2} - e^{-4}$ c) $2e^{-6} - e^{-8}$ d) e^{-2} e) $e^{-4} + e^{-6}$

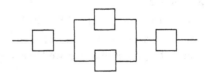

Figure 3.19. System in Multiple choice question no. 51.

Question no. 52

Let X be a random variable whose distribution function is

$$F_X(x) = \begin{cases} 0 & \text{if } x < 0, \\ \frac{1}{4}(x+1) & \text{if } 0 \le x < 1, \\ 1 & \text{if } x \ge 1. \end{cases}$$

Calculate $P\left[0 < X \le \frac{1}{2}\right] + P[X = 1]$.

a) 1/2   b) 5/8   c) 11/16   d) 3/4   e) 7/8

Question no. 53

A continuous random variable X has the probability density function

$$f_X(x) = \frac{1}{2}e^{-|x|} \quad \text{for } -\infty < x < \infty.$$

Calculate $P[-1 \le X < 1] + P[X = 2]$.

a) e^{-1} b) $2e^{-1}$ c) $1 - e^{-1}$ d) $e^{-1} + \frac{1}{2}e^{-2}$ e) $1 - e^{-1} + \frac{1}{2}e^{-2}$

Question no. 54

We consider a random variable X whose probability density function is defined by

$$f_X(x) = \begin{cases} 1 + x & \text{if } -1 \le x < 0, \\ 1 - x & \text{if } 0 \le x \le 1, \\ 0 & \text{elsewhere.} \end{cases}$$

A) Find $F_X(3/4)$.

a) 3/4   b) 13/16   c) 7/8   d) 5/6   e) 31/32

B) Calculate $E[X \mid X > 0]$.

a) 1/6   b) 1/4   c) 1/3   d) 1/2   e) 3/4

C) Let $Y := |X|$. Find $f_Y(y)$ for $0 \le y \le 1$.

a) 1   b) $\frac{3}{2} - y$ c) $\frac{1}{2} + y$ d) $2y$ e) $2 - 2y$

Question no. 55

Let X be a random variable having a uniform distribution on the interval $[0, 1]$. We define $Y = \max\left\{X, \frac{1}{2}\right\}$.

A) Find $F_Y(y)$ for $\frac{1}{2} \le y \le 1$. What type of random variable is Y?

a) $\frac{1}{2}$; discrete   b) y; continuous   c) $2y - 1$; continuous   d) y; mixed   e) $2y - 1$; mixed

B) Calculate $E[Y]$.

a) 9/16   b) 5/8   c) 11/16   d) 3/4   e) 13/16

Question no. 56

We want to find a model for the number N of cars crossing a certain intersection during a one-minute period. According to the data collected, the most frequent value of N is five cars. It also happens, although not frequently, that the traffic is much heavier. Which of the following models seems to be the most appropriate?

a) $\text{Poi}\left(\frac{11}{2}\right)$ b) $B\left(10,\frac{1}{2}\right)$ c) $\text{Geom}\left(\frac{1}{5}\right)$

d) $p_N(n) = (\frac{5}{6})^n(\frac{1}{6})$ for $n = 0, 1, \ldots$ e) $p_N(n) = \frac{1}{10}$ for $n = 1, \ldots, 10$

Question no. 57

We look for a model, as realistic as possible, for a continuous random variable X that represents the lifetime of a machine, and whose mean and variance are equal to 1 and 3, respectively. Which of the following distributions can be acceptable?

1) uniform   2) exponential   3) gamma   4) Gaussian
5) the square of a Gaussian $N(1, 3)$ distribution

a) all except 4)   b) 2), 3) and 5) only   c) 3) and 5) only
d) 3) only   e) 4) only

Question no. 58

A fabrication process for producing one-inch diameter screws is such that the diameter (in inches) of a screw can be considered as being the value taken by a random variable X having (approximately) a Gaussian $N(1, 625 \times 10^{-4})$ distribution, independently from one screw to another.

A) What is the probability that a screw taken at random has a diameter inaccurate by more than 0.05 inch?

Hint. We have: $Q(0.1) \simeq 0.4602$ and $Q(0.2) \simeq 0.4207$.

a) 0.8414   b) 0.8514   c) 0.8614   d) 0.8714   e) 0.8814

B) Let p be the answer in A) and $q := 1 - p$. We take six screws at random (and without replacement). What is the probability that less than two of these screws have a diameter inaccurate by more than 0.05 inch?

a) $p^6 + p^5q$ b) $p^6 + 6p^5q$ c) $p^6 + 5p^5q$ d) $q^6 + pq^5$ e) $q^6 + 6pq^5$

Question no. 59

The average lifetime of a certain type of tire is equal to three years, with a standard deviation of 0.3 year. What can be said, with as much accuracy as possible, concerning the probability p that a tire of this type lasts more than 54 months or less than 18 months?

a) $p \le 1/25$ b) $p \ge 1/25$ c) $p > 1/25$ d) $p \le 24/25$ e) $p = 12/25$

Question no. 60

The characteristic function of a discrete random variable X is given by

$$\phi_X(\omega) = \frac{1}{4}\left(1 + e^{j\omega} + 2e^{2j\omega}\right).$$

Calculate $P[X = 1]$.

a) 0   b) 1/4   c) 1/2   d) 3/8   e) 5/8

Question no. 61

A certain system is made up of two components placed in parallel and operating independently. The lifetime of the first component has a uniform distribution on the interval $(0, 1)$, while the lifetime of the second component has an exponential distribution with parameter 2. Calculate the reliability function of the system at time $1/2$.

a) 0.1839   b) 0.3161   c) 0.6839   d) 0.8161   e) 0.8679

Question no. 62

The electric current X (in amperes) that passes through a resistance of r ohms is a random variable having the probability density function

$$f_X(x) = \begin{cases} \frac{1}{2} & \text{if } 4 < x < 6, \\ 0 & \text{elsewhere.} \end{cases}$$

Furthermore, the power Y (in watts) is given by $Y = rX^2$.

A) Find the value of the distribution function $F_X(5 \mid X \geq 4.5)$.

a) 1/6   b) 1/4   c) 1/3   d) 1/2   e) 5/6

B) Calculate VAR$[X \mid X \geq 4.5]$.

a) 1/16   b) 1/12   c) 1/8   d) 3/16   e) 1/4

C) An observation of the random variable Y is generated. In which interval, among the following, is the observation most likely to be found if $r = 25$?

a) $(400, 401)$ b) $(500, 501)$ c) $(700, 701)$ d) $(899, 900)$
e) the four preceding intervals are equally likely

D) Calculate VAR$[Y]$ if $r = 30$.

a) 30,040   b) 30,050   c) 30,060   d) 30,070   e) 30,080

Question no. 63

Each day a student spends a random time T (in minutes) in front of her computer to do some work. We estimate that $E[T] = 40$ and VAR$[T] = 1600$.

A) Let d be a duration that at least 90% of the work sessions do not exceed. According to Bienaymé–Chebyshev's inequality, what is the minimum value of d?

a) 50.98   b) 72.53   c) 102.45   d) 166.49   e) 227.68

B) Suppose that T has an exponential distribution.

i) If the student has already spent 30 minutes in front of her computer, what is the probability that the work session will last at least 60 more minutes?

a) 0.1245   b) 0.2231   c) 0.2548   d) 0.3679   e) 0.6321

ii) How many work sessions does it take, on average, to obtain a first session that lasts more than 90 minutes?

a) 9.19   b) 9.29   c) 9.39   d) 9.49   e) 9.59

Question no. 64

A person subscribes to a service that allows him to connect to the Internet for 30 hours per month. We suppose that the time X (in hours) that this person spends surfing the Internet during a given day has a gamma distribution with parameters $\alpha = 3$ and $\lambda = 4$, independently from day to day.

A) What is the probability that the person spends more than 90 minutes on the Internet during a given day?

a) 0.022   b) 0.032   c) 0.042   d) 0.052   e) 0.062

B) Suppose that the answer in A), for other values of α and λ, is equal to 0.05. What is the probability that, during a given week, the person spends more than 90 minutes on the Internet on at least three days?

a) 0.0018   b) 0.0028   c) 0.0038   d) 0.0048   e) 0.0058

C) We can show that the total time (in hours) during which the person wants to connect to the Internet, over a 30-day month, has approximately a Gaussian distribution with parameters $\mu = 22.5$ and $\sigma^2 = 5.625$. What is the probability that the person is disconnected because he reached the time limit imposed?

a) $Q(-3.16)$ b) $Q(3.16)$ c) $Q(3.16) + \frac{1}{2}$ d) $\frac{1}{2} - Q(3.16)$

e) $\frac{1}{\sqrt{2\pi}} \exp\{-(3.16)^2/2\}$

Question no. 65

Let

$$f_X(x) = (n+1)x^n \quad \text{for } 0 \le x \le 1,$$

where $n \in \{1, 2, \dots\}$.

A) Calculate the 50th percentile (that is, the median) of X if $n = 3$.

a) 0.8209   b) 0.8309   c) 0.8409   d) 0.8509   e) 0.8609

B) We define $Y = X^n$. Find $f_Y(1)$.

a) 1   b) $n+1$ c) $(n+1)^n$ d) $n/(n+1)$ e) $(n+1)/n$

C) Calculate the characteristic function $\phi_X(\omega)$ of X at $\omega = 2$, if $n = 1$.

a) $\frac{1}{2}\left[e^{2j}(1 - 2j) - 1\right]$ b) $\frac{1}{2}\left[e^{2j}(2j - 1) - 1\right]$ c) $\frac{1}{2}\left[e^{2j}(1 - 2j) + 1\right]$

d) $\frac{1}{2}\left[e^{2j}(2j + 1) - 1\right]$ e) $\frac{1}{2}\left[e^{2j}(2j + 1) + 1\right]$

D) Calculate the failure rate at time $x = 1/2$ of a machine whose lifetime X has the above probability density function f_X with $n = 2$.

a) 5/7   b) 6/7   c) 1   d) 7/6   e) 7/5

Question no. 66

Let X be a random variable of mixed type whose distribution function is

$$F_X(x) = \begin{cases} 0 & \text{if } x < -1, \\ \dfrac{1 - x^2}{2} & \text{if } -1 \le x < b, \\ 1 & \text{if } x \ge b. \end{cases}$$

A) What are the possible values of the constant b?

a) $-1 < b \leq 1$ b) $-1 < b < 1$ c) $-1 < b \leq 0$ d) $0 < b < 1$
e) $0 \leq b \leq 1$

B) Calculate $P[X = (b-1)/2] + P[X \geq b]$.

a) 0   b) $\frac{1}{2} + \frac{b^2}{2}$ c) $\frac{3}{8} + \frac{b}{2} - \frac{b^2}{8}$ d) $\frac{11}{8} + \frac{b}{2} - \frac{b^2}{8}$ e) $\frac{7}{8} + \frac{b}{2} + \frac{3b^2}{8}$

C) Calculate $E[X \mid -1 \leq X < b]$.

a) $\dfrac{b-1}{2}$ b) $-\dfrac{b^3+1}{3}$ c) $\dfrac{b^3-1}{3}$ d) $\dfrac{2}{3}\left(\dfrac{b^3+1}{b^2-1}\right)$ e) $-\dfrac{b^3+1}{b^2+2}$

Question no. 67

A point X is taken at random in the interval $[0, 1]$. Let D be the distance between X and the endpoint of the interval closer to X. That is,

$$D = \begin{cases} X & \text{if } 0 \leq X \leq \frac{1}{2}, \\ 1 - X & \text{if } \frac{1}{2} < X \leq 1. \end{cases}$$

A) Find $f_X(1/4 \mid X^2 < 1/4)$.

a) 0   b) 1/4   c) 1/2   d) 1   e) 2

B) Calculate $E[X^{1/2} + X^{-1/2}]$.

a) 2/3   b) 4/3   c) 5/3   d) 8/3   e) ∞

C) Find $F_D(d)$ for $0 \leq d \leq 1/2$.

a) $\sqrt{2d}$ b) $2d$ c) $d + \frac{1}{2}$ d) $4d^2$ e) $8d^3$

D) What is the 25th percentile of D?

a) 1/32   b) 1/16   c) 1/8   d) 1/6   e) 1/4

Question no. 68

A machine is made up of four components that operate independently from one another. The machine functions if at least three of its components function. The state of the components is checked at the end of each workday and the components that failed are repaired. If the machine fails during the course of the day, it will remain down until the end of that day. We estimate that a given component has a 90% probability of functioning during an entire workday, whether it is new or not, and this independently from day to day.

A) What is the probability that the machine functions during an entire workday?

a) 0.9077   b) 0.9177   c) 0.9277   d) 0.9377   e) 0.9477

B) What is the probability that the machine functions during exactly four entire workdays over a given (seven-day) week?

a) 0.0040   b) 0.0080   c) 0.0120   d) 0.0160   e) 0.0200

C) Calculate the probability that the tenth day of a certain month is the third day during which a given component failed.

a) 0.0132   b) 0.0142   c) 0.0152   d) 0.0162   e) 0.0172

D) Suppose that the number N of failures of the machine that last more than a half day, over a six-month period, has (approximately) a Poisson distribution with parameter $\alpha = 3$. Calculate VAR$[N \mid N \geq 1]$.

a) 2.33   b) 2.66   c) 3   d) 3.33   e) 3.66

Question no. 69

A system has n components connected in parallel and functioning independently from one another. The components function one at a time, starting with component no. 1, so that the system functions as long as there is at least one component that is not down (standby redundancy). We suppose that the lifetime T_k of component k has an exponential distribution with parameter $\lambda = 1$ for $k = 1, 2, \ldots, n$. Let T be the total lifetime of the system.

A) Let $S_k := \ln T_k$. Calculate the moment generating function of S_k, defined by $M_{S_k}(s) = E[e^{-sS_k}]$, at $s = -2$.

a) $\ln 2$ b) $\ln 4$ c) $1/2$ d) 2   e) 4

B) If $n = 2$, we can show that T has a gamma distribution with parameters $\alpha = 2$ and $\lambda = 1$. That is, $f_T(t) = te^{-t}$ for $t \geq 0$. Find the failure rate $r(t)$ of the system for $t > 0$.

a) $\frac{t}{t+1}$ b) $1 + \frac{1}{t}$ c) $\frac{1}{t}$ d) $\frac{t^2}{t+1}$ e) t

C) If $n = 25$, we can show that T has approximately a Gaussian distribution with parameters $\mu = 25$ and $\sigma^2 = 25$. Calculate $P[T \geq 30 \mid T > 25]$.

a) $2Q(1)$ b) $Q(1)$ c) $Q(1)/2$ d) $1 - Q(1)$ e) $1 - 2Q(1)$

Question no. 70

Let X be a random variable whose probability density function is

$$f_X(x) = \begin{cases} ax + b & \text{if } c \leq x \leq d, \\ 0 & \text{elsewhere,} \end{cases}$$

where a, b, c and d are constants.

A) What are all possible values of the constants a and b, if $c = -1$ and $d = 1$?

a) $a > 0; b = 0$ b) $a = \frac{1}{2}; b = \frac{1}{2}$ c) $a = -\frac{1}{2}; b = \frac{1}{2}$

d) $a \in [-\frac{1}{2}, \frac{1}{2}]; b = \frac{1}{2}$ e) $a \in [-\frac{1}{2}, \frac{1}{2}]; b = -\frac{1}{2}$

B) What is the value of a that maximizes the variance of X, if $c = -1$ and $d = 1$?

a) 0   b) $\frac{1}{2}$ c) $-\frac{1}{2}$ d) $\pm\frac{1}{2}$ e) $\pm\frac{1}{4}$

C) Let $Y := aX + b$. Find $f_Y(y)$ if $a > 0, c = 0$ and $d = 1$.

a) $2y$ if $y \in [0, 1]$ b) $\frac{2y}{a^2}$ if $y \in [0, a]$ c) $\frac{2y}{a}$ if $y \in [0, \sqrt{a}]$

d) $\frac{y}{a}$ if $y \in [1 - \frac{a}{2}, 1 + \frac{a}{2}]$ e) $-\frac{y}{a}$ if $y \in [-1 - \frac{a}{2}, -1 + \frac{a}{2}]$

D) Calculate the characteristic function of X, if $a = 2, c = 0$ and $d = 1$.

a) $\frac{2}{j\omega}e^{j\omega}$ b) $\frac{2}{(j\omega)^2}(e^{j\omega} - 1)$ c) $\frac{2}{j\omega}e^{j\omega} - \frac{2}{(j\omega)^2}(e^{j\omega} - 1)$

d) $\frac{2}{(j\omega)^2}e^{j\omega} - \frac{2}{j\omega}(e^{j\omega} - 1)$ e) $\frac{2}{(j\omega)^2}e^{j\omega} - \frac{2}{j\omega}e^{j\omega}$

Question no. 71

A particular case of the Weibull distribution, which is often used in reliability, is the one when the probability density function of the random variable T is given by

$$f_T(t) = \begin{cases} 3t^2 e^{-t^3} & \text{if } t > 0, \\ 0 & \text{elsewhere.} \end{cases}$$

A) Calculate the failure rate $r(t)$ of a system whose lifetime T has the above probability density function.

a) $3te^{-t^3}$ b) $\dfrac{3t^2}{1 - e^{-t^3}}$ c) $\dfrac{3t^2 e^{-t^3}}{1 - e^{-t^3}}$ d) $3t$ e) $3t^2$

B) Find the median t_m of T, as well as the *mode* of T, that is, the value of t that maximizes the probability density function $f_T(t)$.

a) $t_m = (\ln 2)^{1/3}$; mode $= (2/3)^{1/3}$ b) $t_m = (\ln 2)^{1/3}$; mode $= (2/3)^{2/3}$
c) $t_m = (\ln 2)^{2/3}$; mode $= (2/3)^{1/3}$ d) $t_m = (\ln 2)^{1/3}$; mode $= (\ln 2)^{1/3}$
e) $t_m = (2/3)^{1/3}$; mode $= (2/3)^{1/3}$

C) Calculate $E\left[T^{-3/2}\right]$.

a) $\pi/2$ b) π c) $\sqrt{\pi/2}$ d) $\sqrt{\pi}$ e) $\sqrt{2\pi}$

D) A generalization of the probability density function given above is the function

$$f_T(t) = \begin{cases} \beta t^{\beta-1} e^{-t^\beta} & \text{if } t > 0, \\ 0 & \text{elsewhere,} \end{cases}$$

where β is a positive constant. We can show that $E[T] = \Gamma\left(1 + \frac{1}{\beta}\right)$. For what value(s) of β is the probability density function $f_T(t)$ symmetric with respect to the mean of T?

a) 1 only   b) 2 only   c) 1, 3, 5, ...   d) 2, 4, 6, ...   e) none

Question no. 72

We suppose that the number of calls received at a service counter during a one-hour period has a Poisson distribution, and that 30 calls are received per hour, on average. Moreover, we suppose that the number of calls received during a given hour is independent of the number of calls received during the previous hours.

A) What is the probability that exactly 30 calls are received during a given hour, and then at least 30 calls during the next hour?

Hint. We have: $F_X(30) \simeq 0.5484$ and $p_X(30) \simeq 0.0726$ if X has a Poisson distribution with parameter 30.

a) 0.0281   b) 0.0381   c) 0.0481   d) 0.0581   e) 0.0681

B) What is the probability that at most 30 calls are received during each of exactly 20 of the 24 hours of a given day?

a) 0.0017   b) 0.0027   c) 0.0037   d) 0.0047   e) 0.0057

C) What is the average number of hours required to obtain a first hour during which at most 30 calls are received, given that during at least one of the first three hours considered at most 30 calls were received?

a) 1.32   b) 1.42   c) 1.52   d) 1.62   e) 1.72

Question no. 73

A box holds 10 brand A components and 20 brand B components. The lifetime X_A (in years) of the brand A components has an exponential distribution with parameter $\lambda = 1/2$, while the lifetime X_B (in years) of those of brand B has a gamma distribution with parameter $\alpha = 3$ and $\lambda = 2$, so that

$$f_{X_B}(x) = \begin{cases} 4x^2 e^{-2x} & \text{if } x > 0, \\ 0 & \text{elsewhere.} \end{cases}$$

A) A component is taken at random from the box. Let X be its lifetime.

i) Calculate $E[X]$.

a) $\dfrac{19}{12}$ b) $\dfrac{5}{3}$ c) $\dfrac{7}{4}$ d) $\dfrac{11}{6}$ e) $\dfrac{23}{12}$

ii) What is the probability that the component lasts more than one year?

a) 0.5533   b) 0.6033   c) 0.6533   d) 0.7033   e) 0.7533

B) We can show that the total lifetime T of the 30 components has approximately a Gaussian N(50, 55) distribution. Let $Z := (T - 50)/\sqrt{55}$. What is the value of z_0 for which $P\left[Z^2 < z_0\right] \simeq 0.95$?

Hint. We have: $Q(1.645) \simeq 0.05$ and $Q(1.960) \simeq 0.025$.

a) 1.28   b) 1.40   c) 1.92   d) 2.71   e) 3.84

Question no. 74

The distribution function of the random variable X is given by

$$F_X(x) = \begin{cases} 0 & \text{if } x < c, \\ 1 - \dfrac{1}{x^2} & \text{if } x \geq c, \end{cases}$$

where c is a constant.

A) What is the set of possible values of the constant c?

a) $(-\infty, \infty)$ b) $(0, \infty)$ c) $[1, \infty)$ d) $\{1, 2, \dots\}$ e) $(-\infty, -1] \cup [1, \infty)$

B) Let $Y := 1/X$. Find $F_Y(1/4)$ if $c = 2$.

a) 1/16   b) 1/8   c) 1/4   d) 3/8   e) 1/2

C) Calculate $E[X]$ if $c = 2$.

a) 3/2   b) 2   c) 5/2   d) 3   e) 7/2

Question no. 75

We say that a continuous random variable X has a beta distribution with parameters $\alpha > 0$ and $\beta > 0$ if

$$f_X(x) = \begin{cases} \dfrac{\Gamma(\alpha+\beta)}{\Gamma(\alpha)\Gamma(\beta)} x^{\alpha-1}(1-x)^{\beta-1} & \text{for } 0 < x < 1, \\ 0 & \text{elsewhere.} \end{cases}$$

This distribution generalizes the $U(0, 1)$ distribution and is notably useful in reliability.

A) Calculate $F_X(1/2)$, if $\alpha = 1$ and $\beta = 2$.

a) 9/16   b) 5/8   c) 3/4   d) 7/8   e) 15/16

B) Find $f_X\left(\frac{1}{4} \mid X < \frac{1}{2}\right)$, if $\alpha = 3$ and $\beta = 1$.

a) 3/4   b) 1   c) 5/4   d) 3/2   e) 2

C) Let $Y := -\ln X$. Obtain the characteristic function of Y, if $\alpha = \beta = 1$.

a) $\dfrac{1}{1-j\omega}$ b) $\dfrac{1}{1+j\omega}$ c) $\dfrac{j\omega}{1-j\omega}$ d) $\dfrac{j\omega}{1+j\omega}$ e) $\dfrac{1}{(1-j\omega)^2}$

D) The lifetime X of a certain machine has a beta distribution with parameters $\alpha = \beta = 3$. Calculate the failure rate of this machine at time $x = 1/2$.

a) 7/2   b) 15/4   c) 4   d) 17/4   e) 9/2

Question no. 76

In a certain region, there are 1000 pylons (supporting an electric line), numbered from 1 to 1000. We estimate that when a thunderstorm occurs, the probability that a given pylon is hit by lightning is equal to 10^{-4}, independently from one pylon to another.

A) What is the probability that at least two pylons are hit by lightning during a thunderstorm?

a) 0.0027   b) 0.0037   c) 0.0047   d) 0.0057   e) 0.0067

B) Use a Poisson distribution to calculate (approximately) the probability that exactly two pylons are hit by lightning during a thunderstorm.

a) 0.004524   b) 0.005524   c) 0.006524   d) 0.007524   e) 0.008524

C) What is the variance of the number of thunderstorms until pylon no. 1 or pylon no. 2 is hit by lightning for the first time, from a given time instant?

a) 24,997,199.94   b) 24,997,299.94   c) 24,997,399.94   d) 24,997,499.94   e) 24,997,599.94

D) Suppose that the number of thunderstorms that occur over a one-year period is a random variable having a Poisson distribution with parameter $\alpha = 50$. What is the probability that a given pylon is hit by lightning at least once during a certain year?

a) 0.0030   b) 0.0035   c) 0.0040   d) 0.0045   e) 0.0050

Question no. 77

We are interested in the temperature (in degrees Celsius) in a certain city over the months of July, September and December.

A) First, we suppose that in July, when the temperature exceeds 30 °C, the number of degrees above 30 that the temperature reaches has an exponential distribution with mean 3. What is the probability that the temperature is higher than 35 degrees, knowing that it is higher than 31 degrees?

a) 0.2236   b) 0.2336   c) 0.2436   d) 0.2536   e) 0.2636

B) Next, we suppose that, during the month of September, the temperature has a Gaussian distribution with parameters $\mu = 15$ and $\sigma^2 = 25$. What is the temperature that is exceeded only 1% of the time (over a long period) in this city in September?

Hint. We have: $Q^{-1}(0.02) \simeq 2.054$, $Q^{-1}(0.01) \simeq 2.326$ and $Q^{-1}(0.005) \simeq 2.576$.

a) 25.63   b) 26.63   c) 27.63   d) 28.63   e) 29.63

C) Finally, we seek a model for the temperature T in December, as well as a model for T when $T \geq 0$. We consider the following distributions:

1) U [0, 10]; 2) U $\left[5 - 5\sqrt{3}, 5 + 5\sqrt{3}\right]$; 3) Exp(1/5); 4) G(5, 1); 5) N(5, 25).

Among these models, which one seems the most appropriate for
   i) T if $E[T] = 5$ and STD[T] = 5?
   ii) $U := T \mid \{T \geq 0\}$ if, instead, we have: $E[T \mid T \geq 0] = $ STD[$T \mid T \geq 0$] = 5?

a) i) 2; ii) 1   b) i) 2; ii) 3   c) i) 3; ii) 5   d) i) 5; ii) 3   e) i) 5; ii) 4

Question no. 78
   Let X be a continuous random variable taking its values in the interval $[0, \infty)$. We say that X has a Pareto distribution with parameter $\theta > 0$ if its probability density function is of the form

$$f_X(x) = \begin{cases} \dfrac{\theta}{(1+x)^{\theta+1}} & \text{if } x \geq 0, \\ 0 & \text{elsewhere.} \end{cases}$$

In economics, the Pareto distribution is used to represent the (poor) distribution of wealth. Suppose that, in a certain country, the wealth X of an individual (in thousands of dollars) has a Pareto distribution with parameter $\theta = 1.2$.

A) Calculate $f_X(2 \mid 1 < X \leq 3)$.

a) 0.3754   b) 0.3954   c) 0.4154   d) 0.4354   e) 0.4554

B) What is the median wealth in this country?

a) $781.80   b) $1781.80   c) $2781.80   d) $3781.80   e) $4781.80

C) We find that about 11.65% of the population has a personal fortune of at least $5000, which is the average wealth in this population. What percentage of the total wealth of this country do these 11.65% of the population own?

a) 75.5%   b) 77.5%   c) 79.5%   d) 81.5%   e) 83.5%

Question no. 79

Let X be a discrete random variable with probability mass function given by

$$p_X(k) = \begin{cases} \dfrac{1}{2^{k+1}} & \text{for } k = 0, 1, \ldots, \\ 0 & \text{otherwise.} \end{cases}$$

A) Calculate the probability that X takes on a value which is a multiple of the number 3, that is, a value in the set $\{3, 6, 9, \ldots\}$.

a) 1/14   b) 1/12   c) 1/10   d) 1/8   e) 1/6

B) We generate (independent) random numbers according to the distribution of the random variable X. Let M be the number of random numbers that are greater than 1, among the first ten numbers generated.

i) Calculate $P[M = 2]$.

a) 0.2416   b) 0.2616   c) 0.2816   d) 0.3016   e) 0.3216

ii) Suppose that we approximate the probability $P[M = k]$ by $P[N = k]$, for $k = 0, 1, \ldots, 10$, where N is a random variable that has a Poisson distribution with parameter $\alpha = 2.5$. What can we assert concerning $P[N = k]$ with respect to $P[M = k]$ for any value k taken from the set $\{0, 1, \ldots, 10\}$?

a) $P[N = k]$ is strictly less than $P[M = k]$;
b) $P[N = k]$ can only be less than or equal to $P[M = k]$;
c) $P[N = k]$ is strictly greater than $P[M = k]$;
d) $P[N = k]$ can only be greater than or equal to $P[M = k]$;
e) we cannot assert anything.

Question no. 80

An electroluminescent diode emits light at a wavelength X (in microns) that is a continuous random variable taking its values in the interval $(0, \infty)$.

A) What is the distribution of X, if the quantile of order p of X is given by the formula

$$x_p = \frac{1}{2}(1 + p),$$

where $0 < p < 1$?

a) U(0, 1/2)   b) U(0,1)   c) U(1/2, 1)   d) U(1/2, 3/2)   e) U(1, 3/2)

B) Suppose that the probability density function of X is strictly decreasing and is bounded (above) by a positive constant k. Which one, among the models below, seems the most appropriate?

a) G(1/2, 1/2)   b) G(1, 1/2)   c) G(3/2, 1/2)   d) G(3/2, 1)   e) G(2, 1)

C) Suppose that X has a Gaussian distribution, with parameters $\mu = 1$ and $\sigma^2 = 1/16$, to which its negative part and the point 0 have been subtracted, so that

$$f_X(x) = \frac{c}{\sqrt{2\pi}} \exp\left\{-8(x-1)^2\right\} \quad \text{for } x > 0,$$

where c is a constant such that f_X is a valid probability density function.

i) Find the constant c.

a) $1/\Phi(1)$ b) $2/\Phi(2)$ c) $4/\Phi(2)$ d) $2/\Phi(4)$ e) $4/\Phi(4)$

ii) Suppose that the diode is considered defective if the value taken by the random variable X is such that $|X - 1| > 1/4$. Calculate, in terms of c, the probability that a diode is defective.

Hint. We have: $\Phi(1) \simeq 0.8413$ and $\Phi(2) \simeq 0.9772$.

a) $0.17c$ b) $0.24c$ c) $1 - 0.68c$ d) $1 - 0.24c$ e) $1 - 0.17c$

Question no. 81

Let X be a continuous random variable having the probability density function

$$f_X(x) = \begin{cases} \dfrac{x}{\theta^2} \exp\left\{-\dfrac{x^2}{2\theta^2}\right\} & \text{if } x > 0, \\ 0 & \text{elsewhere.} \end{cases}$$

We say that X has a Rayleigh distribution with parameter $\theta > 0$. We can show that $E[X] = \theta\sqrt{\pi/2}$ and $\mathrm{VAR}[X] = \theta^2 [2 - (\pi/2)]$.

A) Let $Y := \ln X$, where X has a Rayleigh distribution with parameter $\theta = 1$.

i) Find $f_Y(1)$.

a) 0.1637 b) 0.1837 c) 0.2037 d) 0.2237 e) 0.2437

ii) Calculate the mathematical expectation $E[e^{j\omega Y}]$ at $\omega = -2j$.

a) $1/2$ b) 1 c) $3/2$ d) 2 e) $5/2$

B) We define $Z = 1/X$. Calculate the mathematical expectation of Z, if $\theta = 1$ as in A).

a) $\sqrt{1/\pi}$ b) $\sqrt{2/\pi}$ c) $\sqrt{\pi/2}$ d) $\sqrt{\pi}$ e) ∞

C) What is the value of the reliability function of a system, whose lifetime has a Rayleigh distribution with parameter $\theta = 2$, at the moment that corresponds to its average lifetime?

a) $e^{-\pi/8}$ b) $e^{-\pi/4}$ c) $e^{-\pi/2}$ d) $e^{-\pi}$ e) $e^{-2\pi}$

4

Random Vectors

4.1 Introduction

Definition 4.1.1. *A function* **X** *that assigns a vector* $(X_1(s), \ldots, X_n(s))$ *of real numbers to each outcome s in a sample space S associated with a random experiment E is called a* **random vector** *of dimension n. Each component of the random vector is a random variable. The set of all possible values of* **X**, *denoted by* $S_\mathbf{X}$, *is a subset of* $\mathbb{R}^n$ *(see Fig. 4.1).*

Example 4.1.1 Let E be the random experiment that consists in observing the length of a program submitted for execution to a computer, as well as the execution time. Then, the elements of the sample space S are of the form $s = (n, t)$, where n can be the number of lines in the program and t the execution time in seconds. Now, let $\mathbf{X} := (X_1, X_2)$, where $X_1(s) = n$ and $X_2(s) = t$. The function $\mathbf{X}$, which here

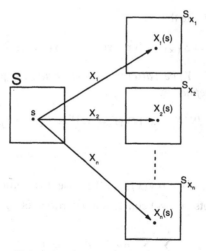

Figure 4.1. Notion of random vector.

is the identity function, is a random vector of dimension 2. In this example, X_1 is a discrete random variable, whereas X_2 is (theoretically) continuous. In general, the components X_k of a random vector are random variables of the same type.

Next, let E^* be the random experiment that consists in observing the execution time only. Then, if $s = t$, where t is in seconds, we can nevertheless define a random vector $\mathbf{X} = (X_1(s), \ldots, X_n(s))$. For instance, we can have: $X_1(s) = s \ (= t)$, $X_2(s) = s^2$, $X_3(s) = \min\{s, 10\}$, $X_4(s) = -s^3$, etc. Thus, there is no relationship between the dimension of the elements s of S and the dimension of the random vector $\mathbf{X}$.

Remarks. i) In reality, the variable X_2 above is a random variable of mixed type because the execution time can be equal to zero (when there is a fatal error in the program, for example), with a strictly positive probability, or it can take on any real positive value. In practice, the execution time is rounded off, so that X_2 becomes a random variable of discrete type.

ii) For simplicity, we will not consider in this book the case when one (or many) random variable(s) in the random vector is (are) of mixed type.

4.2 Random Vectors of Dimension 2

I) Discrete Case

Definition 4.2.1. *Let* $\mathbf{Z} := (X, Y)$ *be a random vector of dimension 2. If* $S_\mathbf{Z}$ *is a finite or denumerably infinite set of points in the plane, that is, if we can write* $S_\mathbf{Z}$ *in the form*

$$S_\mathbf{Z} = \{(x_j, y_k), j = 1, 2, \ldots ; k = 1, 2, \ldots\},$$

then we say that $\mathbf{Z}$ *is a* **discrete** *random vector (of dimension 2).*

Remark. We may also write that

$$S_\mathbf{Z} = S_{X \times Y} = \{(x_j, y_k): x_j \in S_X, y_k \in S_Y\}.$$

Definition 4.2.2. *Let* (X, Y) *be a discrete random vector. The* **joint probability mass function** *of* (X, Y) *is defined by*

$$p_{X,Y}(x_j, y_k) = P[\{X = x_j\} \cap \{Y = y_k\}] \equiv P[X = x_j, Y = y_k] \qquad (4.1)$$

for $j, k = 1, 2, \ldots$.

Properties. i) $p_{X,Y}(x_j, y_k) \geq 0 \ \forall (x_j, y_k)$ (because it is a probability).

ii) Since the set of points (x_j, y_k) constitutes a partition of $S_\mathbf{Z}$, we have:

$$\sum_{j=1}^{\infty} \sum_{k=1}^{\infty} p_{X,Y}(x_j, y_k) = 1. \qquad (4.2)$$

Let A be an event with respect to $S_{\mathbf{Z}}$. That is, $A \subset S_{\mathbf{Z}}$. The probability of A is given by the formula

$$P[A] = \sum_{(x_j, y_k) \in A} p_{X,Y}(x_j, y_k). \tag{4.3}$$

Now, making use of the *total probability rule*, we can write that

$$p_X(x_j) \equiv P[X = x_j]$$
$$= \sum_{k=1}^{\infty} P[\{X = x_j\} \cap \{Y = y_k\}] = \sum_{k=1}^{\infty} p_{X,Y}(x_j, y_k). \tag{4.4}$$

Likewise, we have:

$$p_Y(y_k) \equiv P[Y = y_k]$$
$$= \sum_{j=1}^{\infty} P[\{X = x_j\} \cap \{Y = y_k\}] = \sum_{j=1}^{\infty} p_{X,Y}(x_j, y_k). \tag{4.5}$$

Definition 4.2.3. *The function obtained by summing the function $p_{X,Y}$ over all possible values of either of the two random variables is called a* **marginal probability mass function**.

Example 4.2.1 A box contains six transistors, including one of brand A and one of brand B. Two transistors are taken at random and without replacement. Let X (respectively Y) be the number of brand A (resp. B) transistors among the two picked. Then, the joint probability mass function $p_{X,Y}$ of the random vector $\mathbf{Z} := (X, Y)$ can be given in the form of a table in two dimensions as follows:

$y \backslash x$	0	1	2
0	16/36	8/36	1/36
1	8/36	2/36	0
2	1/36	0	0

For instance, we find that $p_{X,Y}(0,0) = 16/36$, because there are 4 chances in 6 that the first transistor picked is neither a brand A nor a brand B transistor, and the two draws are independent. Moreover, there is 1 chance in 6 that the first transistor picked is of brand A; it follows that

$$p_{X,Y}(1,0) = 2 \times (1/6)(4/6) = 8/36$$

(by symmetry), etc. We can check that $p_{X,Y}(j, k) \geq 0$ for $j, k = 0, 1, 2$, and that the sum of all the fractions in the table is equal to 1. Note that the random variables X and Y both have a binomial distribution with parameters $n = 2$ and $p = 1/6$.

Next, let, for example, $A = \{X + Y \geq 1\}$. We have then:

$$P[A] = 1 - P[A^c] = 1 - p_{X,Y}(0,0) = 1 - 16/36 = 20/36 = 5/9,$$

because there is a single element of $S_{\mathbf{Z}}$ in A^c, namely $(0, 0)$. Similarly, we find that

$$P[X = Y] = \sum_{j=0}^{2} p_{X,Y}(j, j) = 16/36 + 2/36 + 0 = 1/2.$$

Finally, in such a case the marginal probability mass functions are obtained by adding the elements of the lines and of the columns in the table. For example, adding the three elements of each column in the table, we deduce that

j	0	1	2	Σ
$p_X(j)$	25/36	10/36	1/36	1

Furthermore, by symmetry, we can write that $p_Y(k) = p_X(k)$ for $k = 0, 1, 2$.

II) Continuous Case

Definition 4.2.4. *Let* $\mathbf{Z} := (X, Y)$ *be a random vector of dimension 2. If* $S_{\mathbf{Z}}$ *is a non-denumerably infinite subset of the plane, we say that* $\mathbf{Z}$ *is a* **continuous** *random vector. (We assume that* X *and* Y *are two continuous random variables.)*

Definition 4.2.5. *Let* $\mathbf{Z} := (X, Y)$ *be a continuous random vector, and let* $A \subset S_{\mathbf{Z}}$. *The probability of the event* A *is given by the formula*

$$P[A] \equiv P[\mathbf{Z} \in A] = \iint_A f_{X,Y}(x, y)\,dx\,dy, \tag{4.6}$$

where $f_{X,Y}(x, y)$ *is called the* **joint (probability) density function** *of the pair* (X, Y).

For example, let

$$f_{X,Y}(x, y) = \begin{cases} e^{-x-y} & \text{if } x > 0, y > 0, \\ 0 & \text{elsewhere.} \end{cases}$$

Then, we have (see Fig. 4.2, p. 161):

$$\begin{aligned} P[Y > X^2] &= \int_0^\infty \int_0^{\sqrt{y}} e^{-x-y}\,dx\,dy = \int_0^\infty \int_{x^2}^\infty e^{-x-y}\,dy\,dx \\ &= \int_0^\infty e^{-x-x^2}\,dx = e^{1/4} \int_0^\infty e^{-\left(x+\frac{1}{2}\right)^2}\,dx \\ &= e^{1/4}\sqrt{\pi}\,P\left[N\left(-\tfrac{1}{2}, \tfrac{1}{2}\right) > 0\right] \simeq 0.55. \end{aligned}$$

Properties. i) We have:

$$f_{X,Y}(x, y) \simeq \frac{P[x < X \le x + dx, y < Y \le y + dy]}{dx\,dy} \tag{4.7}$$

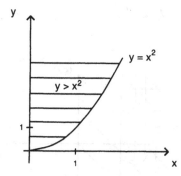

Figure 4.2. Example of computation of probability in two dimensions.

(the equality being obtained by taking the limit as dx and dy decrease to 0). It follows that the function $f_{X,Y}(x, y)$ is *non-negative*.

ii) Since X and Y take on real values, we can write that

$$\int_{-\infty}^{\infty} \int_{-\infty}^{\infty} f_{X,Y}(x, y)\, dx dy = 1. \tag{4.8}$$

Definition 4.2.6. *Let* (X, Y) *be a random vector. The* **joint distribution function** *of the pair* (X, Y) *is defined by*

$$F_{X,Y}(x, y) = P[X \le x, Y \le y]. \tag{4.9}$$

If (X, Y) *is a discrete random vector, then*

$$F_{X,Y}(x, y) = \sum_{x_j \le x} \sum_{y_k \le y} p_{X,Y}(x_j, y_k), \tag{4.10}$$

while when (X, Y) *is continuous,*

$$F_{X,Y}(x, y) = \int_{-\infty}^{x} \int_{-\infty}^{y} f_{X,Y}(u, v)\, dv du. \tag{4.11}$$

Remark. We deduce from Equation (4.11) that

$$f_{X,Y}(x, y) = \frac{\partial^2 F_{X,Y}(x, y)}{\partial x \partial y} \tag{4.12}$$

for all pairs (x, y) where the second mixed partial derivative exists.

Properties. The function $F_{X,Y}$ has properties similar to those of the function F_X.

i) $F_{X,Y}(-\infty, y) = F_{X,Y}(x, -\infty) = 0$ (impossible events);

ii) $F_{X,Y}(\infty, \infty) = 1$ (certain event);

iii) $F_{X,Y}(x_1, y_1) \le F_{X,Y}(x_2, y_2)$ if $x_1 \le x_2$ *and* $y_1 \le y_2$ (because, in this case, $\{X \le x_1, Y \le y_1\} \subset \{X \le x_2, Y \le y_2\}$; see Fig. 4.3, p. 162).

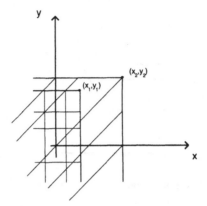

Figure 4.3. Non-decreasingness of the joint distribution function.

iv) We can show that

$$\lim_{\delta \downarrow 0} F_{X,Y}(x + \delta, y) = \lim_{\delta \downarrow 0} F_{X,Y}(x, y + \delta) = F_{X,Y}(x, y).$$

Remark. We can also check, graphically, that

$$P[a < X \leq b, c < Y \leq d] = F_{X,Y}(b, d) - F_{X,Y}(b, c) - F_{X,Y}(a, d) + F_{X,Y}(a, c).$$
$$(4.13)$$

Now, we have:

$$F_X(x) \equiv P[X \leq x] = P[X \leq x, Y < \infty] = F_{X,Y}(x, \infty). \qquad (4.14)$$

Likewise, we obtain: $F_Y(y) = F_{X,Y}(\infty, y)$.

Definition 4.2.7. *The function obtained from $F_{X,Y}(x, y)$ by replacing y (respectively x) by infinity is called the* **marginal distribution function** *of X (resp. Y).*

Finally, we can write (see Remark i) below) that

$$f_X(x) = \frac{d}{dx} F_X(x) = \frac{d}{dx} F_{X,Y}(x, \infty)$$
$$= \frac{d}{dx} \left\{ \int_{-\infty}^{x} \int_{-\infty}^{\infty} f_{X,Y}(u, v) \, dv \, du \right\} = \int_{-\infty}^{\infty} f_{X,Y}(x, v) \, dv.$$

That is,

$$f_X(x) = \int_{-\infty}^{\infty} f_{X,Y}(x, y) \, dy. \qquad (4.15)$$

Similarly, the function $f_Y(y)$ is obtained by integrating $f_{X,Y}(x, y)$ with respect to x, from $-\infty$ to ∞.

Remarks. i) In Equation (4.15), we must actually integrate over all possible values of Y, given x, a *fixed* value of X.

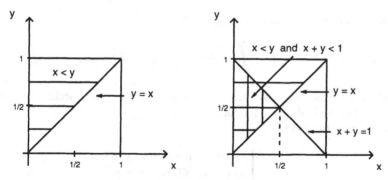

Figure 4.4. Figures for Example 4.2.2.

ii) Let

$$I(y) := \int_{u_1(y)}^{u_2(y)} g(x, y)\, dx.$$

Then, under certain conditions, we have the following formula (due to Leibniz):[1]

$$\frac{dI(y)}{dy} = \int_{u_1(y)}^{u_2(y)} \frac{\partial g(x, y)}{\partial y}\, dx + g(u_2(y), y)\frac{du_2(y)}{dy} - g(u_1(y), y)\frac{du_1(y)}{dy}.$$

Definition 4.2.8. *The functions f_X and f_Y obtained from $f_{X,Y}(x, y)$ as above are called the* **marginal (probability) density functions** *of X and Y, respectively.*

Example 4.2.2 Let

$$f_{X,Y}(x, y) = \begin{cases} 2 & \text{if } 0 \le x \le y \le 1, \\ 0 & \text{elsewhere} \end{cases}$$

(see Fig. 4.4). Since $f_{X,Y}(x, y) \equiv 2$ and the area of the triangle on which the function is positive is equal to $1/2$, the double integral of $f_{X,Y}(x, y)$ over $S_{X \times Y}$ is equal to 1, as required. We have:

$$f_X(x) = \int_x^1 2\, dy = 2(1 - x) \quad \text{if } 0 \le x \le 1 \quad (\text{and } f_X(x) \equiv 0 \text{ elsewhere}),$$

$$f_Y(y) = \int_0^y 2\, dx = 2y \quad \text{if } 0 \le y \le 1 \quad (\text{and } f_Y(y) \equiv 0 \text{ elsewhere}).$$

---

[1] Gottfried Wilhelm von Leibniz, 1646–1716, was born and died in Germany. He was at once mathematician, philosopher and theologian. With Newton, he is the inventor of infinitesimal calculus. He introduced the notation for the derivative and the integral. Among his great achievements in mathematics, we count the development of the binary system and his work on determinants. He also contributed to mechanics, more precisely to dynamics. He corresponded with more than 600 European scientists. In the field of religion, he set himself the objective of reuniting the various branches of the Christian church.

Moreover, we can write (see Fig. 4.4, p. 163) that

$$P[X + Y < 1] = \int_0^{1/2} \int_x^{1-x} 2\,dy\,dx = \int_0^{1/2} 2(1 - 2x)\,dx = 1/2.$$

Remark. In this case, we could have computed the area of the checkered region in Fig. 4.4 and multiplied it by 2 instead.

Finally, we also find that

$$F_{X,Y}(x, y) = \begin{cases} 0 & \text{if } x < 0 \text{ or } y < 0, \\ 2xy - x^2 & \text{if } 0 \le x \le y \le 1, \\ 2x - x^2 & \text{if } 0 \le x \le 1 \text{ and } y > 1, \\ y^2 & \text{if } 0 \le y \le 1 \text{ and } x > y, \\ 1 & \text{if } x > 1 \text{ and } y > 1. \end{cases}$$

Indeed, for any point (x_0, y_0) in the region $S_{X \times Y}$, the area of the checkered region in Fig. 4.5, is given by $\frac{1}{2}x_0^2 + (y_0 - x_0)x_0$. Thus, we do have:

$$F_{X,Y}(x, y) = 2 \times \left(\frac{1}{2}x^2 + (y - x)x \right) = 2xy - x^2 \quad \text{if } 0 \le x \le y \le 1.$$

The value of the function $F_{X,Y}(x, y)$ can then be obtained by reasoning for any point $(x, y) \in \mathbb{R}^2$. First, if $0 \le x \le 1$ but $y > 1$, we only have to replace y by 1 in the above formula. Next, if $0 \le y \le 1$ but $x > y$, we replace x by its maximum value, namely y, in the above formula. The other two cases are elementary.

We will complete this section with the notion of independence applied to random variables.

Definition 4.2.9. *Let (X, Y) be a random vector. We say that X and Y are **independent** random variables if*

$$F_{X,Y}(x, y) = F_X(x)F_Y(y) \quad \forall(x, y). \tag{4.16}$$

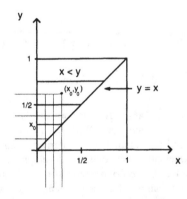

Figure 4.5. Joint distribution function in Example 4.2.2.

Equation (4.16) is equivalent to

$$p_{X,Y}(x_j, y_k) \equiv p_X(x_j)p_Y(y_k) \ \textit{if } (X, Y) \textit{ is a discrete random vector,}$$
$$f_{X,Y}(x, y) \equiv f_X(x)f_Y(y) \quad \textit{if } (X, Y) \textit{ is a continuous random vector.}$$

Remark. We can say, more generally, that X and Y are independent if

$$P[X \in A, Y \in B] = P[X \in A]P[Y \in B],$$

where A (respectively B) is an arbitrary event that involves only X (resp. Y). Definition 4.2.9 then becomes a proposition that we can easily prove.

Proposition 4.2.1. *If X and Y are two independent random variables, then so are $g(X) \equiv X^*$ and $h(Y) \equiv Y^*$.*

Proof. We have:

$$F_{X^*,Y^*}(x^*, y^*) \equiv P[X^* \le x^*, Y^* \le y^*] = P[X \in A, Y \in B],$$

where $\{X \in A\}$ is the event in S_X that is equivalent to $\{X^* \le x^*\}$, and $\{Y \in B\} \subset S_Y$ is equivalent to $\{Y^* \le y^*\}$. Then, given that X and Y are independent, we can write that

$$F_{X^*,Y^*}(x^*, y^*) = P[X \in A]P[Y \in B] = P[X^* \le x^*]P[Y^* \le y^*]. \qquad \square$$

Example 4.2.3 The two random variables in Example 4.2.1 are *not* independent, because, in particular,

$$p_{X,Y}(2, 2) = 0 \neq (1/36)(1/36) = p_X(2)p_Y(2).$$

Remark. In such a case, if there is one or more 0's in the table, then the random variables are not independent. We can also notice that we must have: $0 \le X + Y \le 2$, which directly implies that X and Y are not independent since, for instance, $X = 2 \Rightarrow Y = 0$ (because $Y \ge 0$).

Example 4.2.4 Consider once again the function $f_{X,Y}$ in Example 4.2.2. We already found that $f_X(x) = 2(1 - x), 0 \le x \le 1$, and $f_Y(y) = 2y, 0 \le y \le 1$. Since

$$f_X(0)f_Y(0) = 2 \times 0 = 0 \neq 2 = f_{X,Y}(0, 0),$$

X and Y are *not* independent random variables.

Remark. Given that there is a relationship between x and y in the definition of $S_{X \times Y}$, namely $x \le y$, X and Y could not be independent.

We can show the following proposition.

Proposition 4.2.2. *Let (X, Y) be a continuous random vector. Suppose that $S_{X \times Y}$ is of the form*

$$S_{X \times Y} = \{(x, y): a < x < b, c < y < d\},$$

where a, b, c and d are constants. Then, X and Y are independent random variables if and only if (iff) we can write that

$$f_{X,Y}(x, y) \equiv g(x)h(y), \tag{4.17}$$

where $g(x) > 0$ iff $a < x < b$ and $h(y) > 0$ iff $c < y < d$.

Remark. An analogous result exists in the discrete case.

4.3 Conditionals

Conditional Distribution, Probability Mass and Density Functions

We discussed the notion of conditional distribution function in Section 3.2. If Y is a discrete random variable, then we can write, directly, that the **conditional distribution function** of X, given that $Y = y_k$, is

$$F_{X|Y}(x \mid y_k) = \frac{P[X \leq x, Y = y_k]}{P[Y = y_k]} = \frac{\sum_{x_j \leq x} p_{X,Y}(x_j, y_k)}{p_Y(y_k)} \quad \text{if } p_Y(y_k) > 0. \tag{4.18}$$

Remark. Some authors use the notation $F_X(x \mid Y = y_k)$, as $F_X(x \mid A)$ in Chapter 3, rather than $F_{X|Y}(x \mid y_k)$.

Similarly, if (X, Y) is a discrete random vector, we define the **conditional probability mass function** of X, given that $Y = y_k$, by

$$p_{X|Y}(x_j \mid y_k) = \frac{p_{X,Y}(x_j, y_k)}{p_Y(y_k)} \quad \text{if } p_Y(y_k) > 0. \tag{4.19}$$

Remark. This function has the same properties as the (marginal) probability mass functions. Furthermore, the conditional probability mass function of Y, given that $X = x_j$, is defined in a similar way.

If Y is a continuous random variable, then $P[Y = y] = 0 \; \forall y$. We must therefore consider the event $\{y < Y \leq y + dy\}$ in (4.18) and take the limit as dy decreases to zero. If X also is a continuous random variable, we obtain the proposition that follows.

Proposition 4.3.1. *Let (X, Y) be a continuous random vector. The **conditional distribution function** of X, given that $Y = y$, is given by*

$$F_{X|Y}(x \mid y) = \frac{\int_{-\infty}^{x} f_{X,Y}(u, y) \, du}{f_Y(y)} \quad \text{if } f_Y(y) > 0. \tag{4.20}$$

Corollary 4.3.1. *Let (X, Y) be a continuous random vector. The* **conditional (probability) density function** *of X, given that $Y = y$, is obtained by differentiating the function $F_{X|Y}(x \mid y)$ above with respect to x:*

$$f_{X|Y}(x \mid y) = \frac{f_{X,Y}(x, y)}{f_Y(y)} \quad \text{if } f_Y(y) > 0, \tag{4.21}$$

for any point x where $F_{X|Y}(x \mid y)$ is differentiable.

Proposition 4.3.2. *Let (X, Y) be a random vector. X and Y are independent random variables if and only if*

$$F_{X|Y}(x \mid y) = F_X(x) \quad \forall \, (x, y) \quad (\text{or } F_{Y|X}(y \mid x) \equiv F_Y(y)). \tag{4.22}$$

If (X, Y) is a discrete random vector, this condition is equivalent to

$$p_{X|Y}(x_j \mid y_k) = p_X(x_j) \quad \forall \, (x_j, y_k), \tag{4.23}$$

while Equation (4.22) is equivalent to

$$f_{X|Y}(x \mid y) = f_X(x) \quad \forall \, (x, y) \tag{4.24}$$

when (X, Y) is a continuous random vector.

Proof. This follows directly from Equations (4.18)–(4.21), and from the definition of independence. $\qquad\qquad\qquad\square$

Example 4.3.1 Consider once again the function $f_{X,Y}$ in Example 4.2.2:

$$f_{X,Y}(x, y) = 2 \quad \text{if } 0 \le x \le y \le 1 \quad (= 0 \text{ elsewhere}).$$

The conditional probability density function of X, given that $Y = y$, is

$$f_{X|Y}(x \mid y) = \frac{f_{X,Y}(x, y)}{f_Y(y)} = \frac{2}{2y} = \frac{1}{y} \quad \text{if } 0 \le x \le y \, (\le 1).$$

That is, $X \mid \{Y = y\}$ has a $U[0, y]$ distribution, where $y \in [0, 1]$.

Conditional Expectation

As we mentioned in Section 3.7, the concept of mathematical expectation can be generalized by considering the **conditional** expectation of a random variable.

Definition 4.3.1. *Let (X, Y) be a pair of random variables. The* **conditional expectation** *of X, given that $Y = y$, is given by*

$$E[X \mid Y = y] = \begin{cases} \displaystyle\sum_{j=1}^{\infty} x_j \, p_{X|Y}(x_j \mid y) & \text{if } (X, Y) \text{ is discrete,} \\[2mm] \displaystyle\int_{-\infty}^{\infty} x \, f_{X|Y}(x \mid y) \, dx & \text{if } (X, Y) \text{ is continuous.} \end{cases} \tag{4.25}$$

Now, let $E[X \mid Y]$ be a function of the random variable Y that takes on the value $E[X \mid Y = y]$ when $Y = y$. Then, $E[X \mid Y]$ also is a random variable. We could denote this random variable by $g(Y)$, so that $g(y) = E[X \mid Y = y]$, as in Section 3.6 (p. 92). The following result is important in probability.

Proposition 4.3.3. *Let (X, Y) be a pair of random variables. We have:*

$$E[X] = E[E[X \mid Y]]. \tag{4.26}$$

Proof. Consider the case when (X, Y) is a continuous random vector. We have:

$$E[E[X \mid Y]] = \int_{-\infty}^{\infty} E[X \mid Y = y] f_Y(y)\, dy$$

$$= \int_{-\infty}^{\infty} \left[\int_{-\infty}^{\infty} x f_{X|Y}(x \mid y)\, dx \right] f_Y(y)\, dy.$$

Now, $f_{X|Y}(x \mid y) f_Y(y) = f_{X,Y}(x, y)$. Thus, we can write (changing the order of integration) that

$$E[E[X \mid Y]] = \int_{-\infty}^{\infty} x \left[\int_{-\infty}^{\infty} f_{X,Y}(x, y)\, dy \right] dx = \int_{-\infty}^{\infty} x f_X(x)\, dx = E[X].$$

The proof in the other cases is analogous. $\qquad\square$

Remarks. i) The previous proposition is also true when we replace X by a function $h(X)$. That is,

$$E[E[h(X) \mid Y]] = E[h(X)]. \tag{4.27}$$

ii) In the case when Y is a discrete random variable, Equation (4.26) becomes

$$E[X] = E[E[X \mid Y]] = \sum_{k=1}^{\infty} E[X \mid Y = y_k] p_Y(y_k). \tag{4.28}$$

iii) We can also write that

$$\mathrm{VAR}[X] = E[X^2] - (E[X])^2 = E[E[X^2 \mid Y]] - (E[E[X \mid Y]])^2. \tag{4.29}$$

Finally, let A be an event and W be the indicator variable of the event $\{X \in A\}$. That is,

$$W = \begin{cases} 1 & \text{if } X \in A, \\ 0 & \text{otherwise.} \end{cases} \tag{4.30}$$

Since $E[W] = P[X \in A]$, we deduce from Proposition 4.3.3 the following result, which is the equivalent of the total probability rule for random variables.

Corollary 4.3.2. a) *Let Y be a discrete random variable. Then, we can write that*

$$P[X \in A] = \sum_{k=1}^{\infty} P[X \in A \mid Y = y_k] p_Y(y_k). \tag{4.31}$$

b) *If Y is continuous, Equation (4.31) becomes*

$$P[X \in A] = \int_{-\infty}^{\infty} P[X \in A \mid Y = y] f_Y(y) \, dy. \qquad (4.32)$$

Remark. We also define the **conditional variance** of X, given the random variable Y, by

$$\text{VAR}[X \mid Y] = E[(X - E[X \mid Y])^2 \mid Y] = E[X^2 \mid Y] - (E[X \mid Y])^2. \qquad (4.33)$$

We can show that

$$\text{VAR}[X] = E[\text{VAR}[X \mid Y]] + \text{VAR}[E[X \mid Y]]. \qquad (4.34)$$

Example 4.3.2 Consider one last time the function $f_{X,Y}$ in Example 4.2.2:

$$f_{X,Y}(x, y) = 2 \quad \text{if } 0 \le x \le y \le 1 \quad (= 0 \text{ elsewhere}).$$

Making use of Example 4.3.1, we can write that the conditional expectation of X, given that $Y = y$, is

$$E[X \mid Y = y] = y/2,$$

because $X \mid \{Y = y\} \sim \text{U}[0, y]$. Moreover, we have:

$$E[E[X \mid Y]] = E[Y/2] = \int_0^1 \frac{y}{2} \cdot 2y \, dy = 1/3,$$

which is indeed equal to

$$E[X] = \int_0^1 x \cdot 2(1 - x) \, dx = x^2 - \frac{2}{3}x^3 \Big|_0^1 = 1/3.$$

Example 4.3.3 Suppose that the number X of customers who visit a car dealer during a one-day period has a Poisson distribution with parameter $\alpha = 20$. Furthermore, suppose that the probability that a customer buys a car (on a given visit) is equal to $1/10$, independently from one person to another. Let Y be the number of cars sold during one day. We can write that $Y \mid X \sim \text{B}(n = X, p = 0.1)$. It follows that

$$E[Y] = E[E[Y \mid X]] = E[0.1X] = 0.1 \cdot 20 = 2.$$

Moreover, since $\text{VAR}[Y \mid X] = X (0.1)(0.9)$, we deduce from Equation (4.34) that

$$\text{VAR}[Y] = E[(0.09)X] + \text{VAR}[(0.1)X] = (0.09)(20) + (0.01)(20) = 2.$$

Finally, we can calculate

$$P[Y \ge 1] = 1 - P[Y = 0] = 1 - \sum_{k=0}^{\infty} P[Y = 0 \mid X = k] P[X = k]$$

$$= 1 - \sum_{k=0}^{\infty} P[\text{B}(n = k, p = 0.1) = 0] p_X(k)$$

$$= 1 - \sum_{k=0}^{\infty}(0.9)^k e^{-20}\frac{20^k}{k!}$$

$$= 1 - e^{-20}\sum_{k=0}^{\infty}\frac{18^k}{k!} = 1 - e^{-2} \simeq 0.8647.$$

Example 4.3.4 A number X is taken at random in the interval $[0, 1]$, then a number Y is taken at random in the interval $[0, X]$. Thus, we can write that $X \sim U[0, 1]$ and $Y \mid X \sim U[0, X]$. Then, we obtain that

$$E[Y] = E[E[Y \mid X]] = E[X/2] = (1/2)(1/2) = 1/4.$$

Next, in this case we have: $\text{VAR}[Y \mid X] = \frac{1}{12}X^2$, so that

$$\text{VAR}[Y] = E[X^2/12] + \text{VAR}[X/2] = \frac{1}{12}\int_0^1 x^2 \cdot 1\, dx + \frac{1}{4}\underbrace{\text{VAR}[X]}_{\frac{1}{12}}$$

$$= \frac{1}{12}\cdot\frac{1}{3} + \frac{1}{4}\cdot\frac{1}{12} = \frac{7}{144}.$$

Finally, we compute

$$P[Y > 1/2] = \int_{-\infty}^{\infty} P[Y > 1/2 \mid X = x]f_X(x)\, dx$$

$$= \int_0^1 P[Y > 1/2 \mid X = x] \cdot 1\, dx$$

$$= \int_{1/2}^1 \frac{x - \frac{1}{2}}{x} \cdot 1\, dx = x - \frac{1}{2}\ln x\Big|_{1/2}^1 = \frac{1}{2} - \frac{\ln 2}{2} \simeq 0.1534.$$

Remark. We have:

$$f_{X,Y}(x, y) = f_{Y\mid X}(y \mid x)f_X(x) = \frac{1}{x}\cdot 1 = \frac{1}{x} \quad \text{if } 0 \le y \le x \le 1.$$

4.4 Random Vectors of Dimension $n > 2$

All the definitions that we have seen in the case of pairs of random variables can easily be generalized to the case of random vectors of dimension n, where $n > 2$. For instance, the **joint probability density function** of a continuous random vector $\mathbf{X} = (X_1, \ldots, X_n)$ is a non-negative function $f_{\mathbf{X}}(x_1, \ldots, x_n)$ defined on a non-denumerably infinite subset $S_{\mathbf{X}}$ of $\mathbb{R}^n$, such that

$$\int\int\cdots\int_{S_{\mathbf{X}}} f_{\mathbf{X}}(x_1, \ldots, x_n)dx_1\ldots dx_n = 1.$$

The **marginal probability density function** of X_k is obtained by integrating the function $f_{\mathbf{X}}$ with respect to all variables, except X_k, for $k = 1, \ldots, n$:

$$f_{X_k}(x_k) = \underbrace{\int_{-\infty}^{\infty} \cdots \int_{-\infty}^{\infty}}_{(n-1) \text{ times}} f_{\mathbf{X}}(x_1, \ldots, x_n)\, dx_1 \ldots dx_{k-1} dx_{k+1} \ldots dx_n.$$

Likewise, we have:

$$f_{X_1, X_2}(x_1, x_2) = \underbrace{\int_{-\infty}^{\infty} \cdots \int_{-\infty}^{\infty}}_{(n-2) \text{ times}} f_{\mathbf{X}}(x_1, \ldots, x_n)\, dx_3 \ldots dx_n,$$

etc. That is, we integrate the function $f_{\mathbf{X}}$ with respect to all the variables that we want to remove from the random vector $\mathbf{X}$.

Finally, we say that the random variables $X_1, \ldots, X_n$ are (globally) **independent** if and only if

$$P[X_1 \in A_1, \ldots, X_n \in A_n] = \prod_{k=1}^{n} P[X_k \in A_k], \qquad (4.35)$$

where A_k is an event that involves only X_k, for $k = 1, \ldots, n$. If $\mathbf{X} = (X_1, \ldots, X_n)$ is a continuous random vector, this condition implies that

$$f_{\mathbf{X}}(x_1, \ldots, x_n) \equiv \prod_{k=1}^{n} f_{X_k}(x_k).$$

Moreover, as in the case of (random) events, some variables in a random vector may be independent and others not, as in the following example.

Example 4.4.1 Let

$$f_{\mathbf{W}}(x, y, z) = \begin{cases} k(x+y)e^{-z} & \text{if } 0 < x < 1, 0 < y < 2, z > 0, \\ 0 & \text{elsewhere} \end{cases}$$

be the joint probability density function of the continuous random vector $\mathbf{W} = (X, Y, Z)$. Using the fact that the triple integral of $f_{\mathbf{W}}$ over $S_{\mathbf{W}}$ must be equal to 1, we find that the constant k is equal to $1/3$. In addition, we can calculate, for instance, the probability

$$P[X < Y, Z > 1] = \int_1^{\infty} \left[\int_0^1 \int_x^2 \frac{1}{3}(x+y)\, dy dx \right] e^{-z}\, dz = \frac{5}{6}e^{-1} \simeq 0.3066.$$

Remark. Here, if we integrate with respect to the variable x before the variable y, then we must divide the double integral into two parts:

$$P[X < Y, Z > 1] = \int_1^\infty \left[\int_0^1 \int_0^y \frac{1}{3}(x+y)\,dx\,dy + \int_1^2 \int_0^1 \frac{1}{3}(x+y)\,dx\,dy \right] e^{-z}\,dz.$$

Furthermore, we can assert that the random variable Z is in fact independent of the variables X and Y (or, equivalently, of the random vector (X, Y)).

Example 4.4.2 Let S be a sample space associated with a random experiment E, and let $B_1, \dots, B_4$ be a partition of S. Suppose that we perform m independent repetitions of the experiment E and that the probability $p_k := P[B_k]$ is constant from one repetition to another. Then, if X_k denotes the number of times that the event B_k has occurred among the m repetitions, for $k = 1, \dots, 4$, we can write that

$$P[X_1 = x_1, X_2 = x_2, X_3 = x_3]$$
$$:= p_{\mathbf{X}}(x_1, x_2, x_3) = \left[\frac{m!}{x_1!x_2!x_3!(m - x_1 - x_2 - x_3)!} \right] p_1^{x_1} p_2^{x_2} p_3^{x_3} p_4^{m-x_1-x_2-x_3}$$

where $x_1, x_2, x_3 \in \{0, 1, \dots, m\}$ and $x_1 + x_2 + x_3 \le m$. This three-dimensional probability distribution, called the *quadrinomial distribution* with parameters $p_1, \dots, p_4$, can easily be generalized to the case when the partition of S is made up of n events. In this case, we say that the random vector $(X_1, \dots, X_n)$ has a *multinomial distribution* with parameters $p_1, \dots, p_n$ (where $0 \le p_k \le 1$ $\forall k$ and $p_1 + \cdots + p_n = 1$). It is actually a distribution of dimension $n - 1$ (because the sum of the X_k's is fixed). The binomial distribution is thus the particular case when n is equal to 2.

Remark. The term
$$\frac{m!}{x_1!x_2!x_3!(m - x_1 - x_2 - x_3)!}$$
in the function $p_{\mathbf{X}}(x_1, x_2, x_3)$ above is what we called the *multinomial coefficients* in Chapter 2. In the case of the binomial distribution, the corresponding term is expressed in terms of the *combinations* of k objects taken (without replacement) among n distinct objects. Note, however, that if we have n objects, of which k are of type I and $n - k$ of type II, then the number of permutations of the n objects taken all at once is given by

$$\frac{n!}{k!(n - k)!} = C_k^n.$$

Thus, the number of permutations of n objects of two types, of which k are of type I, is the same as the number of combinations of k objects taken (without replacement) among n distinct objects.

An n-dimensional probability distribution that is very important for the applications is the *multinormal distribution*, which will be seen in Section 4.7 (p. 179).

4.5 Transformations of Random Vectors

In the case when $\mathbf{X}$ is a discrete random vector, we can proceed as in Section 3.6 to find the joint probability mass function of the new random vector $\mathbf{Y} := g(\mathbf{X})$. That is, we apply the transformation g to each vector in $S_{\mathbf{X}}$ and we add the probabilities of the vectors $(x_1, \ldots, x_n)$ that correspond to the same vector $(y_1, \ldots, y_m)$. Note that the dimension of the random vector $\mathbf{Y}$ is not necessarily the same as that of $\mathbf{X}$.

For example, consider the random vector (X_1, X_2) in Example 4.2.1, p. 159. Let $\mathbf{Y} = (Y_1, Y_2) := (X_1 + X_2, X_1 - X_2)$. Then, we find that the joint probability mass function of (Y_1, Y_2) is given by the following table:

$y_2 \backslash y_1$	0	1	2
-2	0	0	1/36
-1	0	8/36	0
0	16/36	0	2/36
1	0	8/36	0
2	0	0	1/36

Indeed, in this case the six pairs in $S_{X_1 \times X_2}$ are transformed as follows:

$$(0,0) \to (0,0), \quad (0,1) \to (1,-1), \quad (0,2) \to (2,-2),$$

$$(1,0) \to (1,1), \quad (1,1) \to (2,0), \quad (2,0) \to (2,2).$$

Thus, here the transformation g is bijective.

Now, suppose that $\mathbf{X}$ is a continuous random vector. To obtain the distribution function of $Z := g(X_1, \ldots, X_n)$, we can find the event $A = \{(x_1, \ldots, x_n): g(x_1, \ldots, x_n) \le z\} \subset S_{\mathbf{X}}$ which is equivalent to the event $\{Z \le z\}$. Then, we write:

$$F_Z(z) = \int \int \cdots \int_A f_{\mathbf{X}}(x_1, \ldots, x_n) \, dx_1 \cdots dx_n. \tag{4.36}$$

Similarly, if $Z_k := g_k(X_1, \ldots, X_n)$ for $k = 1, \ldots, n$, then to obtain the joint distribution function of the new random vector $(Z_1, \ldots, Z_n)$, we can find the event in $S_{\mathbf{X}}$ that is equivalent to $\{Z_1 \le z_1, \ldots, Z_n \le z_n\}$ and integrate the function $f_{\mathbf{X}}$ over this event.

We can also show the following proposition.

Proposition 4.5.1. *Let X and Y be two continuous random variables, and let $V := g_1(X, Y)$ and $W := g_2(X, Y)$. Suppose that*

1) *the system*

$$\left[\begin{matrix} v = g_1(x, y), \\ w = g_2(x, y) \end{matrix} \right.$$

has a unique solution $x = h_1(v, w)$, $y = h_2(v, w)$;

2) *the functions g_1 and g_2 have continuous partial derivatives $\forall (x, y)$, and the Jacobian of the transformation:*

$$J(x, y) := \left| \begin{matrix} \partial g_1/\partial x & \partial g_1/\partial y \\ \partial g_2/\partial x & \partial g_2/\partial y \end{matrix} \right|$$

is different from zero $\forall(x, y)$. *Then, we can write that*

$$f_{V,W}(v, w) = f_{X,Y}(x, y) \mid J(x, y) \mid^{-1}, \tag{4.37}$$

where $x = h_1(v, w)$ *and* $y = h_2(v, w)$.

Remarks. i) In the particular case when $v = x + y$ and $w = x - y$, we find that $x = (v + w)/2$ and $y = (v - w)/2$. Moreover, in the case of such a linear transformation, the partial derivatives of the functions g_1 and g_2 are constants. Consequently, they are continuous for any pair $(x, y) \in \mathbb{R}^2$.

ii) The Jacobian of the inverse transformation is given by

$$J^*(v, w) := \begin{vmatrix} \partial h_1/\partial v & \partial h_1/\partial w \\ \partial h_2/\partial v & \partial h_2/\partial w \end{vmatrix}.$$

We can show that $|J^*(v, w)| = |J(h_1(v, w), h_2(v, w))|^{-1}$. Then, we can also write that

$$f_{V,W}(v, w) = f_{X,Y}(h_1(v, w), h_2(v, w))|J^*(v, w)|, \tag{4.38}$$

provided that $J^*(v, w)$ is not identical to zero.

iii) When the system $v = g_1(x, y)$, $w = g_2(x, y)$ does *not* have a unique solution, we could apply the proposition on disjoint parts of the plane (where the solution is unique) and add all the terms obtained, as in Section 3.6.

iv) The result in Proposition 4.5.1 can easily be generalized to the case of n functions of n continuous random variables $X_1, \dots, X_n$.

v) In the case when g_k is a *linear* transformation $\forall k$, we can write that

$$\mathbf{Z} := (g_1(\mathbf{X}), \dots, g_n(\mathbf{X})) = \mathbf{XA},$$

where $\mathbf{A}$ is an $n \times n$ matrix. If $\mathbf{A}$ is invertible, we find that

$$f_{\mathbf{Z}}(z_1, \dots, z_n) = f_{\mathbf{X}}(x_1, \dots, x_n) \mid \det \mathbf{A} \mid^{-1}, \tag{4.39}$$

where $(x_1, \dots, x_n) = (z_1, \dots, z_n)\mathbf{A}^{-1}$. Note that we consider, in what precedes, that $\mathbf{X}$ (and thus $\mathbf{Z}$) is a row vector.

Example 4.5.1 Let $X \sim \text{Exp}(\lambda)$ and $Y \sim \text{Exp}(\lambda)$ be two independent random variables, and let $V := X + Y$. To obtain the probability density function of V, we can make use of Proposition 4.5.1. We define an *auxiliary* variable: $W = X$. Then, the system

$$\begin{bmatrix} v = x + y, \\ w = \quad x \end{bmatrix}$$

has the unique solution $x = w$, $y = v - w$. Moreover, the partial derivatives of the functions g_1 and g_2 are continuous $\forall(x, y) \in \mathbb{R}^2$, and the Jacobian

$$J(x, y) = \begin{vmatrix} 1 & 1 \\ 1 & 0 \end{vmatrix} = -1$$

is different from zero $\forall (x, y)$. Then, we can write that

$$f_{V,W}(v, w) = f_{X,Y}(w, v - w) \mid -1 \mid^{-1} \overset{\text{ind.}}{=} f_X(w) f_Y(v - w)$$

$$\Rightarrow \quad f_{V,W}(v, w) = \lambda e^{-\lambda w} \lambda e^{-\lambda(v-w)} = \lambda^2 e^{-\lambda v}, \quad \text{if } w \geq 0 \text{ and } v \geq w.$$

Finally, we have:

$$f_V(v) = \int_0^v \lambda^2 e^{-\lambda v} \, dw = \lambda^2 v e^{-\lambda v} \quad \text{if } v \geq 0.$$

Remarks. i) We have, in fact: $V := X + Y \sim G(\alpha = 2, \lambda)$.

ii) Since the transformation considered is *linear*, we could have used (4.39) to obtain $f_{V,W}(v, w)$.

iii) We must not forget to give the set of possible values of the new random vector (V, W), that is, $S_{V \times W}$. For instance, if

$$S_{X \times Y} = \{(x, y): 0 \leq x \leq y \leq 1\}$$

and if we consider the transformation

$$\begin{bmatrix} v = x + y, \\ w = x - y, \end{bmatrix}$$

then we find (see Fig. 4.6) that

$$S_{V \times W} = \{(v, w): -v \leq w \leq 0, 0 \leq v \leq 1 \text{ or } v - 2 \leq w \leq 0, 1 \leq v \leq 2\}.$$

Indeed, we simply have to consider each of the three line segments that define $S_{X \times Y}$ and find out what they become in the (v, w)-plane. For example, the line segment defined by $x = 0, 0 \leq y \leq 1$ becomes the line segment

$$v = y, w = -y \Leftrightarrow v = -w \quad \text{for } 0 \leq v \leq 1$$

in the (v, w)-plane.

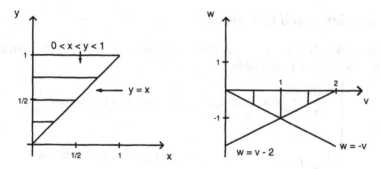

Figure 4.6. Example of transformation of a random vector.

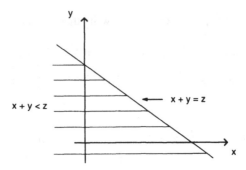

Figure 4.7. Distribution function of the sum of two random variables.

Remark. Let X and Y be two *independent* random variables, and let $Z := X + Y$. Then, if X and Y are *continuous*, we can write (see Fig. 4.7) that

$$F_Z(z) = P[X + Y \leq z] = \int_{-\infty}^{\infty} \int_{-\infty}^{z-u} f_X(u) f_Y(v) \, dv \, du.$$

It follows that

$$f_Z(z) = \int_{-\infty}^{\infty} f_X(u) f_Y(z - u) \, du. \tag{4.40}$$

Thus, we can state that the probability density function of the sum of two independent continuous random variables, X and Y, is the *convolution product* of the probability density functions of X and Y. We can use this result to calculate the probability density function of V in Example 4.5.1. We have:

$$f_V(v) = \int_{-\infty}^{\infty} f_X(u) f_Y(v - u) \, du = \int_0^v \lambda e^{-\lambda u} \lambda e^{-\lambda(v-u)} \, du$$

$$= \lambda^2 \int_0^v e^{-\lambda v} \, du = \lambda^2 v e^{-\lambda v} \quad \text{if } v \geq 0.$$

We will see yet another way of obtaining this same result (and its generalization) in Section 4.9 (p. 187).

4.6 Covariance and Correlation

Definition 4.6.1. *Let (X, Y) be a random vector and $Z := g(X, Y)$. The **mathematical expectation** of Z is defined by*

$$E[Z] = \begin{cases} \sum_{j=1}^{\infty} \sum_{k=1}^{\infty} g(x_j, y_k) p_{X,Y}(x_j, y_k) & \text{if } (X, Y) \text{ is discrete,} \\ \\ \int_{-\infty}^{\infty} \int_{-\infty}^{\infty} g(x, y) f_{X,Y}(x, y) \, dx \, dy & \text{if } (X, Y) \text{ is continuous.} \end{cases} \tag{4.41}$$

Remarks. i) The above definition of the expected value of Z is actually a theorem that we could prove.

ii) We find, in particular, that

$$E[aX + bY] = aE[X] + bE[Y] \tag{4.42}$$

for all constants a and b and for all random variables X and Y, and that

$$E[g_1(X)g_2(Y)] = E[g_1(X)]E[g_2(Y)] \tag{4.43}$$

if X and Y are *independent* random variables.

iii) The mathematical expectation $E[XY]$ is called the **correlation** of X and Y. If $E[XY] = 0$, we say that X and Y are **orthogonal**.

Definition 4.6.2. *Let (X, Y) be a random vector. The* **covariance** *of X and Y is defined by*

$$\text{COV}[X, Y] \equiv \sigma_{X,Y} = E[(X - E[X])(Y - E[Y])]. \tag{4.44}$$

Remarks. i) We have: $\text{COV}[X, X] = \text{VAR}[X]$. Thus, the covariance generalizes the variance. However, the covariance $\text{COV}[X, Y]$ can be *negative*.

ii) We easily show that

$$\text{COV}[X, Y] = E[XY] - E[X]E[Y]. \tag{4.45}$$

It follows that if X and Y are *independent*, then $\text{COV}[X, Y] = 0$.

Definition 4.6.3. *Let (X, Y) be a random vector. The* **correlation coefficient** *of X and Y is defined by*

$$\text{CORR}[X, Y] \equiv \rho_{X,Y} = \frac{\text{COV}[X, Y]}{\text{STD}[X]\text{STD}[Y]}. \tag{4.46}$$

Remarks. i) To avoid any confusion between the correlation and the correlation coefficient, we will generally use the symbol $\rho_{X,Y}$, rather than the operator $\text{CORR}[X, Y]$.

ii) The correlation coefficient is a unitless (or dimensionless) measure of the *linear* relationship between X and Y. We can show that $-1 \leq \rho_{X,Y} \leq 1$. Moreover, if $Y = aX + b$, then $\rho_{X,Y} = 1$ if $a > 0$ and $\rho_{X,Y} = -1$ if $a < 0$.

iii) If X and Y are *independent*, then $\rho_{X,Y} = 0$. However, the converse is not always true. If $\rho_{X,Y} = 0$, but X and Y are *not* independent, we say that they are simply **uncorrelated**. For example, let $X \sim U[-1, 1]$ and $Y := X^2$. We have then:

$$\rho_{X,Y} = \frac{E[XY] - E[X]E[Y]}{\text{STD}[X]\text{STD}[Y]} = \frac{E[X^3] - 0 \cdot (1/3)}{\sqrt{1/3} \cdot \text{STD}[Y]} = \frac{\sqrt{3}(0 - 0)}{\text{STD}[Y]} = 0,$$

because all odd moments of X are equal to zero, and the standard deviation $\text{STD}[Y]$ is strictly positive (actually, $\text{STD}[Y] = 2/\sqrt{45}$). However, X and Y are certainly not independent, since the variable Y is expressed in terms of X: $Y = g(X) = X^2$. In particular, we can write that $X = 0 \Rightarrow Y = 0$. Likewise, we can show that if $X \sim$ N(0, 1), then $\rho_{X, X^2} = 0$.

Note that $\rho_{X,Y} = 0$ is the only case that does not allow us to conclude about the independence (or not) of X and Y, since, if $\rho_{X,Y} \neq 0$, then we can state that X and Y are *not* independent.

iv) An important particular case is the one when X and Y both have a Gaussian distribution and $\rho_{X,Y} = 0$. Indeed, we can then show (see Section 4.7, p. 179) that this implies that X and Y are independent. Thus, in the case when X and Y have Gaussian distributions, the random variables are independent *if and only if* $\rho_{X,Y} = 0$.

In practice, we compute the *point estimate* (see Chapter 6, p. 253) $r_{X,Y}$ of $\rho_{X,Y}$. If $r_{X,Y}$, called the *sample correlation coefficient*, is very small (in absolute value), and if the assumption of normality of the random variables seems reasonable, then we can assume the independence of X and Y when we analyze data, for example. We will see in Chapter 6 how to test, in particular, the hypothesis that a given random variable has (approximately) a Gaussian distribution.

Example 4.6.1 Let

$$f_{X,Y}(x, y) = \begin{cases} 1 \text{ if } -y < x < y, 0 < y < 1, \\ 0 \text{ elsewhere.} \end{cases}$$

We find (see Fig. 4.8) that

$$f_X(x) = \int_{|x|}^{1} 1 \, dy = 1 - |x| \quad \text{if } -1 < x < 1,$$

$$f_Y(y) = \int_{-y}^{y} 1 \, dx = 2y \quad \text{if } 0 < y < 1.$$

Since $f_X(x)$ is symmetric with respect to 0 (and $S_X = (-1, 1)$, which is finite), we can write that $E[X] = 0$. Moreover, we have:

$$E[XY] = \int_{0}^{1} \int_{-y}^{y} xy \cdot 1 \, dx \, dy = 0.$$

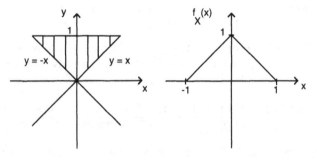

Figure 4.8. Figures for Example 4.6.1.

Then, given that $|E[Y]| < \infty$ (because $S_Y = (0,1)$) and $\text{STD}[X]\text{STD}[Y] > 0$ (because X and Y are not degenerate random variables), we can write that $\rho_{X,Y} = 0$. However, X and Y are not independent, since

$$f_X(x)f_Y(y) = (1 - |x|)2y \neq 1 = f_{X,Y}(x,y)$$

for almost all pairs (x, y). In fact, we could have stated from the outset that X and Y are not independent random variables, since there is the relationship $-y < x < y$ between x and y in $S_{X \times Y}$.

Remark. To obtain the correlation coefficient of the random variables X and Y, we must theoretically calculate five mathematical expectations: $E[X]$, $E[Y]$, $E[X^2]$, $E[Y^2]$ and $E[XY]$. However, as can be seen in the above example, it is preferable to start by calculating the mean of XY and that of X or of Y. Indeed, if the mean $E[XY]$ is equal to zero, as well as $E[X]$ or $E[Y]$, then it is not necessary to calculate the means of the squared variables (in the case when all expected values are real numbers and the random variables are not degenerate).

4.7 Multinormal Distribution

Definition 4.7.1. *Let (X, Y) be a continuous random vector. We say that the random variables X and Y have a* **bivariate normal distribution** *(or* **binormal distribution***) if their joint probability density function is of the form*

$$f_{X,Y}(x,y) = C \exp\left\{ -\frac{1}{2(1-\rho^2)}\left[\left(\frac{x-\mu_X}{\sigma_X}\right)^2 + \left(\frac{y-\mu_Y}{\sigma_Y}\right)^2 \right.\right.$$
$$\left.\left. - 2\rho\frac{(x-\mu_X)(y-\mu_Y)}{\sigma_X\sigma_Y}\right]\right\} \tag{4.47}$$

for $-\infty < x < \infty$, $-\infty < y < \infty$, where $\mu_X \in \mathbb{R}$, $\mu_Y \in \mathbb{R}$, $\sigma_X > 0$, $\sigma_Y > 0$, $-1 < \rho < 1$, and

$$C := \left[2\pi\sigma_X\sigma_Y(1-\rho^2)^{1/2}\right]^{-1} \tag{4.48}$$

(see Fig. 4.9, p. 180). We write: $(X, Y) \sim N(\mu_X, \mu_Y; \sigma_X^2, \sigma_Y^2; \rho)$.

Remarks. i) We can show that the parameter ρ in the function $f_{X,Y}$ above is (indeed) the *correlation coefficient* of X and Y.

ii) We find, integrating the function $f_{X,Y}(x, y)$ with respect to y, that X has a Gaussian distribution with parameters μ_X and σ_X^2. Likewise, we have: $Y \sim N(\mu_Y, \sigma_Y^2)$. We then easily obtain that

$$X \mid \{Y = y\} \sim N\left(\mu_X + \rho(\sigma_X/\sigma_Y)(y - \mu_Y), \sigma_X^2(1-\rho^2)\right). \tag{4.49}$$

We see that if $\rho = 0$, then $f_{X|Y}(x \mid y) \equiv f_X(x)$, from which we deduce the following important result (see Section 4.6, p. 176): in the case of *bivariate normal*

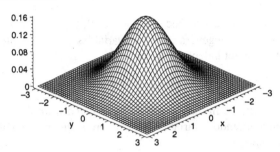

Figure 4.9. Joint probability density function of a random vector having a bivariate normal distribution with $\mu_X = \mu_Y = 0, \sigma_X = \sigma_Y = 1$ and $\rho = 0$.

distributions, if the correlation coefficient of X and Y is equal to zero, then X and Y are *independent* random variables.

iii) We also deduce from (4.49) that $E[X \mid Y = y]$ is of the form $ay + b$, where a and b are constants. We will see in Section 4.8 (p. 182) that this implies that, in the case of the bivariate normal distribution, the best *linear* estimator and the best *non-linear* estimator of X in terms of Y, namely $E[X \mid Y]$, are identical.

Example 4.7.1 Suppose that the lifetime X (in kilometers) and the thickness Y (in millimeters) of the disc brakes of a certain company have a bivariate normal distribution with parameters $\mu_X = 20{,}000, \mu_Y = 10, \sigma_X = 1000, \sigma_Y = 1$ and $\rho = 0.85$. The engineer responsible for quality control wants to determine the lifetime of the brakes by measuring the thickness of the discs. If the thickness of a disc is equal to 9 mm, what is the probability that the brake will last at least 19,500 km?

Solution. We seek

$$P[X \geq 19{,}500 \mid Y = 9] = P[Z \geq 19{,}500], \quad \text{where } Z \sim N(19{,}150; 277{,}500)$$

$$\simeq Q(0.66) = 1 - \Phi(0.66) \overset{\text{Tab. 3.3}}{\simeq} 0.25.$$

Note that the probability of the event $\{Y = 9\}$ is zero, because Y is a continuous random variable. Therefore, we must not try to compute the above probability by making use of the definition of the conditional probability $P[A \mid B]$ given in Chapter 2.

Definition 4.7.2. *We say that the random variables $X_1, X_2, \ldots, X_n$ have a* **multinormal distribution** *if they are a linear combination of the independent $N(0, 1)$ random variables $Z_1, \ldots, Z_m$, that is, if*

$$X_k = \mu_k + \sum_{j=1}^{m} c_{kj} Z_j \quad for \; k = 1, \ldots, n, \tag{4.50}$$

where μ_k is a real constant for all values of k.

Now, we can show that the joint probability density function of the vector $\mathbf{X} = (X_1, \ldots, X_n)$ is completely determined by the vector of means $\mathbf{m}$ and the covariance matrix $\mathbf{K}$, where $\mathbf{m} := (\mu_1, \ldots, \mu_n)$ and

$$
\mathbf{K} := \begin{bmatrix}
\text{VAR}[X_1] & \text{COV}[X_1, X_2] & \ldots & \text{COV}[X_1, X_n] \\
\text{COV}[X_2, X_1] & \text{VAR}[X_2] & \ldots & \text{COV}[X_2, X_n] \\
\ldots & \ldots & \ldots & \ldots \\
\text{COV}[X_n, X_1] & \text{COV}[X_n, X_2] & \ldots & \text{VAR}[X_n]
\end{bmatrix}.
\tag{4.51}
$$

Remark. The matrix $\mathbf{K}$ is symmetric, because $\text{COV}[X, Y] = \text{COV}[Y, X]$. Moreover, it is non-negative definite. That is,

$$
\sum_{i=1}^{n} \sum_{k=1}^{n} c_i c_k \underbrace{\text{COV}[X_i, X_k]}_{\sigma_{ik}} \geq 0 \quad \forall c_i, c_k \in \mathbb{R}.
$$

In fact, we can also show that if $\mathbf{K}$ is a non-singular matrix, then

$$
f_{\mathbf{X}}(\mathbf{x}) = (2\pi)^{-n/2} (\det \mathbf{K})^{-1/2} \exp\left[(-1/2)(\mathbf{x} - \mathbf{m})\mathbf{K}^{-1}(\mathbf{x}^T - \mathbf{m}^T) \right]
\tag{4.52}
$$

for $x_k \in \mathbb{R} \; \forall k$, where $\mathbf{x} := (x_1, \ldots, x_n)$ and ^{T} denotes the transpose of the vector (or the matrix).

Notation. We write: $\mathbf{X} \sim N(\mathbf{m}, \mathbf{K})$.

The joint characteristic function of the vector $\mathbf{X}$, that is,

$$
\phi_{\mathbf{X}}(\omega_1, \ldots, \omega_n) := E\left[\exp[j(\omega_1 X_1 + \cdots + \omega_n X_n)] \right],
\tag{4.53}
$$

is given by

$$
\exp\left[j \sum_{i=1}^{n} \mu_i \omega_i - \frac{1}{2} \sum_{i=1}^{n} \sum_{k=1}^{n} \sigma_{ik} \omega_i \omega_k \right].
$$

We may write that

$$
\phi_{\mathbf{X}}(\omega_1, \ldots, \omega_n) = \exp\left[j\mathbf{m}\omega^T - (1/2)\omega \mathbf{K}\omega^T \right],
\tag{4.54}
$$

where $\omega := (\omega_1, \ldots, \omega_n)$.

Remark. To show this result, we can use the fact (see Section 4.9, p. 187) that

$$
Z := \omega_1 X_1 + \cdots + \omega_n X_n \sim N(\mu_Z, \sigma_Z^2),
$$

where

$$
\mu_Z := \sum_{i=1}^{n} \omega_i \mu_i
$$

and

$$
\sigma_Z^2 := \sum_{i=1}^{n} \sum_{k=1}^{n} \omega_i \omega_k \sigma_{ik},
$$

and that if $X \sim N(\mu, \sigma^2)$, then $E[e^{jX}] = \phi_X(1) = \exp(j\mu - (1/2)\sigma^2)$.

Properties. i) If the random vector **X** has a multinormal distribution and if $\sigma_{ik} = 0$ $\forall i \neq k$, then the random variables $X_1, \ldots, X_n$ are *independent*.

ii) If the random variables $X_1, \ldots, X_n$ have a multinormal distribution, then any set $Y_1, \ldots, Y_k$ of random variables formed by linear combinations of the X_i's also has a multinormal distribution. More precisely, if $\mathbf{Y} = \mathbf{XA}^T$, where A is a $k \times n$ matrix of rank $k \leq n$ (that is, the k rows of A are linearly independent), then $\mathbf{Y} \sim N(\mathbf{mA}^T, \mathbf{AKA}^T)$.

iii) If $\mathbf{X} \sim N(\mathbf{m}, \mathbf{K})$, where $\mathbf{K}$ is of rank n, then there exists a matrix $\mathbf{D}$ such that

$$\mathbf{X} = \mathbf{YD} + \mathbf{m}, \quad \text{where } \mathbf{Y} \sim N(\mathbf{0}, \mathbf{I}_n) \tag{4.55}$$

($\mathbf{I}_n$ being the identity matrix of order n).

Example 4.7.2 Let $\mathbf{X} = (X_1, \ldots, X_n)$ be a random vector having a multinormal distribution $N(\mathbf{0}, \mathbf{I}_n)$. Then, the mathematical expectation of the square of the distance of the vector $\mathbf{X}$ from the origin is given by

$$E[X_1^2 + X_2^2 + \cdots + X_n^2] = \sum_{k=1}^{n} E[X_k^2] \overset{\text{i.d.}}{=} \sum_{k=1}^{n} 1 = n,$$

because all the random variables X_k have a standard Gaussian distribution. In addition, they are (globally) independent, since $\sigma_{ij} = 0\ \forall i \neq j$. Then, given that X_k^2 has a chi-square distribution with 1 degree of freedom (see Example 3.6.3, p. 94), we can show (see Section 4.9, p. 187) that the squared distance

$$D^2 := X_1^2 + X_2^2 + \cdots + X_n^2$$

also has a chi-square distribution, with n degrees of freedom. That is, $D^2 \sim G\left(\alpha = \frac{n}{2}, \lambda = \frac{1}{2}\right)$. It follows that if n is even, then

$$P[D > n] = P[D^2 > n^2] = P[Y < n/2],$$

where $Y \sim \text{Poi}\left(\frac{1}{2}n^2\right)$ (see Section 3.5, p. 87). For example, if $n = 4$, we have:

$$P[D > 4] = P[\text{Poi}(8) < 2] = 9e^{-8} \simeq 0.0030.$$

4.8 Estimation of a Random Variable

We will consider in Chapter 6 the problem of estimating the unknown parameters of a probability density (or mass) function. For example, we will see how to estimate the parameter λ of a random variable X having an $\text{Exp}(\lambda)$ distribution, based on *observations* of this variable.

Here, we are interested in a different type of estimation problem: let Y be a random variable. Suppose that we seek to estimate Y by a function $g(X)$ of the random variable X. We will consider three cases for the type of function $g(X)$ that is admissible.

1) Case When $g(X)$ Is Constant

If we suppose first that $g(X)$ is a *constant*, we easily find that the constant a that minimizes the **mean square error**

$$e(a) := E[(Y - a)^2] \tag{4.56}$$

is $\hat{a} = E[Y]$, so that the *minimum error* e_{min} is equal to VAR[Y]. Indeed, we have:

$$E[(Y - a)^2] = E[Y^2] - 2aE[Y] + a^2.$$

Differentiating with respect to a and setting the derivative equal to 0, we obtain that $a = E[Y]$ does minimize the error $e(a)$:

$$\frac{d}{da}e(a) = -2E[Y] + 2a = 0 \Leftrightarrow a = E[Y]$$

and

$$\frac{d^2}{da^2}e(a) = 2 > 0.$$

Thus,

$$e_{min} = E[(Y - E[Y])^2] \equiv \text{VAR}[Y].$$

2) Linear Case

We now suppose that we look for a function g of the form $g(X) = aX + b$ to estimate Y.

Proposition 4.8.1. *The constants a and b that minimize the mean square error*

$$e(a, b) := E[(Y - (aX + b))^2] \tag{4.57}$$

are given by

$$\hat{a} = \rho_{X,Y} \frac{\text{STD}[Y]}{\text{STD}[X]} \quad \text{and} \quad \hat{b} = E[Y] - \hat{a}E[X]. \tag{4.58}$$

Moreover, we find that the minimum error e_{min} is

$$e_{min} = \text{VAR}[Y](1 - \rho_{X,Y}^2). \tag{4.59}$$

(Idea of the) **Proof.** Let $Z := Y - aX$. Then,

$$e(a, b) := E[(Y - (aX + b))^2] = E[(Z - b)^2].$$

Making use of the result in Case 1, we can write that

$$\hat{b} = E[Z] = E[Y] - aE[X].$$

We then find $\hat{a}$ by solving the equation

$$\frac{d}{da} E[(Y - E[Y] - a(X - E[X]))^2] = 0.$$

Finally, replacing a and b by $\hat{a}$ and $\hat{b}$ in $e(a, b)$, and computing the mathematical expectation, we obtain Equation (4.59). $\square$

3) General Case

Finally, in the general case when $g(X)$ is any random variable, we can show the following proposition.

Proposition 4.8.2. *The (general) function $g(X)$ that minimizes the mean square error*

$$e(g(X)) := E[(Y - g(X))^2] \tag{4.60}$$

is given by $g(X) = E[Y \mid X]$.

Particular cases. i) If $Y = h(X)$, then we have: $g(X) = E[Y \mid X] = E[h(X) \mid X] = h(X)$, and it follows that the minimum error e_{min} is equal to 0.

ii) If X and Y are *independent* random variables, then $E[Y \mid X] = E[Y]$. In this case, we have: $e_{min} = \text{VAR}[Y]$.

Remarks. i) The function $\hat{a}X + \hat{b}$ is the best *linear* estimator of Y in terms of X, while $g(X) = E[Y \mid X]$ is the best *non-linear* estimator of Y in terms of X. Note however that the best "non-linear" estimator of Y in terms of X may be linear, as in the next remark. Therefore, $E[Y \mid X]$ is in fact (simply) the best estimator of Y in terms of X.

ii) If X and Y have Gaussian distributions, then the two estimators of Y in terms of X coincide. Indeed, we then have (see Section 4.7, p. 179):

$$E[Y \mid X = x] = \mu_Y + \rho_{X,Y} \left(\frac{\sigma_Y}{\sigma_X} \right) (x - \mu_X) := a^* x + b^*,$$

which is linear in x. Furthermore,

$$\text{VAR}[Y \mid X = x] = \sigma_Y^2 (1 - \rho_{X,Y}^2),$$

which corresponds to e_{min} in Equation (4.59).

iii) We find, differentiating the mean square error $e(a, b)$ defined in (4.57) with respect to a and setting $\partial e / \partial a$ equal to 0, that

$$E[(Y - \hat{a}X - \hat{b})X] = 0.$$

This equation is known as the *orthogonality condition*. This condition can be stated as follows: the best estimator of the form $aX + b$ of Y is such that the estimation error $Y - (aX + b)$ and the observation X are *orthogonal*.

Example 4.8.1 If $X \sim N(0, 1)$ and $Y := X^2$, then the best non-linear estimator of Y in terms of X is directly $g(X) = X^2$, while the best linear estimator of Y in terms of X is given by $\hat{a}X + \hat{b}$ with (see Section 4.6, p. 178)

$$\hat{a} = 0 \quad \text{and} \quad \hat{b} = E[Y] = E[X^2] = 1.$$

Thus, the best linear estimator of Y in terms of X is simply the constant 1.

4.9 Linear Combinations

An important particular case of transformation of random vectors is the one when $Z := g(X_1, \ldots, X_n)$ is a *linear combination* of the random variables $X_1, \ldots, X_n$. That is, we can write that

$$Z := a_0 + a_1 X_1 + \cdots + a_n X_n, \tag{4.61}$$

where the a_k's are real constants $\forall k$. Then, making use of the (generalized) definition of $E[g(\mathbf{X})]$ (see Section 4.6, p. 176), we easily show that

$$E[Z] = a_0 + \sum_{k=1}^{n} a_k E[X_k]. \tag{4.62}$$

Next, let $Z := a_1 X_1 + a_2 X_2$. We have:

$$
\begin{aligned}
\text{VAR}[Z] &= E[Z^2] - (E[Z])^2 \\
&= E[a_1^2 X_1^2 + 2a_1 a_2 X_1 X_2 + a_2^2 X_2^2] - (a_1 E[X_1] + a_2 E[X_2])^2 \\
&= \sum_{k=1}^{2} a_k^2 \{E[X_k^2] - (E[X_k])^2\} + 2a_1 a_2 \{E[X_1 X_2] - E[X_1]E[X_2]\}.
\end{aligned}
$$

That is,

$$\text{VAR}[Z] = a_1^2 \text{VAR}[X_1] + a_2^2 \text{VAR}[X_2] + 2a_1 a_2 \text{COV}[X_1, X_2]. \tag{4.63}$$

Using mathematical induction, we obtain the following proposition.

Proposition 4.9.1. *Let $X_1, \ldots, X_n$ be random variables, and let Z be a linear combination of the X_k's: $Z = a_0 + \sum_{k=1}^{n} a_k X_k$. Then, we can write that*

$$\text{VAR}[Z] = \sum_{k=1}^{n} a_k^2 \text{VAR}[X_k] + 2 \underbrace{\sum_{i=1}^{n} \sum_{k=1}^{n}}_{i < k} a_i a_k \text{COV}[X_i, X_k]. \tag{4.64}$$

Remarks. i) The constant a_0 does not affect the variance of Z.

ii) We can also write Equation (4.64) in the following form:

$$\text{VAR}[Z] = \sum_{i=1}^{n} \sum_{k=1}^{n} a_i a_k \text{COV}[X_i, X_k].$$

Particular cases. i) If the random variables are *independent*, Equation (4.64) becomes

$$\text{VAR}[Z] = \sum_{k=1}^{n} a_k^2 \text{VAR}[X_k]. \tag{4.65}$$

ii) Suppose that the random variables X_k are independent and *identically distributed* (that is, they have the same distribution function). We write that the X_k's are *i.i.d.* variables. Then, if

$$S_n := X_1 + \cdots + X_n, \tag{4.66}$$

we can write that $E[S_n] = nE[X_1]$ and $\text{VAR}[S_n] = n\text{VAR}[X_1]$.

Remark. In summary, if the mathematical expectations of X and Y exist, we can write that

$$E[X + Y] = E[X] + E[Y].$$

If, in addition, the random variables X and Y are independent, then

$$E[XY] = E[X]E[Y].$$

Likewise,

$$\text{VAR}[X + Y] \stackrel{\text{ind.}}{=} \text{VAR}[X] + \text{VAR}[Y]$$

and

$$\text{VAR}[XY] = E[X^2 Y^2] - (E[XY])^2 \stackrel{\text{ind.}}{=} E[X^2]E[Y^2] - (E[X]E[Y])^2.$$

Note that $\text{STD}[X + Y] \neq \text{STD}[X] + \text{STD}[Y]$ (in general), even if X and Y are independent, because the square root of a sum is not equal to the sum of the square roots.

Reproductive Properties

To obtain the probability density function of the sum of two independent continuous random variables, X and Y, we can consider the characteristic function of their sum Z:

$$\phi_Z(\omega) := E[e^{j\omega Z}] = E[e^{j\omega(X+Y)}] \stackrel{\text{ind.}}{=} E[e^{j\omega X}]E[e^{j\omega Y}] = \phi_X(\omega)\phi_Y(\omega). \tag{4.67}$$

Hence, we deduce the result that we have already mentioned in Section 4.5 (p. 176): the probability density function of Z is given by the convolution product of the density functions of X and Y. That is,

$$f_Z(z) = F^{-1}\{\phi_X(\omega)\phi_Y(\omega)\} = f_X(x) * f_Y(y), \qquad (4.68)$$

where F^{-1} denotes the inverse Fourier transform.

The result (4.67) can be directly generalized to the case when we add n *independent* random variables. Making use of this generalization, we easily show the following *reproductive properties*.

a) If Z is a linear combination of independent Gaussian distributions, then Z has a Gaussian distribution whose mean and variance are given by (4.62) and (4.65), respectively (actually, if the random variables are not independent, Z still has a Gaussian distribution, whose variance is given by Equation (4.64)).

b) The sum of n independent exponential distributions with the same parameter λ (thus, the sum of n independent gamma distributions with parameters 1 and λ) has a gamma distribution with parameters $\alpha = n$ and λ. Indeed, the characteristic function of $X \sim \text{Exp}(\lambda)$ is given by

$$\phi_X(\omega) = \frac{\lambda}{\lambda - j\omega}.$$

Then, for instance, if $X_1 \sim \text{Exp}(\lambda)$ and $X_2 \sim \text{Exp}(\lambda)$ are independent, we can write that

$$\phi_{X_1+X_2}(\omega) \stackrel{\text{ind.}}{=} \phi_{X_1}(\omega)\phi_{X_2}(\omega) = \left(\frac{\lambda}{\lambda - j\omega}\right)^2.$$

Now, if $Y \sim G(\alpha = 2, \lambda)$, we have:

$$\phi_Y(\omega) = \frac{\lambda^2}{(\lambda - j\omega)^2}.$$

Thus, by uniqueness, we may conclude that $X_1 + X_2 \sim G(\alpha = 2, \lambda)$.

More generally, the sum of n independent gamma distributions with parameters α_k, for $k = 1, 2, \ldots, n$, and λ has a gamma distribution with parameters $\alpha := \sum_{k=1}^{n} \alpha_k$ and λ. In particular, since a gamma $G(\frac{m}{2}, \frac{1}{2})$ distribution is a chi-square distribution with m degrees of freedom (see Section 3.5, p. 86), the sum of n independent chi-square distributions with m_k degrees of freedom, for $k = 1, \ldots, n$, also has a chi-square distribution, with $m_1 + \cdots + m_n$ degrees of freedom.

c) The sum of independent Poisson distributions with parameters α_k, for $k = 1, \ldots, n$, also has a Poisson distribution, whose parameter is given by $\alpha := \alpha_1 + \cdots + \alpha_n$.

d) If $X_1 \sim B(n_1, p)$ and $X_2 \sim B(n_2, p)$ are independent, then $X_1 + X_2 \sim B(n_1 + n_2, p)$, which actually follows directly from the definition of a binomial distribution.

Remark. Since the generating function G_X and the moment generating function M_X can be expressed in terms of the characteristic function ϕ_X, Equation (4.67) above, and its generalization, are valid for G_X and M_X as well:

$$G_{X_1+X_2}(z) \overset{\text{ind.}}{=} G_{X_1}(z)G_{X_2}(z) \quad \text{and} \quad M_{X_1+X_2}(s) \overset{\text{ind.}}{=} M_{X_1}(s)M_{X_2}(s).$$

Example 4.9.1 If $X \sim N(0, 1)$ and $Y \sim N(-1, 2)$ are independent random variables, then the random variable $Z := X - 2Y$ has a Gaussian distribution with parameters $\mu_Z = 0 - 2 \times (-1) = 2$ and $\sigma_Z^2 = 1 + 4 \times (2) = 9$. We must not forget to square the coefficients of the random variables when we calculate the variance. Note that otherwise here we would obtain a negative variance!

4.10 The Laws of Large Numbers

We can show the two limit theorems that follow.

Theorem 4.10.1 (Weak law of large numbers). *Let* $X_1, X_2, \ldots$ *be an infinite sequence of i.i.d. random variables such that* $E[X_1] = \mu \in \mathbb{R}$. *Let* S_n *be the sum of the first n variables in the sequence:* $S_n := \sum_{k=1}^{n} X_k$. *Then, for any constant* $c > 0$, *we have:*

$$\lim_{n \to \infty} P\left[\left|\frac{S_n}{n} - \mu\right| < c\right] = 1. \tag{4.69}$$

Remark. In statistics, we say that μ is the mean of the *population*, while S_n/n is the mean of a *random sample* (of size n) of the population. We deduce from this theorem that the sample mean converges toward the population mean. In practice, if the mean μ is unknown, we can therefore estimate it by using the mean of a sample of the population. The larger the size of the sample is, the more accurate the approximation of μ by the numerical value taken by S_n/n should be.

Theorem 4.10.2 (Strong law of large numbers). *Let* $X_1, X_2, \ldots$ *be an infinite sequence of i.i.d. random variables such that* $E[X_1] = \mu \in \mathbb{R}$ *and* $\text{VAR}[X_1] < \infty$, *and let* $S_n := \sum_{k=1}^{n} X_k$. *Then,*

$$P\left[\lim_{n \to \infty} \frac{S_n}{n} = \mu\right] = 1. \tag{4.70}$$

Remarks. i) We must talk about *measure theory* to explain the difference between the two laws. The weak law of large numbers was first proved by Khintchin[2] in 1929.

---

[2] Aleksandr Yakovlevich Khintchin, 1894–1959, was born and died in Russia. His father was an engineer. His first mathematical works were on number theory and probability. He was a professor at Moscow University. This is where he contributed in a very important way to the development of the theory of *stochastic processes*. Next, he became interested in statistical mechanics and in information theory. In addition to his mathematical research, he had a passion for theater and poetry.

ii) We say that S_n/n converges *in probability* (respectively *almost everywhere*) to μ in the case of the weak (resp. strong) law of large numbers.

iii) If we apply the strong law of large numbers to relative frequencies (see Chapter 1, p. 4), we find that the relative frequency of an event A converges to the probability of A with probability 1. This follows from the fact that the relative frequency of an event A is actually the mean of a sequence of independent Bernoulli random variables with parameter p, where p is at once the probability of A and the mean of the Bernoulli distributions.

Example 4.10.1 Let $X_1, X_2, \ldots$ be independent random variables, all having an exponential distribution with parameter $\lambda = 1$. We define the indicator variable

$$I_k = \begin{cases} 1 \text{ if } X_k > 1, \\ 0 \text{ otherwise} \end{cases}$$

for all k. Since I_k has a Bernoulli distribution with parameter $p = P[X_k > 1] = e^{-1}$, we deduce from the strong law of large numbers that

$$\lim_{n \to \infty} \sum_{k=1}^{n} \frac{I_k}{n} = E[I_k] = p = e^{-1},$$

with probability 1.

4.11 The Central Limit Theorem

Theorem 4.11.1 (Central limit theorem). *Let $X_1, \ldots, X_n$ be n i.i.d. random variables with finite mean μ and finite variance σ^2 ($\sigma > 0$). Let S_n be the sum of the n random variables and*

$$Z_n := \frac{S_n - n\mu}{n^{1/2}\sigma}. \tag{4.71}$$

Then, the distribution function of Z_n tends toward that of a Gaussian $N(0, 1)$ distribution.

Remarks. i) We can also write the following result:

$$S_n \approx N(n\mu, n\sigma^2) \quad \text{if } n \text{ is large.} \tag{4.72}$$

Similarly,

$$\frac{S_n}{n} \approx N(\mu, \sigma^2/n) \quad \text{if } n \text{ is large.} \tag{4.73}$$

ii) In general, if $n \geq 30$ we can use the Gaussian distribution to approximate the exact distribution of Z_n. However, if the distribution of the X_k's is *symmetric*, then fewer variables are needed in the sum S_n to obtain a good approximation (and vice versa).

iii) To prove the central limit theorem as stated above, we can use characteristic functions. We show that the characteristic function of Z_n tends toward that of a standard Gaussian distribution, namely $e^{-\omega^2/2}$. Then, since there is a bijective relationship between probability density functions and characteristic functions, we obtain the result we seek. The theorem was first proved by J. W. Lindeberg in 1922. It had been demonstrated previously, among others by Lyapunov,[3] but under more stringent restrictions. The theorem is sometimes called the Lindeberg–Feller[4] central limit theorem, or the Lévy[5] central limit theorem. We can nowadays prove it in a few steps by making use of advanced methods (see reference [13]).

iv) We can generalize the central limit theorem to the following case: let $X_1, \ldots, X_n$ be n independent random variables. Then, under some conditions, S_n/n has approximately (when n is large enough) a Gaussian distribution with parameters

$$\mu = \frac{1}{n} \sum_{k=1}^{n} E[X_k] \quad \text{and} \quad \sigma^2 = \frac{1}{n^2} \sum_{k=1}^{n} \text{VAR}[X_k]. \tag{4.74}$$

So, the random variables X_k do not need to be identically distributed. Note that the expressions for the mean and the variance of S_n/n follow at once from Equations (4.62) and (4.65) of Section 4.9. What is difficult to prove is the fact that the distribution of S_n/n is approximately Gaussian.

Example 4.11.1 Let $X_1, \ldots, X_n$ be independent random variables distributed as the discrete random variable X, whose probability mass function is given by the

---

[3] Aleksandr Mikhailovich Lyapunov, 1857–1918, was born and died in Russia. He was a mathematician and mechanical engineer. He was a school friend of Markov and a student of Chebyshev. He worked, in particular, in the field of differential equations, his methods enabling us to determine the stability of systems of ordinary differential equations, and on probability theory. Moreover, he invented important approximation methods. His doctoral thesis on the stability of motion, which was of great scientific value, was published in English in 1966. He met a violent death (he shot himself when his wife died) soon after the October Revolution of 1917.

[4] William Feller, 1906–1970, was born in Croatia and died in the United States. After having been educated by tutors, he obtained university degrees in Croatia and in Germany. He taught at many renowned universities in the United States. Although he made important contributions to the field of *diffusion processes* in general, and to the theory of *Brownian motion* in particular, he is most famous for his book *An Introduction to Probability Theory and its Applications*, published in two volumes, which is a real classic. The first volume is cited by almost all authors of books on probability theory. The second volume is of a higher mathematical level.

[5] Paul Pierre Lévy, 1886–1971, was born and died in France. Many members of his family were mathematicians. He studied at the École Polytechnique de Paris, where he graduated first in his class. Next, he obtained his doctorate from the École des Mines de Paris. His thesis was on *functional analysis*. He taught at the École des Mines and at the École Polytechnique. In addition to functional analysis, he was interested in probability theory, as well as in partial differential equations and in geometry. He is considered as one of the founders of the rigorous theory of probability, because he published many important papers and books on this subject.

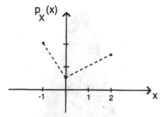

Figure 4.10. Probability mass function of the random variable in Example 4.11.1.

following table:

x	-1	0	2	Σ
$p_X(x)$	1/2	1/8	3/8	1

(see Fig. 4.10). The *exact* distribution of S_n can be obtained by performing what is sometimes called the "convolution summation" of $X_1, \ldots, X_n$. For instance, the distribution of S_2 is the following:

x	-2	-1	0	1	2	4	Σ
$p_{S_2}(x)$	1/4	1/8	1/64	3/8	3/32	9/64	1

Indeed, there is a single pair in $S_{X_1 \times X_2}$ that corresponds to -2, namely $(-1, -1)$, and we have:

$$P[X_1 = -1, X_2 = -1] \overset{\text{ind.}}{=} P[X_1 = -1]P[X_2 = -1] = (1/2)^2 = 1/4,$$

etc.

Figures 4.11 and 4.12 (p. 192) present the probability mass function of S_n for n = 16 and n = 32. We see that when n is equal to 32, the curve looks a lot like that of a Gaussian distribution.

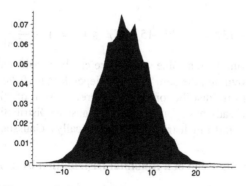

Figure 4.11. Probability mass function of the random variable obtained by adding 16 independent copies of the random variable in Example 4.11.1.

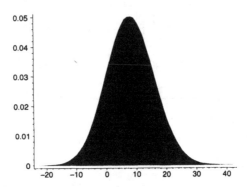

Figure 4.12. Probability mass function of the random variable obtained by adding 32 independent copies of the random variable in Example 4.11.1.

Example 4.11.2 A computer, in adding numbers, rounds each number to the nearest integer. Suppose that the rounding errors are *independent* and have a *uniform* distribution on the interval $(-\frac{1}{2}, \frac{1}{2})$. If 1500 numbers are added, what is the probability that the total error, in absolute value, exceeds 15?

Solution. Let E_T be the total error caused by rounding the 1500 numbers. We can write that $E_T = E_1 + E_2 + \cdots + E_{1500}$, where E_k is the error committed in rounding the kth number $\forall k$. Since $E_k \sim U(-\frac{1}{2}, \frac{1}{2})$ $\forall k$ and the E_k's are independent, we have:

$$E_T \sim N(1500(0), 1500(1/12)) \quad (appr.).$$

Indeed, we have: $E[E_k] = 0$ and $VAR[E_k] = \left[\frac{1}{2} - (-\frac{1}{2})\right]^2 \Big/ 12$ $\forall k$. We seek

$$P[|E_T| > 15] \simeq 2[1 - \Phi(1.34)] \stackrel{\text{Tab. 3.3}}{\simeq} 2(1 - 0.91) = 0.18.$$

Remark. If we had made the error of believing that $E_T \sim U(-750, 750)$, we would have obtained:

$$P[|E_T| > 15] = 1 - P[-15 \le E_T \le 15] = 1 - \frac{30}{1500} = 0.98,$$

which is very far from 0.18, a value which here can be considered as the exact (but rounded) answer. Actually, the sum of two independent uniform random variables does not have a uniform distribution. Furthermore, the sum of 10 independent uniform distributions already has approximately a Gaussian distribution. Thus, the sum of 1500 independent uniform distributions is practically a Gaussian random variable.

Application: Approximation of a Binomial Distribution by a Gaussian Distribution

Let $X \sim B(n, p)$. Since we can represent X as the sum of n i.i.d. Bernoulli random variables, we can use the central limit theorem to approximate the distribution of X.

Indeed, we can write that $X = \sum_{k=1}^{n} X_k$, where X_k is equal to 1 if the kth trial is a success and to 0 otherwise. That is, the binomial distribution counts the number of 1's obtained in n Bernoulli trials.

De Moivre–Laplace Approximation.[6] Let $X \sim B(n, p)$. If n is large enough and p sufficiently close to 1/2, then we can write that

$$p_X(k) \simeq f_Z(k), \tag{4.75}$$

where $Z \sim N(np, npq)$, since $E[X] = np$ and $\text{VAR}[X] = npq$. Note that $p_X(k)$ is a probability, whereas $f_Z(k)$ is not.

Validity condition: the approximation should be good if $\min\{np, nq\}$ is greater than or equal to 5. Thus, if p is equal to 1/2, then $n = 10$ is sufficient to obtain a good approximation, whereas if $p = 1/100$, then n must be at least equal to 500. Actually, if p is very small, or close to 1, we should use the Poisson approximation instead.

Remark. We can also use the distribution function of Z to approximate $F_X(k)$ (or $p_X(k)$). In fact, we generally get more accuracy when we seek to approximate the distribution function, as we already mentioned in Chapter 3 (when we saw Poisson's approximation). In that case, we suggest making a *continuity correction*.

Continuity corrections. To improve the approximation, we generally use the following corrections (where k, a and $b \in \{0, 1, \ldots, n\}$):

a) $P[X = k] = P[k - \frac{1}{2} \leq X \leq k + \frac{1}{2}] \simeq P[k - \frac{1}{2} \leq Z \leq k + \frac{1}{2}]$ (necessary if we want to make use of the distribution function of Z to approximate $p_X(k)$);

b) $P[a \leq X \leq b] = P[a - \frac{1}{2} \leq X \leq b + \frac{1}{2}] \simeq P[a - \frac{1}{2} \leq Z \leq b + \frac{1}{2}]$.

This correction stems from the fact that we replace a *discrete* random variable by a *continuous* random variable (see Fig. 4.13, p. 194).

Remarks. i) If we do not have a closed interval $[a, b]$, we can simply transform the given interval into an interval of this type before making the continuity correction. For instance, we have:

$$P[a < X \leq b] = P[a + 1 \leq X \leq b] = P\left[a + 1 - \frac{1}{2} \leq X \leq b + \frac{1}{2}\right]$$
$$\simeq P\left[a + \frac{1}{2} \leq Z \leq b + \frac{1}{2}\right].$$

---

[6] Abraham de Moivre, 1667–1754, was born in France and died in England. A mathematician of French origin, he emigrated to England in 1685 and made his career there. He was a friend of Newton. His most important works were on analytic geometry and on probability, of which he is one of the pioneers. In 1718, he published his book entitled *The Doctrine of Chance*, in which the definition of statistical independence can be found. The formula attributed to Stirling appeared in a book that he published in 1730. He later used this formula to prove the Gaussian approximation to the binomial distribution. An important formula in *complex analysis* is also due to him. It is said that, like Cardano, he predicted the day on which he would die.

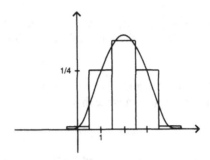

Figure 4.13. Approximation of a binomial distribution by a Gaussian distribution.

ii) When we want to get an approximation for a probability like $P[X \leq b]$ or $P[X \geq a]$, adding the minimum value that X can take, namely 0, or the maximum value, namely n, generally does not change much the numerical value obtained (if the validity condition is respected).

iii) We could also approximate the distribution of a Poisson random variable, with parameter α large enough, by that of a Gaussian random variable with parameters $\mu = \sigma^2 = \alpha$ (see Section 4.9, p. 187).

Example 4.11.3 If 20% of the diodes manufactured by a certain machine are defective, what is the probability that in a batch of 100 (independent) diodes taken at random (and without replacement) among those produced by this machine, there are exactly 15 defectives?

Solution. Let X be the number of defective diodes, among the 100 examined. Then, X has a binomial distribution with parameters $n = 100$ and $p = 0.2$. We seek

$$P[X = 15] = P[14.5 \leq X \leq 15.5]$$
$$\simeq P[14.5 \leq Z \leq 15.5], \qquad \text{where } Z \sim N(20, 16)$$
$$= \Phi(1.375) - \Phi(1.125) \overset{\text{Tab. 3.3}}{\simeq} 0.9155 - 0.8697 = 0.0458.$$

Remarks. i) We obtain practically the same answer by making use of Equation (4.75):

$$f_Z(15) = \frac{1}{\sqrt{2\pi} \cdot 4} \exp\left\{-\frac{1}{2}\frac{(15-20)^2}{16}\right\} \simeq 0.0457.$$

In fact, it is clearly more efficient to use Equation (4.75) when we seek to approximate the value of the probability mass function of X.

ii) The answer obtained by using the binomial distribution is (about) 0.0481. Moreover, if we make use of the Poisson approximation instead, we find that $P[X = 15] \simeq 0.0516$. The probability of a "success," $p = 0.2$, is too large here to obtain a good approximation with a Poisson distribution.

iii) In the preceding example, the validity condition is respected, since we have: $\min\{np, nq\} = \min\{20, 80\} = 20 > 5$. Therefore, we were expecting to get a good approximation with a Gaussian distribution. Note, however, that the percentage of error obtained by using the Gaussian approximation is about 5%, while it is about 7.3% with the Poisson approximation. Thus, here the quality of the two approximations is in fact comparable.

4.12 Exercises, Problems, and Multiple Choice Questions

Solved Exercises

Exercise no. 1 (4.4)

Three companies, X, Y and Z, have a probability equal to $0.4, 0.3$ and 0.3, respectively, of securing an order for a certain product. If three independent orders are placed, what is the probability that each company receives exactly one order?

Solution

Let $\mathbf{W} := (X, Y, Z)$, where X is the number of orders received by the company X, etc. Then, $\mathbf{W}$ has a multinomial (trinomial) distribution with parameters $p_X = 0.4$, $p_Y = 0.3$ and $p_Z = 0.3$. We seek

$$P[X = 1, Y = 1, Z = 1] = \frac{3!}{1!1!1!} (0.4)^1 (0.3)^1 (0.3)^1 = 6(0.036) = 0.216.$$

Exercise no. 2 (4.6)

Let X be a random variable such that $E\left[X^n\right] = \frac{1}{3}$ for $n = 1, 2, \ldots$. We define $Y = X^2$. Calculate $\rho_{X,Y}$.

Solution

We calculate

$$\rho_{X,Y} = \frac{E[XY] - E[X]E[Y]}{\text{STD}[X]\,\text{STD}[Y]} = \frac{E\left[X^3\right] - E[X]E\left[X^2\right]}{\text{STD}[X]\,\text{STD}\left[X^2\right]} = \frac{\frac{1}{3} - \left(\frac{1}{3}\right)\left(\frac{1}{3}\right)}{\frac{2}{9}} = 1$$

because

$$\text{STD}[X] = \left(E[X^2] - (E[X])^2\right)^{\frac{1}{2}} = \left(\frac{1}{3} - \frac{1}{9}\right)^{\frac{1}{2}} = \frac{\sqrt{2}}{3}$$

and

$$\text{STD}[X^2] = \left(E[X^4] - \left(E[X^2]\right)^2\right)^{\frac{1}{2}} = \left(\frac{1}{3} - \frac{1}{9}\right)^{\frac{1}{2}} = \frac{\sqrt{2}}{3}.$$

Exercise no. 3 (4.6)

We define $Z = X + Y$, where X and Y are two independent random variables uniformly distributed on the interval $[a, b]$.
a) Find the probability density function of Z, if $a = 0$ and $b = 1$.
b) Calculate the correlation coefficient of X and Z, if $a = -1$ and $b = 1$.

Hint. We have: $\text{VAR}[X] = 1/3$ if $a = -1$ and $b = 1$.

Solution

a) We can write, by independence, that

$$f_Z(z) = \int_{-\infty}^{\infty} f_X(u) \, f_Y(z - u) \, du = \int_0^1 f_Y(z - u) \, du.$$

Now, we have: $S_Z = [0, 2]$ and

$$f_Y(z - u) = \begin{cases} 1 \text{ if } 0 \le z - u \le 1 \Leftrightarrow z - 1 \le u \le z, \\ 0 \text{ elsewhere.} \end{cases}$$

It follows that

$$f_Z(z) = \begin{cases} \displaystyle\int_0^z 1 \, du = z & \text{if } 0 \le z \le 1, \\ \displaystyle\int_{z-1}^1 1 \, du = 2 - z \text{ if } 1 < z \le 2. \end{cases}$$

Equivalently, if $Z = X + Y$ and $W = X$, then we find that

$$f_{Z,W}(z, w) = 1 \quad \text{if } 0 \le w \le 1, \, 0 \le z \le 2, \, w \le z \le w + 1.$$

Next, we integrate $f_{Z,W}(z, w)$ with respect to w to obtain $f_Z(z)$.
b) We have:

$$\rho_{X,Z} = \frac{E[XZ] - E[X]E[Z]}{\text{STD}[X]\,\text{STD}[Z]} = \frac{\left(\frac{1}{3}\right) - (0)(0)}{\left(\frac{1}{3}\right)^{\frac{1}{2}} \left(\frac{2}{3}\right)^{\frac{1}{2}}} = \frac{\sqrt{2}}{2} \simeq 0.7071,$$

because

$$E[XZ] = E[X^2] + E[XY] \overset{\text{ind.}}{=} \text{VAR}[X] + (E[X])^2 + E[X]E[Y]$$
$$= \text{VAR}[X] + (0)^2 + 0 \times 0 = \text{VAR}[X] = \frac{1}{3}$$

and

$$\text{STD}[Z] = (\text{VAR}[Z])^{\frac{1}{2}} \overset{\text{i.i.d.}}{=} (2\text{VAR}[X])^{\frac{1}{2}} = \left(\frac{2}{3}\right)^{\frac{1}{2}}.$$

Exercise no. 4 (4.6)

    Let

$$f_{X,Y}(x, y) = \begin{cases} 2 - x - y & \text{if } 0 < x < 1, 0 < y < 1, \\ 0 & \text{elsewhere.} \end{cases}$$

Calculate a) $F_{X,Y}(x, y)$; b) $f_X(x)$; c) COV$[X, Y]$; d) $P[X + Y < 1]$.

Solution

a) We calculate

$$F_{X,Y}(x, y) = \int_0^x \int_0^y (2 - u - v)\,dv\,du$$

$$= 2xy - \frac{x^2 y}{2} - \frac{xy^2}{2} \quad \text{if } 0 < x < 1, 0 < y < 1.$$

Hence, we deduce that

$$F_{X,Y}(x, y) = \begin{cases} 0 & \text{if } x \le 0 \text{ or } y \le 0, \\ 2xy - \frac{1}{2}x^2 y - \frac{1}{2}xy^2 & \text{if } 0 < x < 1, 0 < y < 1, \\ \frac{3}{2}y - \frac{1}{2}y^2 & \text{if } x \ge 1, 0 < y < 1, \\ \frac{3}{2}x - \frac{1}{2}x^2 & \text{if } 0 < x < 1, y \ge 1, \\ 1 & \text{if } x \ge 1 \text{ and } y \ge 1. \end{cases}$$

b) We have:

$$f_X(x) = \int_0^1 (2 - x - y)\,dy = \frac{3}{2} - x \quad \text{if } 0 < x < 1.$$

c) First, we calculate

$$E[XY] = \int_0^1 \int_0^1 xy(2 - x - y)\,dx\,dy = \int_0^1 \left(y - \frac{y}{3} - \frac{y^2}{2}\right)dy = \frac{1}{6};$$

$$E[X] \stackrel{b)}{=} \int_0^1 x\left(\frac{3}{2} - x\right)dx = \frac{5}{12}.$$

Since, by symmetry, $E[Y] = E[X]$, we can write that

$$\text{COV}[X, Y] = E[XY] - E[X]\,E[Y] = \frac{1}{6} - \left(\frac{5}{12}\right)^2 = -\frac{1}{144} \simeq -0.0069.$$

d) We can write (see Fig. 4.14, p. 198) that

$$P[X + Y < 1] = \int_0^1 \int_0^{1-x} (2 - x - y)\,dy\,dx$$

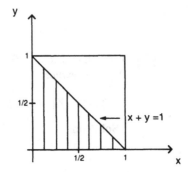

Figure 4.14. Figure for part d) of Exercise no. 4.

$$= \int_0^1 \left[2(1-x) - x(1-x) - \frac{(1-x)^2}{2} \right] dx$$

$$= \int_0^1 \left(\frac{3}{2} - 2x + \frac{x^2}{2} \right) dx = \frac{2}{3}.$$

Exercise no. 5 (4.9)

Let $X \sim \text{Poi}(1)$ and $Y \sim \text{Poi}(2)$ be two independent random variables. Calculate $P\left[3 \le \frac{3}{4}(X+Y) \le 5 \right]$.

Solution

We have:

$$P\left[3 \le \frac{3}{4}(X+Y) \le 5 \right] = P\left[4 \le X+Y \le \frac{20}{3} \right] \stackrel{\text{ind.}}{=} P\left[4 \le \text{Poi}(3) \le 6 \right]$$

$$= e^{-3}\left\{ \frac{3^4}{4!} + \frac{3^5}{5!} + \frac{3^6}{6!} \right\} \simeq 0.3193.$$

Exercise no. 6 (4.11)

Let $X_1, \ldots, X_{50}$ be independent random variables having a Gaussian $N(0, 1)$ distribution. We define $Y = \sum_{k=1}^{50} X_k^2$.
a) Calculate the mean of Y.
b) Use the central limit theorem to calculate $P[Y < 60]$.

Hint. We have: $E[Y^2] = 2600$.

Solution

a) We have:

$$E[X_k^2] = \text{VAR}[X_k] + (E[X_k])^2 = 1 + 0^2 = 1$$

$$\Rightarrow \quad E\,[Y] = \sum_{k=1}^{50} E[X_k^2] = \sum_{k=1}^{50} 1 = 50.$$

b) We can write that $Y \overset{a)}{\sim} N(50, \sigma_Y^2)$ (approximately), where

$$\sigma_Y^2 = E[Y^2] - (E\,[Y])^2 = 2600 - (50)^2 = 100.$$

Then,

$$P\,[Y < 60] \simeq P\left[N\,(0,\,1) < \frac{60 - 50}{10}\right] = \Phi\,(1) \overset{\text{Tab. 3.3}}{\simeq} 0.8413.$$

Exercise no. 7 (4.11)

Let $(X_1, \dots, X_n)$ be a random vector having a multinormal distribution. We suppose that $E\,[X_k] = 0$ and $\text{VAR}\,[X_k] = 1$ for $k = 1, \dots, n$, and that $\text{COV}[X_j, X_k] = 0$ if $j \neq k$. We define $Y = X_1 + X_2$ and $Z = X_3 - X_4$.

a) What is the characteristic function of Y?

Hint. If $X \sim N(\mu, \sigma^2)$, then $\phi_X\,(\omega) = \exp\left(j\mu\omega - \frac{1}{2}\sigma^2\omega^2\right)$.

b) Does the random vector (Y, Z) have a bivariate normal distribution? If so, give its five parameters; if not, justify.

c) Suppose that $n = 50$. We take an observation of each random variable $X_1, \dots, X_{50}$. Use the central limit theorem to calculate (approximately) the probability that exactly 25 observations are positive.

Hint. We have: $\Phi(0.14) \simeq 0.5557$.

Solution

a) $\text{COV}[X_1, X_2] = 0 \overset{\text{here}}{\Rightarrow} X_1$ and X_2 are independent. Then, we can write that $Y \sim N(0 + 0, 1 + 1)$, so that

$$\phi_Y\,(\omega) = \exp\left(j(0)\omega - (1/2)2\omega^2\right) = e^{-\omega^2}.$$

b) Since Y and Z are linear combinations of independent $N(0, 1)$ random variables, the random vector (Y, Z) has a bivariate normal distribution. Its parameters are: $\mu_Y = \mu_Z = 0$, $\sigma_Y^2 = \sigma_Z^2 = 2$ and $\rho_{Y,Z} = 0$ (because Y and Z are independent random variables).

c) Let N be the number of positive observations. We can write that $N \sim B(n = 50, p = 1/2)$. We want $P\,[N = 25] \overset{\text{CLT}}{\simeq} f_Z\,(25)$, where Z has a Gaussian $N(25, 12.5)$ distribution. Therefore,

$$P\,[N = 25] \simeq \frac{1}{\sqrt{2\pi}\sqrt{12.5}} \exp\left\{-\frac{(25 - 25)^2}{2\,(12.5)}\right\} = \frac{1}{\sqrt{25\pi}} \simeq 0.1128.$$

Remark. Or: $P\,[N = 25] \simeq 2\Phi\,(0.14) - 1 \simeq 0.1114.$

Exercise no. 8 (4.11)

The average lifetime of certain electronic components is equal to six months and the standard deviation of the lifetime is equal to two months.

a) We consider two components of this type. Let T_1 and T_2 be their respective lifetimes. We suppose that T_1 and T_2 have (approximately) a Gaussian distribution and that their correlation coefficient is equal to 1/2. Calculate the probability $P[T_2 > 8.5 \mid T_1 = 4]$.

b) Suppose now that the components are independent, but that the distribution of their lifetime is unknown (we only know the mean and the standard deviation of the lifetime). We consider a machine made up of n components of this type placed in parallel. Furthermore, only one component is active at a time (standby redundancy). Use the central limit theorem to find the smallest value of n for which the probability that the machine functions during at least 15 years is greater than 90%.

Hint. We have: $Q^{-1}(0.1) \simeq 1.2815$ and $Q^{-1}(0.9) \simeq -1.2815$.

Solution

a) We can write that

$$T_2 \mid \{T_1 = 4\} \sim N\left(\mu_2 + \rho_{T_2, T_1}\left(\frac{\sigma_2}{\sigma_1}\right)(4 - \mu_1), \sigma_2^2\left(1 - \rho_{T_2, T_1}^2\right)\right) \equiv N(5, 3).$$

Then, $P[T_2 > 8.5 \mid T_1 = 4] = P[N(5, 3) > 8.5] \simeq Q(2.02) \overset{\text{Tab. 3.3}}{\simeq} 1 - 0.9783 \simeq 0.022$.

b) Let $S_n := T_1 + \cdots + T_n$. If n is large enough, then by the central limit theorem we can write that $S_n \approx N(n \cdot 6, n \cdot 4)$. We seek $n_{\min}$ such that

$$P[S_n \geq 180] > 0.90 \Leftrightarrow P[N(6n, 4n) \geq 180] > 0.90$$

$$\Rightarrow Q\left(\frac{180 - 6n}{2\sqrt{n}}\right) > 0.90 \Rightarrow \frac{180 - 6n}{2\sqrt{n}} < -1.2815.$$

We consider the equation $6n - 2.563\sqrt{n} - 180 = 0$. We find that $\sqrt{n} \simeq -5.27$ or 5.69. It follows that $n \simeq 32.4$. Thus, we must take $n = 33$.

Exercise no. 9 (4.11)

Suppose that $X_1, \ldots, X_n$ are independent random variables that are all uniformly distributed on the interval $[0, 1]$.

a) Calculate $P[\mid X_1 - X_2 \mid < 1/2]$.

b) i) Find the characteristic function of $Y := a_1 X_1 + a_2 X_2$, where a_1 and a_2 are real constants ($\neq 0$).

ii) Making use of part i), can we assert that Y has a uniform distribution? If so, on what interval. If not, justify your answer.

Hint. We have: $\phi_X(\omega) = \frac{e^{j\omega b} - e^{j\omega a}}{j\omega(b-a)}$, if $X \sim U[a, b]$.

c) Use the central limit theorem to find the smallest value of n for which $P\left[X_1 + \cdots + X_n < \frac{n}{4}\right] < 10^{-5}$.

Hint. We have: $Q^{-1}(10^{-5}) \simeq 4.2649$.

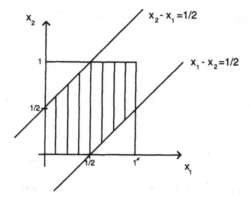

Figure 4.15. Figure for part a) of Exercise no. 9.

Solution

a) We have:

$$f_{X_1, X_2}(x_1, x_2) \overset{\text{ind.}}{=} 1 \times 1 = 1 \quad \text{for } 0 \le x_1 \le 1, 0 \le x_2 \le 1.$$

Then, we can write (see Fig. 4.15) that

$$P\left[|X_1 - X_2| < \frac{1}{2}\right] = 1 - 2\left(\frac{1}{2} \times \frac{1}{2}\right)\frac{1}{2} = \frac{3}{4}.$$

b) i) $\phi_Y(\omega) := E\left[e^{j\omega Y}\right] = E\left[e^{j\omega(a_1 X_1 + a_2 X_2)}\right] \overset{\text{ind.}}{=} E\left[e^{j\omega a_1 X_1}\right] E\left[e^{j\omega a_2 X_2}\right]$

$$= \phi_{X_1}(\omega a_1)\phi_{X_2}(\omega a_2) = \frac{e^{j\omega a_1} - 1}{j\omega a_1} \cdot \frac{e^{j\omega a_2} - 1}{j\omega a_2}$$

$$= \frac{\left(e^{j\omega a_1} - 1\right)\left(e^{j\omega a_2} - 1\right)}{-a_1 a_2 \omega^2}.$$

ii) Since $\phi_Y(\omega) \ne \phi_X(\omega)$ for any uniform random variable X, the random variable Y does not have a uniform distribution.

c)

$$X_1 + \cdots + X_n \overset{\text{CLT}}{\approx} N\left(n \times \frac{1}{2}, n \times \frac{1}{12}\right)$$

$$\Rightarrow P\left[X_1 + \cdots + X_n < \frac{n}{4}\right] \simeq P\left[N\left(\frac{n}{2}, \frac{n}{12}\right) < \frac{n}{4}\right] = P\left[N(0, 1) < \frac{\frac{n}{4} - \frac{n}{2}}{\sqrt{\frac{n}{12}}}\right]$$

$$= \Phi\left(-\frac{\sqrt{n}}{4}\sqrt{12}\right) = Q\left(\frac{\sqrt{n}}{2}\sqrt{3}\right) = 10^{-5}$$

$$\Leftrightarrow \frac{\sqrt{n}}{2}\sqrt{3} \simeq 4.2649.$$

It follows that $n \simeq 24.25$, whence we deduce that $n_{\min} = 25$.

Unsolved Problems

Problem no. 1
　Find $f_X(x)$ if

$$f_{X,Y}(x, y) = \frac{1}{2\pi} \quad \text{if } x^2 + y^2 \leq 2 \quad (= 0 \text{ elsewhere}).$$

Problem no. 2
　Let

$$f_{X,Y}(x, y) = \frac{1}{4} \quad \text{if } 0 < x < 2, \, 0 < y < 2 \quad (= 0 \text{ elsewhere}).$$

Calculate $P[2X^2 > Y]$.

Problem no. 3
　Suppose that

$$p_{X|Y}(x \mid y) = \frac{1}{3} \quad \text{if } x = 0, 1, 2 \text{ and } y = 1, 2,$$

and

$$p_Y(y) = \frac{1}{2} \quad \text{if } y = 1 \text{ or } 2.$$

Find $p_{X,Y}(x, y)$.

Problem no. 4
　Let $X_1 \sim N(\mu, \sigma^2)$ and $X_2 \sim N(\mu, \sigma^2)$ be two independent random variables. We define $Y_1 = X_1 + X_2$ and $Y_2 = X_1 + 2X_2$. Find the joint probability density function $f_{Y_1, Y_2}(y_1, y_2)$.

Problem no. 5
　Let

x	0	1	2
$p_X(x)$	$\frac{1}{4}$	$\frac{1}{4}$	$\frac{1}{2}$

and

y	-1	0	1
$p_Y(y)$	$\frac{1}{3}$	$\frac{1}{2}$	$\frac{1}{6}$

Find $p_Z(z)$, where $Z := XY$, assuming that X and Y are independent random variables.

Problem no. 6
　Calculate VAR $[XY]$ if $X \sim N(-1, 1)$ and $Y \sim N(2, 4)$ are two independent random variables.

Problem no. 7
　Let $X_1 \sim N(0, \sigma^2)$ and $X_2 \sim N(0, \sigma^2)$ be two independent random variables, and let $Y := X_1 + X_2$.

a) Find the best estimator of Y, based on X_1.

b) Find the best estimator of X_1, based on Y.

Problem no. 8

Calculate $P[2X - Y < 8]$, where $X \sim N(0, 1)$ and $Y \sim N(1, 5)$ are two independent random variables.

Problem no. 9

Let X, Y and Z be independent random variables such that $\text{VAR}[X] = 1$, $\text{VAR}[Y] = 4$ and $\text{VAR}[Z] = 11$. We define $W = 3X + 2Y - Z$. Calculate $\text{STD}[W]$.

Problem no. 10

Suppose that X is a random variable such that $E[X] = \text{VAR}[X] = 1$. Use the central limit theorem to calculate (approximately) the probability $P[\sum_{k=1}^{36} X_k < 42]$, where $X_1, \ldots, X_{36}$ are independent random variables having the same distribution function as X.

Problem no. 11

Let $X \sim B(n = 25, p = 1/2)$. Use the Gaussian approximation to the binomial distribution to calculate $P[X < 12]$.

Problem no. 12

Let

$$f_{X,Y}(x, y) = \begin{cases} \frac{3}{2}x & \text{if } 1 \leq x \leq y \leq 2, \\ 0 & \text{elsewhere.} \end{cases}$$

a) Find the marginal density functions f_X and f_Y.

b) Calculate the probability $P[X^2 < Y]$.

Problem no. 13

Let $X_1, \ldots, X_{50}$ be random variables having a geometric distribution with parameter $p = 1/4$, and let $S := X_1 + \cdots + X_{50}$.

a) Calculate $\text{VAR}[S]$ if $\rho_{X_j, X_k} = 1/2 \quad \forall j \neq k$.

b) Suppose now that the X_k's are globally independent. Use the central limit theorem to calculate $P[S < 201]$.

Remark. Do not use a continuity correction.

Problem no. 14

Let

$$f_{X,Y}(x, y) = \begin{cases} (4xy)^{-\frac{1}{2}} & \text{if } 0 < y < x < 1, \\ 0 & \text{elsewhere.} \end{cases}$$

Calculate a) $f_X(x)$ and $f_Y(y)$; b) $E[Y \mid X]$; c) $P[X > \frac{1}{2}, Y > \frac{1}{2}]$.

Problem no. 15

The input to a communication channel is a random variable X having a standard Gaussian distribution. The output Y is given by $Y = X + N$, where N, the noise, has a Gaussian distribution with zero mean and variance σ_N^2. Furthermore, X and N are independent.

a) Calculate $P[Y = 0 \mid X = 0]$.

b) Calculate the correlation coefficient, $\rho_{X,Y}$, of X and Y.

c) If $\sigma_N^2 = 8$, we find that $\rho_{X,Y} = 1/3$. Moreover, the vector (X, Y) has a bivariate normal distribution. What is the best estimator of Y in terms of X?

Problem no. 16

The joint probability mass function of the discrete random vector (X, Y) is given by the following table:

$y \backslash x$	-1	0	1
0	1/9	2/9	1/9
1	1/9	2/9	1/9
2	0	1/9	0

a) Find $p_{Y \mid X}(y \mid 0)$.

b) Calculate $F_{X,Y}(2, 0)$.

c) i) Calculate the correlation coefficient of X and Y. ii) Are X and Y independent random variables? Justify your answer.

d) Let $Z := X + Y$. Find $p_Z(z)$.

Problem no. 17

A computer generates random numbers X according to a Gaussian distribution with parameters 0 and σ^2. Let

$$Y := \begin{cases} X & \text{if } X > 0, \\ 0 & \text{otherwise} \end{cases} \quad \text{and} \quad Z := \begin{cases} X & \text{if } X \leq 0, \\ 0 & \text{otherwise.} \end{cases}$$

a) Does the sum $Y + Z$ have a Gaussian distribution? Justify your answer.

b) Does the pair (Y, Z) have a bivariate normal distribution? Justify.

c) Calculate $E[Y \mid X^2 = 1]$.

Problem no. 18

In a quality control operation, we examine the paint of n brand-new cars taken at random among those produced by a certain company. Let $X_k = 1$ if the paint of the kth car examined has at least one flaw, and $X_k = 0$ otherwise, for $k = 1, 2, \ldots, n$. We assume that the random variables X_k are independent and that the probability that the paint of a brand-new car is flawless is equal to 0.75. Therefore, X_k has a Bernoulli distribution with parameter $p = 0.25$ for all k.

a) Calculate $\lim_{n \to \infty} P\left[\left| \sum_{k=1}^{n} X_k - \frac{n}{4} \right| \geq \frac{n}{2} \right]$.

b) Let $Y := X_1 + X_2$ and $Z := X_1 - X_2$. Calculate COV[Y, Z].

c) Use the central limit theorem to calculate $P\left[\sum_{k=1}^{n} X_k = 10 \right]$, if $n = 40$.

Hint. We have: $\Phi(0.18) \simeq 0.5714$.

Problem no. 19

Let (X, Y) be a discrete random vector whose probability mass function is given by the following table:

$y \setminus x$	0	1	2
-1	1/3	0	1/3
0	0	1/6	0
1	0	1/6	0

a) Are X and Y independent random variables? Justify your answer.
b) Calculate $E[X \mid Y = -1]$.
c) Let $U := X + Y$ and $V := X - Y$. Find $p_{U,V}(u, v)$.

Problem no. 20

Let (X, Y) be a random vector having a bivariate normal distribution with parameters $\mu_X = 0, \mu_Y = 1, \sigma_X^2 = 1, \sigma_Y^2 = 5$ and $\rho_{X,Y}$.

a) Calculate $\mathrm{VAR}[2X - Y]$, if $\rho_{X,Y} = \frac{1}{2}$.
b) Calculate $P[2X - Y > 2]$, if $\rho_{X,Y} = 0$.
Hint. We have: $Q(1) \simeq 1.59 \times 10^{-1}$ and $Q(2) \simeq 2.28 \times 10^{-2}$.
c) Find the best estimator of X in terms of Y, if $\rho_{X,Y} = -1$.

Problem no. 21

Let

$$f_{X,Y}(x, y) = \begin{cases} e^{-y} & \text{if } 0 \leq x \leq y < \infty, \\ 0 & \text{elsewhere.} \end{cases}$$

Calculate
a) the marginal probability density functions of X and Y;
b) the covariance of X and Y;
c) the best estimator of X in terms of Y;
d) $P[X = Y]$.
Hint. We have: $\int_0^\infty y^{\alpha-1} e^{-y} dy = \Gamma(\alpha)$.

Problem no. 22

The joint probability mass function of the discrete random vector (X, Y) is given by the following table:

$y \setminus x$	-1	0	1
0	1/9	1/9	1/9
2	1/9	1/9	1/6
4	1/9	1/18	1/9

a) Calculate i) $p_X(x)$; ii) $P[X + Y < 2]$; iii) $p_{X|Y}(x \mid 2)$; iv) $p_Z(z)$, where $Z := XY$; v) $\mathrm{COV}[X, Y]$.
b) Are X and Y independent random variables? Justify your answer.

Problem no. 23

A source transmits a signal X uniformly distributed on the interval $(0, 2\pi)$. Because of the presence of additive Gaussian noise, the signal received, Y, is a random variable whose conditional probability density function is given by

$$f_{Y|X}(y \mid x) = \frac{1}{\sqrt{2\pi}\sigma} \exp\left\{-\frac{1}{2}\frac{(y-x)^2}{\sigma^2}\right\} \quad \text{for } x \in (0, 2\pi), \ y \in \mathbb{R}.$$

a) Calculate i) $f_{X,Y}(x, y)$; ii) the best non-linear estimator of Y in terms of X; iii) $E[Y]$; iv) $F_{X,Y}(\pi, x)$.

b) We consider 48 signals X_k, where $k = 1, 2, \ldots, 48$, transmitted by the source. What is (approximately) the probability that the average signal $M_{48} := \frac{1}{48}\sum_{k=1}^{48} X_k$ is i) smaller than $\frac{11}{12}\pi$? ii) equal to $\frac{3}{4}\pi$?

Hint. We have: $Q(1) \simeq 0.159$ and $Q(2) \simeq 0.0228$.

Problem no. 24

Let X and Y be two independent random variables having, respectively, a binomial distribution with parameters $n = 2$ and $p = 1/2$, and a uniform distribution on the interval $[0, 1]$. Calculate

a) $E\left[X^2 Y^2\right]$;

b) $P[X + Y < 1]$;

c) the joint distribution function of the pair (X, Y) at the point $(\frac{3}{2}, 2)$;

d) the correlation coefficient of the pair (X, Z), where $Z := X + 1$.

Problem no. 25

Let

$$f_{X,Y}(x, y) = \begin{cases} 1/2 & \text{if } 0 \le x \le 2, \ 0 \le y \le 1, \\ 0 & \text{elsewhere.} \end{cases}$$

a) Find $f_X(x)$ and $f_Y(y)$.

b) Calculate the correlation coefficient of X and Y.

c) Calculate $F_{X,Y}(-1, 1) + F_{X,Y}(1, 2)$.

d) Let $U := X + Y$ and $V := X - Y$;

    i) find $f_{U,V}(u, v)$ and draw the region where $f_{U,V}(u, v) > 0$;

    ii) calculate $E[UV]$.

Problem no. 26

a) Let X, Y and Z be random variables such that $\text{VAR}[X] = 1$, $\text{VAR}[Y] = 4$, $\text{VAR}[Z] = 9$, $\text{COV}[X, Y] = 1/2$, $\text{COV}[X, Z] = 0$ and $\text{COV}[Y, Z] = -1/2$. Calculate $\text{VAR}[X - \frac{1}{2}Y + \frac{1}{3}Z]$.

b) We take 40 independent observations of the random variable X. Let N be the number of observations that are greater than 1. Use the central limit theorem to calculate (approximately) the probability $P[N < 5]$, if $X \sim N(0, 1)$.

Hint. We have: $Q(0.8) \simeq 0.212$, $Q(1) \simeq 0.159$ and $Q(1.2) \simeq 0.115$.

Problem no. 27

The joint conditional probability mass function $p_{X|Y}(x \mid y)$ of the random vector (X, Y) is given by the following table:

$y \backslash x$	0	1	2
−1	1/4	1/4	1/2
0	2/3	0	1/3

So, we have: $1/4 = p_{X|Y}(0 \mid -1)$, etc. Furthermore, we suppose that $p_Y(y) = 1/2$ if $y = -1$ or 0.

a) Are the random variables X and Y independent? Justify your answer.

Calculate

b) the function $p_{X,Y}(x, y)$;

c) the function $p_X(x)$;

d) COV$[X, Y]$.

Problem no. 28

Let

$$f_{X,Y}(x, y) = \begin{cases} ke^{-x^2-y} & \text{if } -\infty < x < \infty, y > 0, \\ 0 & \text{elsewhere,} \end{cases}$$

where $k > 0$ is a constant.

a) Find the constant k.

Hint. We have: $\int_0^\infty e^{-(x^2+ax)} dx = \sqrt{\pi}\, e^{a^2/4} Q(a/\sqrt{2})$.

b) Calculate $P[X > Y]$.

Hint. We have: $Q(0.7071) \simeq 0.24$.

c) Calculate VAR$[2X - Y]$.

d) Let $Z := XY$. Find the joint probability density function of the random vector (Z, Y).

Problem no. 29

Let

$$f_{X,Y}(x, y) = k \cos(x + y) \quad \text{for } 0 \le x \le \tfrac{\pi}{4}, 0 \le y \le \tfrac{\pi}{4}.$$

a) Calculate
   i) the constant k;
   ii) the joint distribution function of the random vector (X, Y);
   iii) the marginal probability density functions of X and Y;
   iv) the probability $P[X < Y]$.

b) Are the random variables X and Y orthogonal? Justify.

Multiple Choice Questions

Question no. 1

Let

$$F_{X,Y}(x, y) = \begin{cases} 0 & \text{if } x \leq 0 \text{ or } y \leq 0, \\ xy/2 & \text{if } 0 < x \leq 2 \text{ and } 0 < y \leq 1, \\ x/2 & \text{if } 0 < x \leq 2 \text{ and } y > 1, \\ y & \text{if } x > 2 \text{ and } 0 < y \leq 1, \\ 1 & \text{if } x > 2 \text{ and } y > 1 \end{cases}$$

be the joint distribution function of the pair (X, Y) of continuous random variables.

A) Calculate $F_X(3) + F_Y(1/2)$.

a) 3/2   b) 2   c) $y + \frac{1}{2}$ d) $\frac{x}{4} + y$ e) $\frac{x}{4} + 1$

B) Are the random variables X and Y orthogonal and independent?

a) not orthogonal and dependent
b) not orthogonal, but independent
c) orthogonal and dependent
d) orthogonal and independent
e) not orthogonal and we cannot tell whether they are independent

C) Calculate $P[X > Y]$.

a) 0   b) 1/4   c) 1/2   d) 3/4   e) 1

D) Calculate VAR$[XY]$.

a) 0   b) 1/36   c) 7/36   d) 4/9   e) 1/2

E) Let $V := X^2$ and $W := 2Y$. Calculate $f_{V,W}(v, w)$.

a) $1/(8\sqrt{v})$ b) $\pm 1/(4\sqrt{v})$ c) $1/(4\sqrt{v})$ d) $4\sqrt{v}$ e) $8\sqrt{v}$

Question no. 2

Let $X_1, X_2, \ldots$ be independent and identically distributed random variables such that $E[X_1] = 0$ and VAR$[X_1] = 3$.

A) Suppose that the X_i's have a Gaussian distribution. Let $Y := X_1 + X_2$ and $Z := X_1 - X_2$. Are the random variables Y and Z orthogonal and independent?

a) not orthogonal and dependent
b) not orthogonal, but independent
c) orthogonal and dependent
d) orthogonal and independent
e) not orthogonal and we cannot tell whether they are independent

B) Under the same assumptions as in A), what is the best estimator of Y in terms of Z?

a) 0   b) Y c) Z d) $\frac{Y+Z}{2}$ e) we cannot calculate it

C) Suppose that the X_i's have a uniform distribution on the interval $[-3, 3]$. What is the approximate distribution of $X_1^2 + \cdots + X_{50}^2$?

a) U$[-150, 150]$ b) U$[0, 450]$ c) N$(0, 1)$ d) N$(3, \frac{36}{5})$ e) N$(150, 360)$

Question no. 3

Let (X, Y) be a continuous random vector with joint distribution function

$$F_{X,Y}(x, y) = (1 - e^{-x})(1 - e^{-y}) \quad \text{if } x > 0, y > 0.$$

Calculate $P[X \geq 1]$.

a) e^{-2} b) e^{-1} c) $(1 - e^{-1})^2$ d) $1 - e^{-1}$ e) $1 - e^{-2}$

Question no. 4

A discrete random vector (X, Y) has the joint probability mass function given by the following table:

y \ x	0	1	2
0	1/8	p_1	1/4
1	1/8	1/8	p_2

What values can the constants p_1 and p_2 take if X and Y are independent?

a) $p_1 = 1/4, p_2 = 1/8$ b) $p_1 = 3/8, p_2 = 0$ c) $p_1 = 1/8, p_2 = 1/4$
d) $p_1 = 3/16, p_2 = 3/16$ e) none of these answers

Question no. 5

Let $X \sim N(0, 1)$. Calculate the correlation coefficient of X and $Y := X^4$.

Hint. We have: $\phi_X(\omega) = e^{-\omega^2/2}$.

a) -1 b) 0   c) 1/2   d) $1/\sqrt{2}$ e) 1

Question no. 6

A random vector (X, Y) has a bivariate normal distribution with parameters $\mu_X = 0, \mu_Y = 0, \sigma_X^2 = 1, \sigma_Y^2 = 4$ and $\rho = 0$. Calculate $P[XY < 0]$.

a) 1/16   b) 1/8   c) 1/4   d) 1/2   e) 3/4

Question no. 7

Let X and Y be two independent random variables, both having a Poisson distribution with parameter $\alpha = 1.5$. Calculate $P[X + Y \geq 1.5]$.

a) 0.05   b) 0.20   c) 0.58   d) 0.80   e) 0.95

Question no. 8

If 10% of the diodes manufactured by a certain machine are defective, what is the probability that in a lot of 100 diodes taken at random (and without replacement) among those produced by this machine, there are less than 15 defectives?

a) $Q(1.67)$ b) $Q(1.5)$ c) $Q(0.05)$ d) $1 - Q(1.67)$ e) $1 - Q(1.5)$

Question no. 9

The joint probability mass function of a discrete random vector (X, Y) is given by

$$p_{X,Y}(x, y) = \binom{2}{x} \frac{e^{-1}}{4y!} \quad \text{for } x = 0, 1, 2; y = 0, 1, 2, \ldots .$$

Calculate $P[X = 2Y]$.

a) 0   b) $\frac{1}{4}e^{-1}$ c) $\frac{1}{2}e^{-1}$ d) e^{-1} e) 1

Question no. 10

Let (X, Y) be a continuous random vector for which the conditional probability density function of X, given that $Y = y$, is given by

$$f_{X|Y}(x \mid y) = \frac{1}{y}e^{-x/y} \quad \text{for } x > 0, 0 < y < 1.$$

Calculate $E[X]$, if $E[Y] = 1/2$.

a) 1/4   b) 1/2   c) 2   d) y e) ∞

Question no. 11

A point (X, Y, Z) is taken at random inside the unit sphere, so that

$$f_{X,Y,Z}(x, y, z) = \frac{3}{4\pi} \quad \text{if } x^2 + y^2 + z^2 \leq 1.$$

Calculate the probability that the distance between the chosen point and the origin is smaller than 1/2.

a) 1/64   b) 1/8   c) 1/4   d) $\pi/8$ e) 1/2

Question no. 12

Let

$$f_{X,Y}(x, y) = 1 \quad \text{if } 0 < x < 1, 0 < y < 1$$

be the joint probability density function of the continuous random vector (X, Y). We define $U = X^2$ and $V = Y^2$. Find $f_{U,V}(u, v)$ for $0 < u < 1, 0 < v < 1$.

a) 1   b) x^2y^2 c) $\frac{1}{4}(uv)^{-1}$ d) $\frac{1}{4}(uv)^{-1/2}$ e) $\frac{9}{4}(uv)^{1/2}$

Question no. 13

Let (X, Y) be a random vector having a bivariate normal distribution with parameters $\mu_X = 0, \mu_Y = 0, \sigma_X^2 = 1, \sigma_Y^2 = 1$ and $\rho = 0$. Calculate the best estimator of X^2 in terms of Y.

a) 0   b) X^2 c) Y d) Y^2 e) none of these answers

Question no. 14

Let $X_1, \ldots, X_{100}$ be a set of independent random variables such that $E[X_k] = 0$ and $\text{VAR}[X_k] = 1$ for $k = 1, 2, \ldots, 100$. Use the central limit theorem to calculate the probability $P[|\sum_{k=1}^{100} X_k| > 10]$.

a) $Q(1)$ b) $2Q(1)$ c) $1 - Q(1)$ d) $1 - 2Q(1)$ e) $2(1 - Q(1))$

Question no. 15

Suppose that $X \sim B(n = 100, p)$. What is the value of p (smaller than or equal to 1/2), if $P[X = 100p] \simeq 0.1330$?

a) 1/10   b) 1/5   c) 3/10   d) 2/5   e) 1/2

Question no. 16

Let X_1, X_2 and X_3 be N(0, 1) random variables such that $\rho_{X_1,X_2} = 0, \rho_{X_1,X_3} = 0$ and $\rho_{X_2,X_3} = -1$.

A) Let $Y := X_1 + X_2$. Calculate ρ_{Y, X_3}.

a) $-1/2$ b) $-1/\sqrt{2}$ c) 0 d) $1/\sqrt{2}$ e) $1/2$

B) Calculate $m_1 := E[X_3 \mid X_2]$ and $m_2 := E[X_3^2 \mid X_2]$.

a) $m_1 = 0, m_2 = 1$ b) $m_1 = -X_2, m_2 = 1$ c) $m_1 = -X_2, m_2 = X_2^2$
d) $m_1 = X_2, m_2 = 1$ e) $m_1 = X_2, m_2 = X_2^2$

C) Let $X_{11}, \ldots, X_{1n}$ be n independent random variables distributed like X_1. We define

$$I_k = \begin{cases} 1 \text{ if } X_{1k} > 0, \\ 0 \text{ otherwise}, \end{cases}$$

for $k = 1, \ldots, n$. According to the strong law of large numbers, what is, with probability 1, the value of the limits

$$l_1 := \lim_{n \to \infty} \sum_{k=1}^{n} \frac{I_k}{n} \quad \text{and} \quad l_2 := \lim_{n \to \infty} \sum_{k=1}^{n} \frac{I_k^2}{n}?$$

a) $l_1 = 0, l_2 = 0$ b) $l_1 = 1/2, l_2 = 1/4$ c) $l_1 = 1/2, l_2 = 1/2$
d) $l_1 = 1, l_2 = 1$ e) $l_1 = \infty, l_2 = \infty$

Question no. 17

Calculate $P[X^2 + Y^2 > 2\ln 4]$ if X and Y are two independent random variables having a Gaussian $N(0, 1)$ distribution.

a) $1/8$ b) $1/4$ c) $1/2$ d) $3/4$ e) $7/8$

Question no. 18

Find the characteristic function of $Z := X + 2Y$, where X and Y are independent random variables having an exponential distribution with parameter α and with parameter β, respectively.

Hint. If $X \sim \text{Exp}(\lambda)$, then $\phi_X(\omega) = \lambda/(\lambda - j\omega)$.

a) $\left(\frac{\alpha}{\alpha - j\omega}\right) + 2\left(\frac{\beta}{\beta - j\omega}\right)$ b) $\left(\frac{\alpha}{\alpha - j\omega}\right)\left(\frac{2\beta}{2\beta - j\omega}\right)$ c) $\left(\frac{\alpha}{\alpha - j\omega}\right)\left(\frac{\beta}{\beta - 2j\omega}\right)$
d) $\left(\frac{\alpha}{\alpha - j\omega}\right)\left(\frac{\beta}{\beta - j\omega}\right)^2$ e) $2\left(\frac{\alpha}{\alpha - j\omega}\right)\left(\frac{\beta}{\beta - j\omega}\right)$

Question no. 19

Let $X_1, X_2, \ldots$ be independent random variables that are all distributed as $X \sim U(-\frac{1}{2}, \frac{1}{2})$, and let $S_n := \sum_{k=1}^{n} X_k$. Use the central limit theorem to calculate $P[S_{1500}^2 \leq 125]$.

a) $Q(1)$ b) $2Q(1)$ c) $1 - 2Q(1)$ d) $2(1 - Q(1))$ e) $(1 - Q(1))^2$

Question no. 20

A number X is taken at random in the interval $(0, 1]$, and then a number Y is taken at random in the interval $(0, X]$. We can show that

$$f_{X,Y}(x, y) = \frac{1}{x} \quad \text{if } 0 < y \leq x \leq 1.$$

Calculate $P[Y < 1/2]$.

a) $\frac{1 - \ln 2}{2}$ b) $\frac{\ln 2}{2}$ c) $\ln 2$ d) $\frac{1 + \ln 2}{2}$ e) none of these answers

Question no. 21

Let

$$F_{X,Y}(x, y) = \begin{cases} 0 & \text{if } x < 0 \text{ or } y < 0, \\ xy & \text{if } 0 \le x \le 1, 0 \le y \le 1, \\ x & \text{if } 0 \le x \le 1, y > 1, \\ y & \text{if } x > 1, 0 \le y \le 1, \\ 1 & \text{if } x > 1, y > 1. \end{cases}$$

Calculate $F_X(1/2) + F_X(3/2)$.

a) 1/2   b) 3/2   c) $\frac{1}{2}y$ d) $1 + \frac{1}{2}y$ e) $\frac{3}{2}y$

Question no. 22

Calculate $E[Y]$, if $Y \mid X \sim N(X, 1)$ and $X \sim \text{Exp}(2)$.

a) 1/2   b) 2   c) $X/2$ d) X e) $2X$

Question no. 23

Suppose that X has a uniform distribution on the interval $(-1, 1)$. We define $Y = X^2$. Are the random variables X and Y orthogonal and independent?

a) not orthogonal and not independent

b) not orthogonal, but independent

c) orthogonal and correlated

d) orthogonal and independent

e) orthogonal, uncorrelated but not independent

Question no. 24

We consider a random vector (X, Y) having a bivariate normal distribution with parameters $\mu_X = 1, \mu_Y = 2, \sigma_X^2 = 1, \sigma_Y^2 = 4$ and $\rho = 1/2$. Calculate $E[XY]$.

a) 1   b) 2   c) 3   d) 4   e) 5

Question no. 25

Use the central limit theorem to calculate $P[(X_1 + \cdots + X_n)^2 > c]$, where the random variables $X_1, \ldots, X_n$ are independent and all have an exponential distribution with parameter 2, and c is a positive constant.

a) $Q\left(\frac{2\sqrt{c}-n}{\sqrt{n}}\right)$ b) $Q\left(\frac{n-2\sqrt{c}}{\sqrt{n}}\right)$ c) $2Q\left(\frac{2\sqrt{c}-n}{\sqrt{n}}\right) - 1$ d) $2Q\left(\frac{n-2\sqrt{c}}{\sqrt{n}}\right) - 1$

e) $1 - 2Q\left(\frac{n-2\sqrt{c}}{\sqrt{n}}\right)$

Question no. 26

Let

$$p_{X,Y}(j, k) = \left(\frac{1}{2}\right)^{j+1} \frac{e^{-1}}{k!} \quad \text{for } j, k = 0, 1, 2, \ldots.$$

Calculate $P[X = Y]$.

a) 0   b) $\frac{1}{2}e^{-1}$ c) $\frac{1}{2}e^{-1/2}$ d) e^{-1} e) $e^{-1/2}$

Question no. 27

Calculate $E[XY]$ if we suppose that Y has a uniform distribution on the interval $(0, 1)$ and that

$$f_{X|Y}(x \mid y) = \frac{2x}{y^2} \quad \text{if } 0 < x < y.$$

a) 2/9   b) 1/3   c) 2/3   d) ∞ e) does not exist

Question no. 28

Let

$$f_{X,Y}(x, y) = \frac{1}{4} \quad \text{if } 0 \le x \le 2, 0 \le y \le 2.$$

Find $f_{V,W}(v, w)$, where $V := X - 2$ and $W := -Y$.

a) $\frac{1}{4}$ if $-2 \le v \le 0, -2 \le w \le 0$ b) $\frac{1}{4}$ if $0 \le v \le 2, -2 \le w \le 0$

c) $\frac{1}{4}vw$ if $-2 \le v \le 0, -2 \le w \le 0$ d) $-\frac{1}{4}$ if $-2 \le v \le 0, 0 \le w \le 2$

e) none of these answers

Question no. 29

Suppose that

$$p_X(x) = 1/2 \quad \text{if } x = 1 \text{ or } 2.$$

We consider n independent random variables, $X_1, \ldots, X_n$, such that $p_{X_i}(x) \equiv p_X(x)$ for $i = 1, 2, \ldots, n$. Calculate the generating function $G_{S_n}(z)$ of the random variable $S_n := X_1 + \cdots + X_n$.

a) $\frac{1}{2}(z + z^2)$ b) $\left[\frac{1}{2}(z + z^2)\right]^{1/n}$ c) $\left[\frac{1}{2}(z + z^2)\right]^n$ d) $\frac{1}{2n}(z + z^2)$ e) $\frac{n}{2}(z + z^2)$

Question no. 30

Suppose that $X \sim N(0, 1)$, $Y \sim N(1, 4)$ and $\rho_{X,Y} = -1/2$. Calculate the best estimator of X^2 in terms of Y.

a) $\frac{3}{4}$ b) $\frac{1-Y}{4}$ c) $\frac{3}{4} + \frac{1-Y}{4}$ d) $\frac{(Y-1)^2}{16}$ e) $\frac{3}{4} + \frac{(Y-1)^2}{16}$

Question no. 31

Calculate $P[X^2 + Y^2 < 1]$, if

$$f_{X,Y}(x, y) = \begin{cases} 2 \text{ if } 0 \le x \le y \le 1, \\ 0 \text{ elsewhere.} \end{cases}$$

a) $\pi/16$ b) $\pi/8$ c) $\pi/4$ d) 1/2   e) 3/4

Question no. 32

Let

$$p_{X,Y}(x, y) = \binom{20}{x}\left(\frac{1}{2}\right)^{y+20} \quad \text{for } x = 0, 1, \ldots, 20; \ y = 1, 2, \ldots.$$

Calculate $F_{X,Y}(25, 5)$.

a) 0   b) 0.03125   c) 0.05   d) 0.95   e) 0.96875

Question no. 33

The joint probability mass function $p_{X,Y}(x, y)$ of a discrete random vector is given by

$$p_{X,Y}(x, y) = \begin{cases} 1/5 & \text{if } (x, y) = (0, 0), (2, 1), \\ 1/10 & \text{if } (x, y) = (0, -1), (0, 1), (1, 0), (1, 1), (2, -1), (2, 0), \\ 0 & \text{otherwise.} \end{cases}$$

Calculate $E[X \mid Y = 0]$.

a) 0   b) 3/10   c) 2/5   d) 3/4   e) 4/5

Question no. 34

Suppose that $X \sim U(0, 1)$ and $Y \sim U(0, 2)$ are independent random variables. Find the function $f_{V,Y}(v, y)$, where $V := X + Y$.

a) $\frac{1}{2}$ if $y < v < y + 1, 0 < y < 2$ b) $\frac{1}{2}$ if $0 < v < 3, 0 < y < 2$

c) $\frac{1}{4}$ if $y \leq v \leq 3, 0 < y < 2$ d) $\frac{1}{2}$ if $0 < v < 1, 0 < y < 2$

e) $\frac{1}{4}$ if $0 < v < y + 1, 0 < y < 2$

Question no. 35

Let $X \sim U(-2, 2)$. We define $Y = X^2$ and $Z = X^3$. Are the random variables Y and Z correlated? independent? orthogonal?

a) orthogonal and independent
b) orthogonal and uncorrelated, but not independent
c) not orthogonal, but independent
d) not orthogonal, uncorrelated, but not independent
e) orthogonal, uncorrelated, but we cannot conclude regarding independence

Question no. 36

The random variables X_1, X_2 and X_3 have a Gaussian $N(0, 2)$ distribution. We suppose that $COV[X_1, X_2] = 0$ and $COV[X_1, X_3] = -2$. Calculate the variance $VAR[X_1 + X_2 + X_3]$.

a) 2   b) 4   c) 6   d) 8   e) 10

Question no. 37

We define $S_{10} = \sum_{k=1}^{10} X_k$, where the X_k's are independent random variables, all having a Poisson distribution with parameter $\alpha = 1$.

i) Calculate $P[S_{10} > 2]$.

ii) Use the central limit theorem to calculate $P[S_{10} > 2]$.

a) i) 0.9972; ii) $\Phi(0.8)$ b) i) 0.9972; ii) $\Phi(2.53)$ c) i) 0.9995; ii) $\Phi(0.7)$
d) i) 0.9995; ii) $\Phi(0.8)$ e) i) 0.9995; ii) $\Phi(2.53)$

Question no. 38

Let

$$f_{X,Y}(x, y) = \frac{3}{4} \quad \text{if } 0 < y < 1, x^2 < y.$$

A) Find $f_X(x)$ and $f_Y(y)$.

a) $f_X(x) = \frac{3}{4}(1 - x^2)$ if $\mid x \mid < 1$; $f_Y(y) = \frac{3}{2}y^{1/2}$ if $0 < y < 1$

b) $f_X(x) = \frac{3}{4}(1 - x^2)$ if $\mid x \mid < 1$; $f_Y(y) = 1$ if $0 < y < 1$

c) $f_X(x) = 2$ if $|x| < 1$; $f_Y(y) = \frac{3}{2}y^{1/2}$ if $0 < y < 1$

d) $f_X(x) = 2$ if $|x| < 1$; $f_Y(y) = \frac{3}{8}$ if $x^2 < y < 1$

e) $f_X(x) = 2$ if $|x| < 1$; $f_Y(y) = 1$ if $0 < y < 1$

B) Calculate $P[Y \geq X]$.

a) 1/8   b) 1/4   c) 1/2   d) 3/4   e) 7/8

C) Calculate $F_{X,Y}(\frac{1}{2}, \frac{1}{9})$.

a) 0   b) 1/27   c) 1/9   d) 1/3   e) 1

D) Are the random variables X and Y independent and orthogonal?

a) correlated, therefore dependent, and not orthogonal
b) dependent, but orthogonal
c) independent, but not orthogonal
d) independent and orthogonal
e) uncorrelated, but dependent, and not orthogonal

E) Find the best linear estimator of Y in terms of X.

a) 0   b) 1/2   c) 3/5   d) X e) $\frac{1+X^2}{2}$

F) We take ten independent observations of the pair (X, Y). i) What is the probability p that they are all located in the first quadrant? ii) What is the approximate value of p, according to the central limit theorem?

a) i) $(\frac{1}{4})^{10}$; ii) $\frac{1}{\sqrt{10\pi}}e^{-2.5}$ b) i) $(\frac{1}{4})^{10}$; ii) $\frac{1}{\sqrt{5\pi}}e^{-5}$ c) i) $(\frac{1}{2})^{10}$; ii) $\frac{\sqrt{2}}{5\sqrt{\pi}}e^{-5}$

d) i) $(\frac{1}{2})^{10}$; ii) $\frac{1}{\sqrt{10\pi}}e^{-2.5}$ e) i) $(\frac{1}{2})^{10}$; ii) $\frac{1}{\sqrt{5\pi}}e^{-5}$

Question no. 39

Let

$$f_{X,Y}(x, y) = \begin{cases} x + xy & \text{if } 0 < x < 1, -1 < y < 1, \\ 0 & \text{elsewhere.} \end{cases}$$

A) Calculate $f_X(\frac{1}{2})$.

a) $\frac{1}{2}$ b) 1   c) $\frac{3}{2}$ d) 2   e) $\frac{1}{2}(1+y)$

B) Calculate $P[XY < 0]$.

a) $\frac{1}{4}$ b) $\frac{1}{3}$ c) $\frac{1}{2}$ d) $\frac{2}{3}$ e) $\frac{3}{4}$

C) Are the random variables X and Y orthogonal? correlated? independent?

a) orthogonal and independent
b) orthogonal and correlated
c) not orthogonal, but independent
d) not orthogonal, but correlated
e) not orthogonal and uncorrelated, but not independent

D) Let $Z := 1/X$. What is the best linear estimator of X in terms of Z?

a) $1/Z$ b) Z c) $-6Z + 6$ d) 2/3   e) 2

E) Let $Y_1, Y_2, \ldots$ be independent random variables having the same distribution function as Y. We define $S_n = \sum_{k=1}^{n} Y_k$.

i) By the weak law of large numbers, toward what value does S_n tend as $n \to \infty$?

a) $-\infty$ b) 0   c) $\frac{1}{3}$ d) ∞ e) none of these answers

ii) According to the central limit theorem, what is (approximately) the characteristic function of S_n?

Hint. If $X \sim N(\mu, \sigma^2)$, then $\phi_X(\omega) = \exp\{j\mu\omega - \frac{1}{2}\sigma^2\omega^2\}$.

a) $\exp\{\frac{1}{3n} j\omega - \frac{1}{9n^2}\omega^2\}$ b) $\exp\{\frac{1}{3}nj\omega - \frac{1}{9}n\omega^2\}$ c) $\exp\{\frac{1}{3} j\omega - \frac{1}{9n}\omega^2\}$

d) $\exp\{\frac{1}{3n} j\omega - \frac{1}{9n}\omega^2\}$ e) $\exp\{-\frac{1}{2}\omega^2\}$

Question no. 40

A point X_1 is taken in the interval $[0, 1]$ according to a uniform distribution. Let x_1 be the value taken by X_1. Next, a point X_2 is taken uniformly in the interval $[x_1, 1]$. We consider the random vector (X_1, X_2).

A) What is the joint probability density function of (X_1, X_2)?

a) 1   if $0 \le x_1 \le 1, 0 \le x_2 \le 1$ b) $\frac{1}{2}$ if $0 \le x_1 + x_2 \le 2$

c) $\frac{1}{x_1}$ if $0 \le x_1 \le x_2 \le 1$ d) $\frac{1}{1-x_1}$ if $0 \le x_1 \le x_2 \le 1$

e) $\frac{1}{(2\ln 2 - 1)(1+x_1)}$ if $0 \le x_1 \le x_2 \le 1$

B) Calculate $E[X_2 \mid X_1 = x_1]$.

a) $\frac{3}{4}$ b) $\frac{1-x_1}{2}$ c) $\frac{1-x_1^2}{2}$ d) $\frac{1+x_1}{2}$ e) $\frac{(1+x_1)^2}{2}$

C) Calculate $E[X_2^2]$.

a) $\frac{5}{9}$ b) $\frac{11}{18}$ c) $\frac{2}{3}$ d) $\frac{13}{18}$ e) $\frac{7}{9}$

Question no. 41

A) Let X be a random variable having a uniform distribution on the interval $[-1, 1]$. We define $Y = X^3$. Calculate $\text{VAR}[X - Y]$.

a) 0.0762   b) 0.1905   c) 0.2762   d) 0.4762   e) 0.8762

B) Let $X_1, \ldots, X_{36}$ be independent random variables such that

$$f_{X_k}(x) = \frac{1}{x} \quad \text{if } 1 \le x \le e,$$

for $k = 1, \ldots, 36$. Use the central limit theorem to calculate the probability $P[\prod_{k=1}^{36} X_k > e^{19}]$.

a) $Q(0.38)$ b) $Q(0.48)$ c) $Q(0.58)$ d) $Q(0.68)$ e) $Q(0.78)$

Question no. 42

Let

$$F_{X,Y}(x, y) = \begin{cases} 0 & \text{if } x < -1 \text{ or } y < 0, \\ \left(\frac{x+1}{2}\right)(1 - e^{-y}) & \text{if } -1 \le x \le 1 \text{ and } y \ge 0, \\ 1 - e^{-y} & \text{if } x > 1 \text{ and } y \ge 0 \end{cases}$$

be the joint distribution function of the continuous random vector (X, Y).

A) Calculate $F_X(1/2)$.

a) 1/8   b) 1/4   c) 1/2   d) 3/4   e) 7/8

B) Calculate $P[X + Y > 0]$.

a) 0.8161   b) 0.8261   c) 0.8361   d) 0.8461   e) 0.8561

C) Are the random variables X and Y independent (or simply uncorrelated)? orthogonal?

a) independent and orthogonal
b) uncorrelated, but not independent, and orthogonal
c) correlated, therefore not independent, but orthogonal
d) independent, but not orthogonal
e) uncorrelated, but not independent, and not orthogonal

Question no. 43

Let $X_1, \ldots, X_{36}$ be independent random variables having an exponential distribution with parameter 2.

A) We define $Y = 2X_1 + X_2$. Calculate the covariance of Y and X_1.

a) 0   b) 1/4   c) 1/2   d) 3/4   e) 1

B) According to the central limit theorem, what is the approximate distribution of $S := X_1^2 + \cdots + X_{36}^2$?

Hint. We have: $\Gamma(x) = \int_0^\infty y^{x-1} e^{-y} \, dy$.

a) N(0, 1)   b) N(18, 9)   c) N(18, 18)   d) N(18, 36)   e) N(18, 45)

Question no. 44

Let (X, Y) be a random vector with joint probability density function

$$f_{X,Y}(x, y) = \begin{cases} x + y & \text{if } 0 < x < 1,\ 0 < y < 1, \\ 0 & \text{elsewhere.} \end{cases}$$

A) Calculate $F_{X,Y}(\frac{1}{2}, \frac{1}{2}) + F_{X,Y}(\frac{1}{2}, \frac{3}{2})$.

a) 1/16   b) 1/8   c) 1/4   d) 3/8   e) 1/2

B) Calculate $P[\frac{X}{Y} < \frac{1}{2}]$.

a) 5/24   b) 1/4   c) 7/24   d) 1/3   e) 3/8

C) Calculate $E[X \mid Y = \frac{1}{4}]$.

a) 11/18   b) 2/3   c) 13/18   d) 7/9   e) 5/6

D) We define the random variables $U = X + Y$ and $V = XY$. We can show that

$$f_U(u) = \begin{cases} u^2 & \text{if } 0 < u \le 1, \\ 2u - u^2 & \text{if } 1 < u < 2, \\ 0 & \text{elsewhere} \end{cases} \quad \text{and} \quad f_V(v) = \begin{cases} 2 - 2v & \text{if } 0 < v < 1, \\ 0 & \text{elsewhere.} \end{cases}$$

Calculate COV$[U, V]$.

a) $-1/18$ b) 0   c) 1/18   d) 1/12   e) 1/9

E) Let $V_1, V_2, \ldots, V_{36}$ be independent random variables having the same probability density function as the random variable $V = XY$.

i) Let $W := V_1 - 2V_2 + 3V_3$. Find the characteristic function $\phi_W(\omega)$ of W in terms of that of V.

a) $2\phi_V(\omega)$ b) $\phi_V(\omega) + \phi_V(-2\omega) + \phi_V(3\omega)$ c) $\phi_V(\omega) + \phi_V^{-2}(\omega) + \phi_V^3(\omega)$

d) $\phi_V(\omega)\phi_V^{-2}(\omega)\phi_V^3(\omega)$ e) $\phi_V(\omega)\phi_V(-2\omega)\phi_V(3\omega)$

ii) Use the central limit theorem to calculate $P[\sum_{k=1}^{36} V_k > 11]$.

a) $1 - Q(0.707)$ b) $Q(0.707)$ c) $1 - Q(1/2)$ d) $Q(1/2)$ e) $1/2$

Question no. 45

We consider the random variables

$$V := \alpha X + (1 - \alpha)Y \quad \text{and} \quad W := (1 - \alpha)X + \alpha Y,$$

where $X \sim N(0, 1)$ and $Y \sim N(-1, 1)$ are independent random variables, and where α is a constant taken from the interval $[0, 1]$.

A) For what value(s) of α are the random variables V and W orthogonal?

a) 0 only   b) 1/2 only   c) 1 only   d) 0 and 1 only   e) 0, 1/2 and 1 only

B) For what value(s) of α are the random variables V and W completely correlated (that is, $\rho_{V,W}^2 = 1$)?

a) 0 only   b) 1/2 only   c) 1 only   d) 0 and 1 only   e) 0, 1/2 and 1 only

C) What is the best linear estimator of V in terms W, if $\alpha = 1/4$?

a) $-\frac{3}{4}$ b) $-\frac{3}{4} + \frac{3}{5}W$ c) $\frac{3}{5}W$ d) $W + \frac{1}{4}$ e) $-\frac{3}{4} + \frac{3}{5}\left(W + \frac{1}{4}\right)$

D) Find the characteristic function of W, if $\alpha = 1/2$.

Hint. If $X \sim N(\mu, \sigma^2)$, then $\phi_X(\omega) = \exp(j\mu\omega - \frac{1}{2}\sigma^2\omega^2)$.

a) $\exp(-\frac{1}{2}j\omega - \frac{1}{4}\omega^2)$ b) $\exp(-\frac{1}{2}j\omega)$ c) $\exp(-\frac{1}{4}\omega^2)$ d) $\exp(-\frac{1}{2}j\omega - \frac{1}{2}\omega^2)$

e) $\exp(-\frac{1}{2}\omega^2)$

E) Suppose that $X_1, \ldots, X_{30}$ and $Y_1, \ldots, Y_{30}$ are independent random variables distributed like X and like Y, respectively. We define $D_i = X_i - Y_i$ for $i = 1, \ldots, 30$. Let M be the number of D_i's that will take on a positive value. Use a Gaussian distribution to calculate (approximately) $P[M = 28]$.

Hint. We have: $\Phi(0.7071) \simeq 0.7602$ and $\Phi(0.8071) \simeq 0.7902$.

a) 0.00045   b) 0.0045   c) 0.0145   d) 0.0245   e) 0.0445

Question no. 46

The joint probability density function of the continuous random vector (X, Y) is given by

$$f_{X,Y}(x, y) = \begin{cases} \frac{1}{2}x^2(2 + 3y^2) & \text{if } -1 < x < 1, 0 < y < 1, \\ 0 & \text{elsewhere.} \end{cases}$$

We can show that

$$E[X] = 0, \quad E[X^2] = \frac{3}{5}, \quad E[Y] = \frac{7}{12} \quad \text{and} \quad E[Y^2] = \frac{19}{45}.$$

A) Are the random variables X and Y independent, or simply uncorrelated, and orthogonal?

a) independent and orthogonal
b) uncorrelated, but not independent, and orthogonal
c) independent, but not orthogonal
d) uncorrelated, but not independent, and not orthogonal
e) not independent and not orthogonal

B) Calculate $P[X > 0, Y < \frac{1}{2}]$.

a) $\frac{1}{16}$ b) $\frac{1}{8}$ c) $\frac{3}{16}$ d) $\frac{1}{4}$ e) $\frac{1}{2}$

C) Let $Z := XY$. What is the best estimator of Z in terms of X?

a) 0   b) X c) $\frac{1}{2}X$ d) $\frac{7}{12}X$ e) XY

D) Let $W := 2X + 3Y$, and let $W_1, \ldots, W_{25}$ be independent random variables having the same distribution function as W. According to the central limit theorem, what is the approximate distribution of the sum $W_1 + \cdots + W_{25}$?

a) $N(41.75, 78.4375)$ b) $N(41.75, 79.4375)$ c) $N(42.75, 77.4375)$
d) $N(43.75, 79.4375)$ e) $N(43.75, 78.4375)$

Question no. 47

The joint probability density function of the continuous random vector (X, Y) is given by

$$f_{X,Y}(x, y) = \begin{cases} \dfrac{k}{\sqrt{xy}} & \text{if } 0 < x < y < 1, \\ 0 & \text{elsewhere.} \end{cases}$$

A) Find the constant k.

a) 1/16   b) 1/8   c) 1/4   d) 1/2   e) 1

B) Calculate $f_X(1/4)$.

a) $k/2$ b) k c) $2k$ d) $4k$ e) $8k$

C) Find the best (non-linear) estimator of Y in terms of X.

a) $\frac{1-X^{3/2}}{3(1-X^{1/2})}$ b) $\frac{1-X^{3/2}}{3}$ c) $\frac{1}{3(1-X^{1/2})}$ d) $\frac{X}{3}$ e) X

Question no. 48

Suppose that $X_1, \ldots, X_{30}$ are independent random variables having a $U(-1, 1)$ distribution. We define $S = \sum_{k=1}^{30} X_k^2$.

A) Calculate the mean of S.

a) 5   b) 10   c) 15   d) 20   e) 30

B) Use the central limit theorem to calculate (approximately) $P[S > 12.5]$.

Hint. We have: $E[S^2] = 924/9$.

a) $Q(1.13)$ b) $Q(1.23)$ c) $Q(1.33)$ d) $Q(1.43)$ e) $Q(1.53)$

5

Stochastic Processes

5.1 Introduction

Definition 5.1.1. *Let E be a random experiment and let S be a sample space associated with E. A* **stochastic process** *(or* **random process***) is a set of random variables $\{X(t, s), t \in T\}$. Thus, with each result $s \in S$, we associate a function $X(t, s)$. The domain of t is a set T of real numbers.*

Remarks. i) As in the case of random variables, we often suppress the argument s, and we denote the stochastic process (sp) by $\{X(t), t \in T\}$. In fact, many authors simply denote the stochastic process by $X(t)$, without specifying the set T. This notation is generally sufficiently clear, because in most cases that are of interest to us the set T is the interval $[0, \infty)$. However, there can be some confusion between the stochastic process itself and the value taken by the process at a (fixed) time t. Consequently, we will use the more complete notation.

ii) The graph of $X(t, s)$ as a function of t, for a fixed s, is called a *realization* or a *trajectory* (or a *sample path*) of the stochastic process.

iii) If T is a denumerably infinite set, then we say that $\{X(t), t \in T\}$ is a *discrete-time* stochastic process. Generally, we take $T = \{0, 1, \ldots\}$. We then have a sequence of random variables that we denote by $X_0, X_1, \ldots$. If T is an interval, or a set of intervals, $\{X(t), t \in T\}$ is called a *continuous-time* stochastic process. Actually, we sometimes encounter the case when the set T is *finite*. However, this case is much less important in practice.

iv) Moreover, we say that $\{X(t), t \in T\}$ is a *discrete-state* (respectively *continuous-state*) stochastic process if the set of possible values of the $X(t)$'s, called the *state space*, is finite or denumerably infinite (resp. non-denumerably infinite).

Now, if t is fixed, then $X(t, s) \equiv X(t)$ is a random variable. We define the following functions.

1) **First-order distribution function:**

$$F(x; t) = P[X(t) \leq x] = P[\{s \in S : X(t, s) \leq x\}]. \tag{5.1}$$

Remark. Note that the function $F(x; t)$ is in fact the distribution function of the random variable $X(t)$, so that $F(x; t) \equiv F_{X(t)}(x)$. Therefore, it is not a new concept.

2) **First-order density function** (for continuous-state sp's):

$$f(x; t) = \frac{\partial}{\partial x} F(x; t) \quad \forall x \text{ where } F(x; t) \text{ is differentiable.} \tag{5.2}$$

3) **Second-order distribution and density functions**:

$$F(x_1, x_2; t_1, t_2) = P[X(t_1) \le x_1, X(t_2) \le x_2] \tag{5.3}$$

and

$$f(x_1, x_2; t_1, t_2) = \frac{\partial^2}{\partial x_1 \partial x_2} F(x_1, x_2; t_1, t_2) \tag{5.4}$$

$\forall (x_1, x_2)$ where the second mixed partial derivative exists.

4) **nth-order distribution function**:

$$F(x_1, \ldots, x_n; t_1, \ldots, t_n) = P[X(t_1) \le x_1, \ldots, X(t_n) \le x_n]. \tag{5.5}$$

Example 5.1.1 A *Bernoulli process* is a sequence $X_1, X_2, \ldots$ of random variables associated with Bernoulli trials. It is therefore a discrete-time and discrete-state stochastic process, since $T = \{1, 2, \ldots\}$ and $S_{X_k} = \{0, 1\} \ \forall k$. Then, by independence, we may write, for instance, that

$$p(x, y; t_1 = 1, t_2 = 2) \equiv P[X_1 = x, X_2 = y] = p^x(1-p)^{1-x} p^y(1-p)^{1-y}$$
$$= p^{x+y}(1-p)^{2-x-y},$$

where x and $y \in \{0, 1\}$. Furthermore, if we say that an *arrival* corresponds to a success, then the number N of trials between two arrivals (including the trial on which a success occurred) has a geometric distribution.

5.2 Characteristics of Stochastic Processes

In order to establish the properties of a stochastic process, we must theoretically know its *n*th-order distribution function. However, we can also (partially) characterize a stochastic process by its moments, with the help of the following functions.

1) **Mean of a stochastic process**:

$$m_X(t) = E[X(t)] = \int_{-\infty}^{\infty} x f(x; t) \, dx \quad \text{(continuous case).} \tag{5.6}$$

2) **Autocorrelation function:**

$$R_X(t_1, t_2) = E[X(t_1)X(t_2)] = \int_{-\infty}^{\infty} \int_{-\infty}^{\infty} x_1 x_2 f(x_1, x_2; t_1, t_2) \, dx_1 dx_2. \quad (5.7)$$

Remark. The value of $R_X(t, t) = E[X^2(t)]$ is sometimes called the *average power* of the stochastic process $\{X(t), t \in T\}$.

3) **Autocovariance function:**

$$C_X(t_1, t_2) = R_X(t_1, t_2) - m_X(t_1)m_X(t_2). \quad (5.8)$$

In particular, the **variance** of $X(t)$ is given by

$$\mathrm{VAR}[X(t)] = C_X(t, t). \quad (5.9)$$

4) **Correlation coefficient:**

$$\rho_X(t_1, t_2) = \frac{C_X(t_1, t_2)}{[C_X(t_1, t_1)C_X(t_2, t_2)]^{1/2}}. \quad (5.10)$$

We have: $-1 \le \rho_X(t_1, t_2) \le 1$ and $\rho_X(t, t) = 1$.

Remark. The functions $R_X(t_1, t_2), C_X(t_1, t_2)$ and $\rho_X(t_1, t_2)$ are non-negative definite. Indeed, we have, for example:

$$\sum_{i=1}^{n} \sum_{k=1}^{n} a_i a_k R_X(t_i, t_k) = E\left[\left\{\sum_{i=1}^{n} a_i X(t_i)\right\}^2\right] \ge 0 \quad \forall a_i \in \mathbb{R}.$$

Definition 5.2.1. *If $m_X(t)$ is a constant and $R_X(t_1, t_2) = R_X(t_2 - t_1) \, \forall t_1, t_2 \in T$, we say that the stochastic process $\{X(t), t \in T\}$ is* **wide-sense stationary** *(WSS).*

Remarks. i) We could replace the function R_X by the function C_X in the definition, since if $m_X(t) \equiv c$, then the functions R_X and C_X only differ by the constant c^2.

ii) It would be more rigorous to write in the definition that we must have $R_X(t_1, t_2) = R_X^*(t_2 - t_1)$, where R_X^* is a function of a single variable. However, the formulation used above is classic.

iii) The concept of stationarity will be seen in detail in Section 5.6 (p. 235).

Definition 5.2.2. *If the increments $X(t_4) - X(t_3)$ and $X(t_2) - X(t_1)$ of the sp $\{X(t), t \in T\}$ are independent (respectively uncorrelated) $\forall t_1 < t_2 \le t_3 < t_4$, we say that $\{X(t), t \in T\}$ is a stochastic process with* **independent increments** *(resp.* **uncorrelated increments***).*

Definition 5.2.3. *If the distribution of $X(t_2+\tau)-X(t_1+\tau)$ and that of $X(t_2)-X(t_1)$ are identical $\forall \tau$, we say that $\{X(t), t \in T\}$ is a stochastic process with* **stationary increments**.

Remark. If two random variables are identically distributed, this means that they have the same distribution function. However, this does *not* mean that they are equal.

Suppose now that we consider two stochastic processes: $\{X(t), t \in T_1\}$ and $\{Y(t), t \in T_2\}$. We then define the following functions.

1) **Cross-correlation function**:

$$R_{X,Y}(t_1, t_2) = E[X(t_1)Y(t_2)] = R_{Y,X}(t_2, t_1). \tag{5.11}$$

2) **Cross-covariance function**:

$$C_{X,Y}(t_1, t_2) = R_{X,Y}(t_1, t_2) - m_X(t_1)m_Y(t_2). \tag{5.12}$$

Definition 5.2.4. *The stochastic processes $\{X(t), t \in T_1\}$ and $\{Y(t), t \in T_2\}$ are said to be*
a) **orthogonal** *if $R_{X,Y}(t_1, t_2) = 0 \ \forall t_1, t_2$;*
b) **uncorrelated** *if $C_{X,Y}(t_1, t_2) = 0 \ \forall t_1, t_2$.*

Definition 5.2.5. *The stochastic processes $\{X(t), t \in T_1\}$ and $\{Y(t), t \in T_2\}$ are called* **independent** *if the random variables $X(t_1), \ldots, X(t_n)$ and $Y(\tau_1), \ldots, Y(\tau_m)$ are (globally) independent $\forall n, m \in \{1, 2, \ldots\}$.*

Example 5.2.1 Let $\{X(t), t > 0\}$ be the random process defined by

$$X(t) = Y^2 t \quad \text{for } t > 0,$$

where Y is a random variable having a uniform distribution on the interval $(0, 1)$.
a) Calculate
   i) the first-order density function of the process;
   ii) the mean $E[X(t)]$ of the process for $t > 0$;
   iii) the autocovariance function $C_X(t, t+s)$ of the process for $s, t > 0$.
b) Is the process WSS?

Solution.
a) i) We can write (see Section 3.6, p. 94) that

$$f(x; t) \equiv f_{X(t)}(x) = f_Y(y) \left| \frac{dy}{dx} \right|, \quad \text{where } y = \sqrt{x/t}$$

$$= 1 \left| \frac{1}{2\sqrt{xt}} \right| = \frac{1}{2\sqrt{xt}} \quad \text{if } x \in (0, t).$$

ii) We have:

$$E[X(t)] = \int_0^1 y^2 t \cdot 1 \, dy = \frac{t}{3} \quad \text{for } t > 0.$$

We can also use part i) and write that

$$E[X(t)] = \int_0^t x \frac{1}{2\sqrt{xt}} \, dx = \frac{1}{2} \int_0^t \frac{x^{1/2}}{t^{1/2}} \, dx = \frac{1}{3} \frac{x^{3/2}}{t^{1/2}} \Big|_0^t = \frac{t}{3} \quad \text{for } t > 0.$$

iii) Since

$$X(t)X(t+s) = Y^2 t \cdot Y^2 (t+s) = Y^4 t(t+s),$$

we first calculate

$$E[X(t)X(t+s)] = E[Y^4 t(t+s)] = \frac{t(t+s)}{5} \quad \text{for } s, t > 0.$$

It follows that

$$C_X(t, t+s) = E[X(t)X(t+s)] - E[X(t)]E[X(t+s)]$$
$$\overset{ii)}{=} \frac{t(t+s)}{5} - \frac{t(t+s)}{9} = \frac{4t(t+s)}{45} \quad \text{for } s, t > 0.$$

b) Since the mean of $X(t)$ depends on t (see part ii) of a)), the stochastic process $\{X(t), t > 0\}$ is *not* WSS. Note that the condition $C_X(t, t+s) = C_X(s)$ is not fulfilled either. Therefore, even if the mean $E[X(t)]$ were constant, the process would not be WSS.

5.3 Markov Chains

The most important discrete-time stochastic process is known as a *Markov chain*.

Definition 5.3.1. *A stochastic process $\{X_n, n = 0, 1, \ldots\}$ whose state space is either finite or denumerably infinite is called a **Markov chain** if*

$$P[X_{n+1} = j \mid X_n = i, X_{n-1} = i_{n-1}, \ldots, X_0 = i_0] \overset{(1)}{=} P[X_{n+1} = j \mid X_n = i]$$

$$\overset{(2)}{=} p_{ij} \tag{5.13}$$

for all states $i_0, \ldots, i_{n-1}, i, j$ and for all $n \geq 0$.

Remarks. i) The equation (1) is called *Markov's property*. It means that the future, given the past and the present, is independent of the past and thus depends only on the present.

ii) If the equation (2) is satisfied, we actually have a *stationary* (or *time-homogeneous*) Markov chain, because $P[X_{n+1} = j \mid X_n = i]$ does not depend on n. In the general case, we can denote this probability by $p_{ij}(n)$. We can also write p_{ij} as follows: $p_{i,j}$. In fact, it is sometimes necessary to separate the states i and j by a comma to avoid any confusion (between the states 1 and 11, for instance).

iii) We generally use the set $\mathbb{N}^0 := \{0, 1, \ldots\}$ as the state space.

Now, we deduce from the definition of the p_{ij}'s that

$$p_{ij} \geq 0 \quad \forall i, j \tag{5.14}$$

and

$$\sum_{j=0}^{\infty} p_{ij} = 1 \quad \forall i. \tag{5.15}$$

Definition 5.3.2. *The matrix* **P** *defined by*

$$\mathbf{P} = \begin{bmatrix} p_{00} & p_{01} & \cdots \\ p_{10} & p_{11} & \cdots \\ \cdots & \cdots & \cdots \end{bmatrix} \tag{5.16}$$

is called the **one-step transition probability matrix** *(or simply the* **transition matrix***) of the Markov chain.*

Remark. The matrix **P** is called a *stochastic* matrix. If, in addition, we have:

$$\sum_{i=0}^{\infty} p_{ij} = 1 \quad \forall j, \tag{5.17}$$

the matrix **P** is said to be *doubly stochastic*.

Definition 5.3.3. *The set* $\{p_i, i = 0, 1, \ldots\}$, *where* p_i *is defined by*

$$p_i = P[X_0 = i], \tag{5.18}$$

is called the **initial distribution** *of the Markov chain.*

Finally, let

$$p_{ij}^n := P[X_{m+n} = j \mid X_m = i], \quad \text{where } m, n, i, j \geq 0. \tag{5.19}$$

That is, p_{ij}^n denotes the probability of moving from state i to state j in n steps. Then, if $\mathbf{P}^{(n)}$ denotes the matrix of the p_{ij}^n's, we can show that

$$\mathbf{P}^{(n)} = \mathbf{P}^n. \tag{5.20}$$

It follows that

$$\mathbf{P}^{(n+m)} = \mathbf{P}^n \mathbf{P}^m. \tag{5.21}$$

The various equations obtained from Equation (5.21) are known as the *Chapman–Kolmogorov*[1] *equations*.

Particular cases

1) A Markov chain whose state space is the set $\mathbf{Z}$ of all integers is called a *random walk* if

$$p_{i,i+1} = p = 1 - p_{i,i-1} \quad \text{for } i = 0, \pm 1, \pm 2, \ldots \quad (5.22)$$

for a certain $p \in (0, 1)$. So, we have (with $q := 1 - p$):

$$\mathbf{P} = \begin{bmatrix} \ddots & \ddots & \ddots & & \\ & q & 0 & p & \\ & & q & 0 & p \\ & & & \ddots & \ddots & \ddots \end{bmatrix}.$$

If the state space is $\{0, 1, \ldots, N\}$ and

$$p_{00} = p_{NN} = 1,$$

the states 0 and N are said to be *absorbing*. If $p_{01} = 1$, the state 0 is called *reflecting*.

2) Let $\{Y_n, n = 0, 1, \ldots\}$ be a set of i.i.d. random variables whose possible values are integers. Let

$$a_k := P[Y_n = k] \quad \forall k.$$

Then, if we define

$$X_n = \sum_{k=0}^{n} Y_k \quad \text{for } n = 0, 1, \ldots,$$

we find that $\{X_n, n = 0, 1, \ldots\}$ is a Markov chain and that

$$p_{ij} = a_{j-i}.$$

We then say that the chain is *homogeneous with respect to the state variable*.

Example 5.3.1 Suppose that a particle moves on the set of all integers and that, when it is in state i, the probability that it jumps from that state to state j depends also on the state it visited before i. More precisely, suppose that

---

[1] Andrei Nikolaevich Kolmogorov, 1903–1987, was born and died in Russia. He was a great mathematician who, before getting his Ph.D., had already published 18 scientific papers, many of which were written during his undergraduate studies. His work on Markov processes in continuous time and with continuous state space is the basis of the theory of *diffusion processes*. His book on theoretical probability, published in 1933, marks the beginning of *modern* probability theory. He also contributed in an important way to many other domains of mathematics, notably to the theory of dynamical systems.

$$P[X_{n+1} = j \mid X_n = i, X_{n-1} = k, X_{n-2} = i_{n-2}, \ldots, X_0 = i_0]$$
$$= P[X_{n+1} = j \mid X_n = i, X_{n-1} = k]$$

for all states $i_0, \ldots, i_{n-2}, i, j, k \in \mathbf{Z}$ and for $n = 1, 2, \ldots$. If the state space of the stochastic process $\{X_n, n = 0, 1, \ldots\}$, where X_n denotes the position of the particle after n transitions, is the set $\mathbf{Z}$ of all integers, then the process is *not* a Markov chain. However, it is possible to satisfy Markov's property by defining a state of the process as the vector (k, i) consisting of the last two states visited by the particle (after at least one transition).

For example, a generalization of the random walk is obtained by writing that the probability of going from (i_1, j_1) to (i_2, j_2) is given by

$$P_{(i_1,j_1),(i_2,j_2)} = \begin{cases} p_1 & \text{if } j_2 = j_1 + 1, i_2 = j_1 \text{ and } i_1 = j_1 - 1, \\ q_1 & \text{if } j_2 = j_1 - 1, i_2 = j_1 \text{ and } i_1 = j_1 - 1, \\ p_2 & \text{if } j_2 = j_1 + 1, i_2 = j_1 \text{ and } i_1 = j_1 + 1, \\ q_2 & \text{if } j_2 = j_1 - 1, i_2 = j_1 \text{ and } i_1 = j_1 + 1, \\ 0 & \text{otherwise,} \end{cases}$$

where $p_1 + q_1 + p_2 + q_2 = 1$.

Similarly, if we must keep track of the last three states visited by the particle, then the states of the Markov chain will be triplets, etc. Thus, even if Markov's property is not satisfied with a given state space, it may be satisfied with another state space.

5.4 The Poisson Process

One of the most important stochastic processes in applied sciences is the *Poisson process*. This process counts the number of events that occur in an interval. It is therefore a particular *counting process*, which we define first.

Definition 5.4.1. *A stochastic process* $\{N(t), t \geq 0\}$ *is called a* **counting process** *if* $N(t)$ *denotes the total number of events that occurred up to time* t.

Properties. i) $N(t) \geq 0 \ \forall t \geq 0$.

ii) $N(t)$ is a random variable whose possible values are $0, 1, \ldots$.

iii) $N(t)$ is non-decreasing.

iv) If $t_1 < t_2$, then $N(t_2) - N(t_1)$ is the number of events that occurred in the interval $(t_1, t_2]$.

Definition 5.4.2. *A counting process* $\{N(t), t \geq 0\}$ *is called a* **Poisson process** *with* **rate** $\lambda \ (> 0)$ *if*

i) $N(0) = 0$;

ii) $\{N(t), t \geq 0\}$ *has independent increments;*

iii) $N(t + \tau) - N(\tau) \sim Poi(\lambda t) \ \forall \tau, t \geq 0$.

Remarks. i) The condition iii) implies that a Poisson process has *stationary incre-ments*. Moreover, we can write that

$$N(t) = N(t+0) - N(0) \Rightarrow N(t) \sim \text{Poi}(\lambda t) \; \forall t \geq 0.$$

ii) We can in fact replace the condition iii) by the following condition: the sp $\{N(t), t \geq 0\}$ has stationary increments and

$$\left. \begin{array}{l} P[N(\delta) = 1] = \lambda\delta + o(\delta), \\ P[N(\delta) = 0] = 1 - \lambda\delta + o(\delta), \end{array} \right] \tag{5.23}$$

where $o(\delta)$ is such that

$$\lim_{\delta \downarrow 0} \frac{o(\delta)}{\delta} = 0. \tag{5.24}$$

This means that, in a short interval, the probability of two or more events occurring is negligible.

Now, we have:

$$\left. \begin{array}{l} E[N(t)] = \lambda t \quad (\text{because } N(t) \sim \text{Poi}(\lambda t)), \\[2mm] E[N^2(t)] = \text{VAR}[N(t)] + (E[N(t)])^2 = \lambda t + \lambda^2 t^2. \end{array} \right] \tag{5.25}$$

Hence, if $t_1 < t_2$, we can write (making use of the fact that the increments of a Poisson process are independent) that

$$\begin{aligned} R_N(t_1, t_2) &:= E[N(t_1)N(t_2)] = E[N(t_1)(N(t_2) - N(t_1))] + E[N^2(t_1)] \\ &\overset{\text{ind.}}{=} E[N(t_1)]E[N(t_2) - N(t_1)] + E[N^2(t_1)] \\ &= \lambda t_1 \lambda(t_2 - t_1) + (\lambda t_1 + \lambda^2 t_1^2) = \lambda^2 t_1 t_2 + \lambda t_1. \end{aligned}$$

Since $R_N(t_1, t_2) = R_N(t_2, t_1)$, it follows that

$$R_N(t_1, t_2) = \begin{cases} \lambda t_1 + \lambda^2 t_1 t_2 & \text{if } t_1 \leq t_2, \\ \lambda t_2 + \lambda^2 t_1 t_2 & \text{if } t_1 > t_2. \end{cases} \tag{5.26}$$

Finally, we deduce from Equation (5.26) that

$$C_N(t_1, t_2) = R_N(t_1, t_2) - (\lambda t_1)(\lambda t_2) = \lambda \min\{t_1, t_2\}. \tag{5.27}$$

Properties. i) Let T_1 be the arrival time of the first event, and T_n be the interarrival time between the $(n-1)$st and the nth event, for $n = 2, 3, \ldots$. We can show that the T_k's are *independent* random variables for $k \geq 1$ and that

$$T_k \sim \text{Exp}(\lambda) \; \forall k.$$

Indeed, we have:

$$P[T_1 > t] = P[N(t) = 0] = e^{-\lambda t}.$$

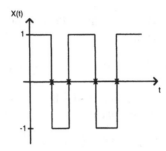

Figure 5.1. Example of trajectory of a telegraph signal process.

It follows that the time required for n events to occur, from any time instant, has a *gamma distribution* with parameters $\alpha = n$ and λ (see Section 4.9, p. 187). If we denote the arrival times of the events by $S_1, S_2, \ldots$, we have: $S_n = \sum_{k=1}^{n} T_k$ for $n = 1, 2, \ldots$.

ii) Given that a single event occurred in the interval $(0, t]$, the time T_a at which it happened has a *uniform distribution* on the interval $(0, t]$.

Remark. The Poisson process is a special case of what is known as *birth and death processes*, which themselves are particular *continuous-time Markov chains*. It is also a particular *renewal process*.

Now, from a Poisson process, we can define a stochastic process called the *telegraph signal process*, as follows:

$$X(t) = \begin{cases} 1 \text{ if } N(t) \text{ is even,} \\ -1 \text{ if } N(t) \text{ is odd} \end{cases} \qquad (5.28)$$

(see Fig. 5.1). We have:

$$P[X(t) = 1] = \sum_{k=0}^{\infty} P[N(t) = 2k] = \sum_{k=0}^{\infty} e^{-\lambda t} \frac{(\lambda t)^{2k}}{(2k)!} = e^{-\lambda t} \cosh(\lambda t), \qquad (5.29)$$

where "cosh" denotes the hyperbolic cosine. Then, we find that

$$P[X(t) = -1] = 1 - e^{-\lambda t} \cosh(\lambda t) = e^{-\lambda t} \sinh(\lambda t). \qquad (5.30)$$

It follows that

$$E[X(t)] = 1 \cdot [e^{-\lambda t} \cosh(\lambda t)] + (-1) \cdot [e^{-\lambda t} \sinh(\lambda t)] = e^{-2\lambda t}. \qquad (5.31)$$

Finally, we also find that

$$R_X(t, t + s) = e^{-2\lambda |s|}. \qquad (5.32)$$

Indeed, if $s > 0$, we can write that

$$R_X(t, t+s) := E[X(t)X(t+s)] = E[(-1)^{N(t)}(-1)^{N(t+s)}]$$
$$= E[(-1)^{2N(t)}(-1)^{N(t+s)-N(t)}] = E[(-1)^{N(s)}]$$
$$= E[X(s)] = e^{-2\lambda s}.$$

Remark. $\{X(t), t \geq 0\}$ is actually called a *semi-random* telegraph signal process. A *random* telegraph signal process is defined by

$$Y(t) = Z \cdot X(t), \tag{5.33}$$

where Z is a random variable independent of $X(t)$ for all t and such that $P[Z = 1] = P[Z = -1] = 1/2$. We have:

$$E[Y(t)] \stackrel{\text{ind.}}{=} E[Z]E[X(t)] = 0 \cdot E[X(t)] = 0$$

and

$$R_Y(t, t+s) = E[ZX(t) \cdot ZX(t+s)] \stackrel{\text{ind.}}{=} E[Z^2]R_X(t, t+s) = 1 \cdot e^{-2\lambda|s|}.$$

Thus, $\{Y(t), t \geq 0\}$ is a WSS stochastic process. Note that Z^2 is actually equal to the constant 1.

Example 5.4.1 Telephone calls arrive at an exchange according to a Poisson process, at the (average) rate of one call per minute.

a) What is the probability that (at least) five minutes have elapsed since i) the last call arrived? ii) the penultimate call arrived?

b) Calculate the probability that no calls arrive during the next two minutes, knowing that at least one call was received over the last two minutes.

Solution. Let $N(t)$ be the number of calls in the interval $[0, t]$. We have:

$$N(t) \sim \text{Poi}(1 \cdot t), \quad \text{where } t \text{ is in minutes.}$$

a) i) We seek

$$P[N(5) = 0] = P[Y = 0], \quad \text{where } Y \sim \text{Poi}(5)$$
$$= e^{-5} \simeq 0.0067.$$

ii) We now seek

$$P[N(5) \leq 1] = P[Y \leq 1] = e^{-5}(5 + 1) \simeq 0.0404.$$

b) Since the Poisson process has independent increments, we simply have to calculate

$$P[N(2) = 0] = e^{-2} \simeq 0.1353.$$

5.5 The Wiener Process

The most important continuous-time and continuous-state stochastic process is the *Wiener*[2] *process*, also known as *Brownian motion*. It is a particular case of the processes called *Gaussian*, which we define first.

Definition 5.5.1. *A stochastic process* $\{X(t), t \geq 0\}$ *is called a* **Gaussian process** *if the random variables* $X(t_1), \ldots, X(t_n)$ *have a multinormal distribution for all* n *and* $\forall t_1, \ldots, t_n$.

Remark. Under some conditions, continuous-time and continuous-state stochastic processes are called *diffusion processes*.

Consider a particle moving at random on the set of all integers. More precisely, consider the *symmetric* random walk, that is, the (discrete-time) Markov chain for which

$$p_{i,i+1} = p_{i,i-1} = 1/2 \quad \text{for } i = 0, \pm 1, \pm 2, \ldots . \tag{5.34}$$

Suppose that we modify the process as follows: instead of moving one unit to the left or to the right at each time unit, the particle moves ϵ unit to the left or to the right at each δ time unit, where ϵ and δ are very small positive quantities. So, we speed up the process: the particle moves very often, but covers very little distance at each displacement.

Let $X(t)$ be the position of the particle at time t. Suppose that it is at the origin at time $t = 0$. Then, the position of the particle after the first n displacements is given by

$$X(n\delta) = (2N - n)\epsilon, \tag{5.35}$$

where N is the number of steps to the right made in the course of these n displacements. Since N has a *binomial distribution* with parameters n and $p = 1/2$, we can write that

$$\left. \begin{array}{l} E[X(n\delta)] = (2(n/2) - n)\epsilon = 0, \\[2mm] \text{VAR}[X(n\delta)] = 4\epsilon^2 \text{VAR}[N] = 4\epsilon^2 (n/4) = n\epsilon^2. \end{array} \right] \tag{5.36}$$

Suppose now that we let δ decrease to 0 in the random walk. We then obtain a continuous-time process. Since

$$\text{VAR}[X(t)] = t\epsilon^2/\delta \tag{5.37}$$

if $t = n\delta$, we must express ϵ in terms of δ. Let

$$\epsilon^2 = \sigma^2 \delta. \tag{5.38}$$

---

[2] Norbert Wiener, 1894–1964, was born in the United States and died in Sweden. He obtained his Ph.D. in philosophy from Harvard University at the age of 18. His research subject was *mathematical logic*. After a stay in Europe to study mathematics, he started working at the Massachusetts Institute of Technology, where he did some research on *Brownian motion*. He contributed, in particular, to communication theory and to control. In fact, he is the inventor of *cybernetics*, which is the "science of communication and control in the animal and the machine."

Then, when δ decreases to zero, the process obtained is also a continuous-state stochastic process and we have:

$$\left.\begin{array}{l} E[X(t)] = 0, \\[2mm] \text{VAR}[X(t)] \to \sigma^2 t \quad \text{as } \delta \downarrow 0. \end{array}\right] \tag{5.39}$$

In fact, the variance of $X(t)$ is equal to $\sigma^2 t$ for any positive value of δ.

The stochastic process $\{W(t), t \geq 0\}$ defined by

$$W(t) = \lim_{\delta \downarrow 0} X(t) \tag{5.40}$$

is called the *Wiener process* or, as mentioned above, a *Brownian motion*, named in honor of the English botanist Robert Brown,[3] who observed in 1827 the movement of microscopic particles suspended in a fluid. This motion seems completely irregular, because of the molecular collisions that occur continuously. The process is also known as the *Wiener–Einstein process* or the *Wiener–Bachelier process*.

Using the Gaussian approximation to the binomial distribution, we can write that

$$P[W(t) \leq w] = \Phi\left(\frac{w}{\sigma t^{1/2}}\right). \tag{5.41}$$

That is, $W(t) \sim N(0, \sigma^2 t)$.

Remark. If we start with a non-symmetric random walk, then we obtain (under some conditions) a Wiener process with non-zero *drift coefficient* μ, so that $W(t) \sim N(\mu t, \sigma^2 t)$.

We now give a more formal definition of a Wiener process.

Definition 5.5.2. *A stochastic process* $\{W(t), t \geq 0\}$ *is called a* **Wiener process** *if*

i) $W(0) = 0$;

ii) $\{W(t), t \geq 0\}$ *is a process with independent and stationary increments;*

iii) $W(t) \sim N(0, \sigma^2 t) \ \forall t > 0$.

Remarks. i) When $\sigma = 1$, the process is called the *standard Brownian motion*. Since the variance of $W(t)/\sigma$ is equal to t, we can always transform an arbitrary Wiener process into a standard Brownian motion.

ii) We can generalize the definition by letting $W(0)$ take on any real value w_0 and then by replacing condition iii) by $W(t) \sim N(w_0, \sigma^2 t) \ \forall t > 0$. Note that the process $\{W^*(t), t \geq 0\}$ defined by $W^*(t) = W(t) - w_0$ then satisfies the preceding definition (if w_0 is a constant). Actually, we could consider the case when $W(0)$ is a random variable W_0 independent of $W(t) \ \forall t > 0$.

iii) We can show that $W(t)$ is a continuous function of t with probability 1 (see Fig. 5.2, p. 234).

---

[3] Robert Brown, 1773–1858, was born in Scotland and died in England. He is considered the greatest British botanist of the nineteenth century. He took part in an expedition to Australia, from which he brought many unknown species to Europe. In 1810, he was elected to the Fellowship of the Royal Society.

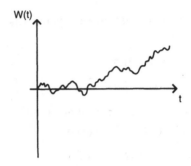

Figure 5.2. Example of trajectory of a Wiener process.

To obtain the nth-order density function of $\{W(t), t \geq 0\}$, we use the fact that the Wiener process has *independent* and *stationary increments*. Then, if $t_1 < t_2 < \cdots < t_n$, we can write that

$$f(w_1, \ldots, w_n; t_1, \ldots, t_n) = f(w_1; t_1) \prod_{k=2}^{n} f(w_k - w_{k-1}; t_k - t_{k-1}), \quad (5.42)$$

where $f(w; t)$ is the probability density function of a Gaussian $N(0, \sigma^2 t)$ distribution.

Finally, if $t < \tau$, we have:

$$\begin{aligned}
C_W(t, \tau) &\equiv \text{COV}[W(t), W(\tau)] = \text{COV}[W(t), W(t) + W(\tau) - W(t)] \\
&= \text{COV}[W(t), W(t)] + \text{COV}[W(t), W(\tau) - W(t)] \\
&\overset{\text{ind.}}{=} \text{COV}[W(t), W(t)] = \text{VAR}[W(t)] = \sigma^2 t,
\end{aligned}$$

where we used the fact that

$$\begin{aligned}
\text{COV}[X, Y + Z] &= E[X(Y + Z)] - E[X]E[Y + Z] \\
&= \{E[XY] - E[X]E[Y]\} + \{E[XZ] - E[X]E[Z]\} \\
&= \text{COV}[X, Y] + \text{COV}[X, Z].
\end{aligned}$$

In the general case, we obtain:

$$C_W(t, \tau) = \sigma^2 \min\{t, \tau\}. \quad (5.43)$$

Example 5.5.1 What is the distribution of $W(t) + W(\tau)$, where $t \leq \tau$?
Solution. We can write that

$$W(t) + W(\tau) = 2W(t) + [W(\tau) - W(t)] := Y + Z,$$

where Y and Z are *independent* random variables having Gaussian distributions with zero means and variances given by

$$\mathrm{VAR}[Y] = 4\mathrm{VAR}[W(t)] = 4\sigma^2 t \quad \text{and} \quad \mathrm{VAR}[Z] = \sigma^2(\tau - t),$$

because the Wiener process has stationary increments. It follows that

$$W(t) + W(\tau) \sim \mathrm{N}(0, \sigma^2(3t + \tau)).$$

Remark. It is important not to forget that the random variables $Z := W(\tau) - W(t)$ and $W(\tau - t)$ have the same *distribution*, but are *not* identical. For example, suppose that $t = 1$ and $\tau = 2$. If $W(2-1) = 0$, then this does not imply that $Z = W(2) - W(1) = 0$, since $P[W(2) = 0] = 0$.

Remark. Let $\{W(t), t \geq 0\}$ be a standard Brownian motion, and let f be a function whose derivative exists and is *continuous* in the interval $[a, b]$. We define the *stochastic integral*

$$\int_a^b f(t)\, dW(t) = f(b)W(b) - f(a)W(a) - \int_a^b W(t)\, df(t). \qquad (5.44)$$

Note that this formula is the one that we obtain by integrating by parts. Moreover, if $f(t) = t^2$, for example, then we have:

$$\int_a^b W(t)\, df(t) = \int_a^b 2t\, W(t)\, dt.$$

The process $\{dW(t), t \geq 0\}$ is called a (Gaussian) *white noise*. In engineering, people often write: $dW(t)/dt = \epsilon \ (= \epsilon(t))$. We can show that the integral defined in (5.44) has a *Gaussian* distribution with zero mean and variance

$$\int_a^b f^2(t)\, dt.$$

5.6 Stationarity

Definition 5.6.1. *A stochastic process* $\{X(t), t \in T\}$ *is said to be* **stationary**, *or* **strict-sense stationary** *(SSS), if*

$$F(x_1, \ldots, x_n; t_1, \ldots, t_n) = F(x_1, \ldots, x_n; t_1 + \tau, \ldots, t_n + \tau) \qquad (5.45)$$

$\forall \tau, n$ *and* $t_1, \ldots, t_n$. *If the process starts at* $t = 0$, *for instance, then* $t_k + \tau$ *must be non-negative* $\forall k$ *in Equation (5.45).*

Likewise, the sp's $\{X(t), t \in T_1\}$ and $\{Y(t), t \in T_2\}$ are **jointly stationary** *if the joint distribution function of the processes remains unchanged if we add the constant* τ *to every value of* t *in the function.*

We deduce from the definition that if $\{X(t), t \in T\}$ is a *continuous* SSS process, then

$$f(x; t) = f(x; t + \tau) \ \forall t, \tau,$$
$$f(x_1, x_2; t_1, t_2) = f(x_1, x_2; t_1 + \tau, t_2 + \tau) \ \forall t_1, t_2, \tau,$$

so that

$$\left. \begin{array}{l} f(x; t) = f(x) \ \forall t, \\[2ex] f(x_1, x_2; t_1, t_2) = f(x_1, x_2; t_2 - t_1) \ \forall t_1, t_2. \end{array} \right] \tag{5.46}$$

These equations lead to the definition of WSS processes, introduced in Section 5.2.

Definition 5.6.2. *A stochastic process* $\{X(t), t \in T\}$ *is said to be* **wide-sense stationary** *(WSS) if*

$$\left. \begin{array}{l} E[X(t)] \equiv m \ (\text{a constant}), \\[2ex] E[X(t)X(t + \tau)] = R_X(\tau) \ (= R_X(-\tau)). \end{array} \right] \tag{5.47}$$

We deduce from (5.47) that

$$E[X^2(t)] = R_X(0), \tag{5.48}$$

so that the *average power* of a WSS stochastic process is independent of t. Furthermore, we can write that

$$C_X(t, t + \tau) = C_X(\tau) = R_X(\tau) - m^2 \tag{5.49}$$

and

$$\rho_X(t, t + \tau) = \rho_X(\tau) = C_X(\tau)/C_X(0). \tag{5.50}$$

Finally, the sp's $\{X(t), t \in T_1\}$ and $\{Y(t), t \in T_2\}$ are said to be **jointly WSS** if $\{X(t), t \in T_1\}$ and $\{Y(t), t \in T_2\}$ are WSS and

$$R_{X,Y}(t + \tau, t) \equiv E[X(t + \tau)Y(t)] = R_{X,Y}(\tau) \ (\neq R_{X,Y}(-\tau), \text{ in general}). \tag{5.51}$$

Similarly, we will have:

$$C_{X,Y}(t + \tau, t) = C_{X,Y}(\tau) = R_{X,Y}(\tau) - m_X m_Y. \tag{5.52}$$

Remark. An SSS stochastic process is also WSS, but the converse is not always true.

Proposition 5.6.1. *If a* **Gaussian** *process* $\{X(t), t \in T\}$ *is* **WSS**, *then it is also* **SSS**.

Proof. This follows from the fact that a Gaussian process is completely determined by its mean and its autocovariance function. Indeed, the *nth-order characteristic* *function* of the process $\{X(t), t \in T\}$ is given by

$$\exp\left\{ jm_X \sum_{k=1}^{n} \omega_k - \frac{1}{2} \sum_{i=1}^{n} \sum_{k=1}^{n} C_X(t_i - t_k)\omega_i \omega_k \right\}. \tag{5.53}$$

Note that this function is *invariant* under any change of origin.   □

Example 5.6.1 The Poisson process is *not* stationary. Indeed, we have:

$$E[N(t)] = \lambda t.$$

Since the mean of $N(t)$ is not constant, $\{N(t), t \geq 0\}$ is not WSS. Consequently, $\{N(t), t \geq 0\}$ is not stationary. However, the random telegraph signal $\{Y(t), t \geq 0\}$ is WSS because (see p. 231)

$$E[Y(t)] = 0$$

and

$$R_Y(t, t + \tau) = e^{-2\lambda\tau},$$

where we assume that $\tau \geq 0$. Actually, we can show that the *random* telegraph signal is (strict-sense) stationary. However, the semi-random telegraph signal is not stationary, not even WSS.

Example 5.6.2 The Wiener process has a mean equal to zero. However, its autocovariance function is given by

$$C_W(t_1, t_2) = \sigma^2 \min\{t_1, t_2\}. \tag{5.54}$$

Since this function does not depend only on the difference between t_2 and t_1, $\{W(t), t \geq 0\}$ is *not* a stationary process. In fact, we can show that

$$\min\{t_1, t_2\} = \frac{1}{2}(t_1 + t_2 - |t_1 - t_2|).$$

Remark. Given that a Gaussian process is completely determined by its *mean* and its *autocovariance function*, and that the Brownian motion is Gaussian, we can also define a Wiener process as a *Gaussian* process, with *zero mean* and *autocovariance function* given by Equation (5.54). In addition, the process must start at the origin.

Example 5.6.3 Let $\{Z(t), t \geq 0\}$ be the process defined by $Z(0) = 0$ and

$$Z(t) = tW(1/t) \text{ if } t > 0,$$

where $\{W(t), t \geq 0\}$ is a Brownian motion. We have:

$$E[Z(t)] = tE[W(1/t)] = t \cdot 0 = 0$$

and

$$
\begin{aligned}
C_Z(t_1, t_2) &= E[Z(t_1)Z(t_2)] - 0^2 \\
&= E[t_1 W(1/t_1) t_2 W(1/t_2)] = t_1 t_2 C_W(1/t_1, 1/t_2) \\
&= t_1 t_2 \sigma^2 \min\{t_1^{-1}, t_2^{-2}\} = \sigma^2 \min\{t_1, t_2\}.
\end{aligned}
$$

Since $\{Z(t), t \geq 0\}$ is a Gaussian process (because it is a *linear* transformation of a Gaussian process) that starts at the origin and whose mean and autocovariance function are identical to those of $\{W(t), t \geq 0\}$, we can state that it is a Brownian motion.

5.7 Ergodicity

To estimate the mean $m_X(t)$ of a stochastic process $\{X(t), t \in T\}$ at time t, we must take a large number of *observations* $X(t, s_k)$ of the process. We write:

$$
m_X(t) \simeq \frac{1}{n} \sum_{k=1}^{n} X(t, s_k). \tag{5.55}
$$

That is, we estimate the mean of the sp at time t by the mean of a *random sample* taken at this time instant. Suppose, however, that we only have a *single* observation, $X(t, s)$, of $X(t)$. Can we then use the *temporal mean*

$$
\langle X(t) \rangle_S := \frac{1}{2S} \int_{-S}^{S} X(t, s)\, dt \tag{5.56}
$$

to estimate $m_X(t)$? A *necessary* condition is that $m_X(t)$ be a *constant*.

Remark. In (5.56), we can also integrate over the interval $[0, 2S]$ (for example if $T = [0, \infty]$).

Definition 5.7.1. *A stochastic process* $\{X(t), t \in T\}$ *is said to be* **ergodic** *if, with probability 1, every characteristic of* $\{X(t), t \in T\}$ *can be obtained from a single realization* $X(t, s)$ *of* $X(t)$.

There exist many types of ergodicity. Here, we will limit ourselves to the case when the sp $\{X(t), t \in T\}$ is *mean ergodic*.

Definition 5.7.2. *A stochastic process* $\{X(t), t \in T\}$, *whose mean is a constant m, is said to be* **mean ergodic** *if, with probability 1, we have:*

$$
\lim_{S \to \infty} \langle X(t) \rangle_S = m. \tag{5.57}
$$

Now, $\langle X(t) \rangle_S$ is a random variable whose mean is given by

$$
E[\langle X(t) \rangle_S] = \frac{1}{2S} \int_{-S}^{S} E[X(t, s)]\, dt = \frac{1}{2S} \int_{-S}^{S} m\, dt = m. \tag{5.58}
$$

Hence, we deduce that the sp $\{X(t), t \in T\}$, for which $m_X(t) \equiv m$, is mean ergodic if and only if the variance of the random variable $\langle X(t) \rangle_S$ decreases to 0 as S tends to infinity.

Remark. For example, if X has a Gaussian $N(\mu, \sigma^2)$ distribution, then the mean $\bar{X}$ of a *random sample* of size n of X has a Gaussian $N(\mu, \sigma^2/n)$ distribution. Since $\text{VAR}[\bar{X}] \downarrow 0$ as n tends to infinity, we can conclude that $\lim_{n \to \infty} \bar{X} = \mu$.

We can show the following proposition.

Proposition 5.7.1. *Let* $\{X(t), t \in T\}$ *be a WSS process. Then,* $\{X(t), t \in T\}$ *is mean ergodic if and only if*

$$\text{VAR}[\langle X(t) \rangle_S] = \frac{1}{2S} \int_{-2S}^{2S} C_X(\tau)\left[1 - \frac{|\tau|}{2S}\right] d\tau \downarrow 0 \quad \text{as } S \to \infty. \tag{5.59}$$

Remarks. i) In the general case when the stochastic process $\{X(t), t \in T\}$ is not necessarily WSS, the criterion (5.59) becomes

$$\lim_{S \to \infty} \text{VAR}[\langle X(t) \rangle_S] = \lim_{S \to \infty} \frac{1}{4S^2} \int_{-S}^{S} \int_{-S}^{S} C_X(t_1, t_2)\, dt_1 dt_2 = 0. \tag{5.60}$$

ii) There exist other (mean) ergodicity criteria. For instance, a WSS process $\{X(t), t \in T\}$ is mean ergodic if

$$C_X(0) < \infty \quad \text{and} \quad \lim_{|\tau| \to \infty} C_X(\tau) = 0. \tag{5.61}$$

Example 5.7.1 Consider once again the random telegraph signal, denoted by $\{Y(t), t \geq 0\}$ in Section 5.4. We already saw that $R_Y(\tau) = e^{-2\lambda|\tau|}$. Since $\{Y(t), t \geq 0\}$ is a process with zero mean, we can write that

$$C_Y(\tau) = e^{-2\lambda|\tau|}.$$

It follows that

$$\text{VAR}[\langle Y(t) \rangle_S] = \frac{2}{2S} \int_0^{2S} e^{-2\lambda\tau}\left[1 - \frac{\tau}{2S}\right] d\tau.$$

Given that the expression between the square brackets is bounded by 0 and 1 (when τ varies from 0 to $2S$), we can replace this term by 1, and we then find that

$$\text{VAR}[\langle Y(t) \rangle_S] < \frac{1 - e^{-4\lambda S}}{2\lambda S}.$$

It is now easy to see that the variance of $\langle Y(t) \rangle_S$ decreases to 0 as S tends to infinity. Thus, by the previous proposition, the random telegraph signal is *mean ergodic*.

Finally, we could have used the criterion (5.61) above. Indeed, we obtain:

$$C_X(0) = 1 < \infty \quad \text{and} \quad \lim_{|\tau| \to \infty} C_X(\tau) = \lim_{|\tau| \to \infty} e^{-2\lambda|\tau|} = 0.$$

Note, however, that this criterion is a *sufficient*, but not a necessary condition.

5.8 Exercises, Problems, and Multiple Choice Questions

Solved Exercises

Exercise no. 1 (5.3)

We suppose that the probability that a certain machine functions without failure today is equal to

0.7 if the machine functioned without failure yesterday and the day before yesterday (state 0);

0.5 if the machine functioned without failure yesterday, but not the day before yesterday (state 1);

0.4 if the machine functioned without failure the day before yesterday, but not yesterday (state 2);

0.2 if the machine did not function without failure, neither yesterday nor the day before yesterday (state 3).

a) Find the one-step transition probability matrix of the Markov chain associated with the functioning state of the machine.

b) Calculate $p_{0,1}^2$, that is, the probability of moving from state 0 to state 1 in two steps.

c) Calculate the average number of days without failure of the machine over the next two days, given that the Markov chain is presently in state 0.

Solution

a) We have, for example:

$$p_{0,2} = P[\text{At least one failure today} \mid \text{state } 0] = 1 - 0.7 = 0.3,$$

etc. We find that

$$P = \begin{bmatrix} 0.7 & 0 & 0.3 & 0 \\ 0.5 & 0 & 0.5 & 0 \\ 0 & 0.4 & 0 & 0.6 \\ 0 & 0.2 & 0 & 0.8 \end{bmatrix}.$$

b) We seek

$$p_{0,1}^2 = p_{0,0}\,p_{0,1} + p_{0,1}\,p_{1,1} + p_{0,2}\,p_{2,1} + p_{0,3}\,p_{3,1} = 0 + 0 + (0.3)(0.4) + 0 = 0.12.$$

c) Let N be the number of days without failure over the two-day period. We have:

$$N = \begin{cases} 2 & \text{if the Markov chain moves from 0 to 0 to 0,} \\ 1 & \text{if the Markov chain moves from 0 to 2 to 1 or from 0 to 0 to 2,} \\ 0 & \text{if the Markov chain moves from 0 to 2 to 3.} \end{cases}$$

It follows that

$$E[N] = 2 \times p_{0,0}p_{0,0} + 1 \times (p_{0,2}p_{2,1} + p_{0,0}p_{0,2}) + 0 \times p_{0,2}p_{2,3}$$
$$= 2(0.7)^2 + (0.3)(0.4) + (0.7)(0.3) + 0 = 1.31.$$

Exercise no. 2 (5.4)

Let $N(t)$ be the number of failures of a computer system in the interval $[0, t]$. We suppose that $\{N(t), t \geq 0\}$ is a Poisson process with rate $\lambda = 1$ per week.

a) Calculate the probability that

   i) the system functions without failure during two consecutive weeks;

   ii) the system has exactly two failures during a given week, knowing that it has functioned without failure during the previous two weeks;

   iii) less than two weeks elapse before the third failure occurs.

b) Let

$$Z(t) := e^{-N(t)} \quad \text{for } t \geq 0.$$

Is the stochastic process $\{Z(t), t \geq 0\}$ wide-sense stationary? Justify your answer.

Hint. We have: $E[e^{-sX}] = \exp[\alpha(e^{-s} - 1)]$, if $X \sim \text{Poi}(\alpha)$.

Solution

a) i) We seek $P[N(2) = 0] = P[\text{Poi}(2) = 0] = e^{-2} \simeq 0.1353$.

   ii) Since the increments of a Poisson process are independent, we look for

$$P[N(1) = 2] = P[\text{Poi}(1) = 2] = e^{-1}\frac{1^2}{2!} \simeq 0.1839.$$

   iii) Let S_3 be the time when the third failure occurs. We have: $S_3 < 2 \Leftrightarrow N(2^-) \geq 3$. We then calculate

$$P[S_3 < 2] = P[N(2) \geq 3] = 1 - P[N(2) \leq 2] = 1 - P[\text{Poi}(2) \leq 2]$$
$$= 1 - \left[e^{-2} + e^{-2}2 + e^{-2}\frac{2^2}{2!}\right] = 1 - 5e^{-2} \simeq 0.3233.$$

Or: $S_3 \sim G(\alpha = 3, \lambda = 1)$, so that

$$P[S_3 < 2] = \int_0^2 \frac{x^2 e^{-x}}{2}\,dx = -\frac{x^2}{2}e^{-x}\Big|_0^2 + \int_0^2 xe^{-x}\,dx$$
$$= -2e^{-2} - xe^{-x}\Big|_0^2 + \int_0^2 e^{-x}\,dx = -4e^{-2} - e^{-x}\Big|_0^2 = 1 - 5e^{-2}.$$

b) We have: $N(t) \sim \text{Poi}(t)$, so that $E[Z(t)] = \exp[t(e^{-1} - 1)]$. Since $E[Z(t)]$ is not a constant, the stochastic process $\{Z(t), t \geq 0\}$ is not WSS.

Unsolved Problems

Problem no. 1

Let $\{N(t), t \geq 0\}$ be a Poisson process with rate λ. We define the stochastic process $\{X(t), 0 \leq t \leq 1\}$ by

$$X(t) = N(t) - tN(1) \quad \text{for } 0 \leq t \leq 1.$$

a) Calculate the mean of $X(t)$.

b) Calculate the autocovariance function of the process $\{X(t), 0 \leq t \leq 1\}$.

Hint. We have: $R_N(t_1, t_2) = \lambda^2 t_1 t_2 + \lambda \min\{t_1, t_2\}$.

c) Is the process $\{X(t), 0 \leq t \leq 1\}$ WSS? Justify.

Problem no. 2

Let $\{W(t), t \geq 0\}$ be a standard Brownian motion. We define

$$X(t) = W^2(t) \quad \text{for } t \geq 0,$$

a) Is the stochastic process $\{X(t), t \geq 0\}$ a Wiener process? Justify.

b) Is $\{X(t), t \geq 0\}$
   i) WSS?
   ii) mean ergodic? Justify your answers.

Problem no. 3

The stochastic process $\{X(t), t > 0\}$ is defined by

$$X(t) = e^{-Yt} \quad \text{for } t > 0.$$

We suppose that Y has a uniform distribution on the interval $(0, 1)$.

a) Calculate $E[X(t)]$.

b) Calculate $C_X(t, t+s)$, where $s, t > 0$.

c) Is the stochastic process $\{X(t), t > 0\}$ WSS? Justify your answer.

d) Find the first-order density function of the process $\{X(t), t > 0\}$.

Problem no. 4

We suppose that customers arrive at a counter in a bank according to a Poisson process with rate $\lambda = 10$ per hour. Let $N(t)$ be the number of customers in the interval $[0, t]$.

a) What is the probability that no customers arrive during a 15-minute period?

b) Knowing that eight customers arrived during a given hour, what is the probability that at most two customers arrive during the following hour?

c) Knowing that a customer arrived during a 15-minute period, what is the probability that he arrived during the first five minutes of the period considered?

d) Let $X(t) := \frac{1}{t}N(t)$. Is the stochastic process $\{X(t), t > 0\}$ WSS? Justify your answer.

Hint. We have: $R_N(t_1, t_2) = 10 \min\{t_1, t_2\} + 100 t_1 t_2$.

Problem no. 5

Let $\{W(t), t \geq 0\}$ be a standard Brownian motion. We define the stochastic process $\{X(t), t \geq 0\}$ by

$$X(t) = e^{-\alpha t} W(e^{2\alpha t}) \quad \text{for } t \geq 0,$$

where α is a positive constant.

a) Calculate the mean of $X(t)$.

b) Calculate the autocovariance function of the process $\{X(t), t \geq 0\}$.

Hint. We have: $C_W(t, \tau) = \min\{t, \tau\}$.

c) Is the stochastic process $\{X(t), t \geq 0\}$ i) a Brownian motion? ii) stationary? iii) mean ergodic? Justify your answers.

Problem no. 6

Let $\{N(t), t \geq 0\}$ be a Poisson process with rate $\lambda > 0$. We define the stochastic process $\{X(t), t \geq 0\}$ by

$$X(t) = \frac{N(t + \delta^2) - N(t)}{\delta} \quad \text{for } t \geq 0,$$

where $\delta > 0$ is a constant.

a) Is the process $\{X(t), t \geq 0\}$ a Poisson process? Justify.

b) Calculate the mean of $X(t)$.

c) Calculate the autocovariance function of the process $\{X(t), t \geq 0\}$ for $t_1 = 1$, $t_2 = 2$ and $\delta = 1$.

Hint. We have: $R_N(t_1, t_2) = \lambda \min\{t_1, t_2\} + \lambda^2 t_1 t_2$.

d) We consider the process $\{Y(t), t \geq 0\}$ defined from $\{X(t), t \geq 0\}$ by taking the limit as δ decreases toward zero. We can show that $E[Y(t)] \equiv 0$ and that

$$C_Y(t_1, t_2) = \begin{cases} \lambda & \text{if } t_1 - t_2 = 0, \\ 0 & \text{otherwise.} \end{cases}$$

Is the process $\{Y(t), t \geq 0\}$ i) WSS? ii) mean ergodic? Justify your answers.

e) Let $Z_n := N(n)$ for $n = 0, 1, 2, \ldots$.

   i) The stochastic process $\{Z_n, n = 0, 1, \ldots\}$ is a Markov chain. Justify this assertion.

   ii) Calculate $p_{i,j}$ for $i, j \in \{0, 1, 2, \ldots\}$.

Problem no. 7

Let $\{W(t), t \geq 0\}$ be a standard Brownian motion. We define the stochastic process $\{X(t), t \geq 0\}$ by $X(t) = W(t) + W(t^2)$ for $t \geq 0$.

a) Calculate the mean of $X(t)$.

b) Calculate $\text{COV}[X(t), X(t + \tau)]$ for $\tau \geq 0$.

Hint. We have: $\text{COV}[W(t_1), W(t_2)] = \min\{t_1, t_2\}$.

c) Is the stochastic process $\{X(t), t \geq 0\}$ i) Gaussian? ii) stationary? iii) a Brownian motion? Justify your answers.

d) Calculate the correlation coefficient of $W(t)$ and $W(t^2)$ for $t > 0$.

Problem no. 8

Let $\{X_n, n = 0, 1, \ldots\}$ be a Markov chain whose state space is the set $\{0, 1\}$ and whose one-step transition probability matrix $\mathbf{P}$ is given by

$$\mathbf{P} = \begin{bmatrix} 1/2 & 1/2 \\ p & 1-p \end{bmatrix},$$

where $0 \leq p \leq 1$.

a) Suppose that $p = 1$ and that $X_0 = 0$. Calculate $E[X_2]$.

b) Suppose that $p = 1/2$ and that $P[X_0 = 0] = P[X_0 = 1] = 1/2$. We define the continuous-time stochastic process $\{Y(t), t \geq 0\}$ by $Y(t) = t X_{[t]}$ for $t \geq 0$, where $[t]$ denotes the integer part of t.

i) Calculate $C_Y(t, t+1)$.

ii) Is the stochastic process $\{Y(t), t \geq 0\}$ WSS? Justify your answer.

iii) Calculate $\lim_{n \to \infty} P[X_n = 0]$.

Multiple Choice Questions

Question no. 1

A man plays independent repetitions of the following game: at each repetition, he throws a dart onto a circular target. Suppose that the distance D (in centimeters) between the impact point of the dart and the center of the target has a U[0, 30] distribution. If $D \leq 5$, the player wins $1; if $5 < D < 25$, the player neither wins nor loses anything; if $D \geq 25$, the player loses $1. The player's initial fortune is equal to $1 and he will stop playing when either he is ruined or his fortune reaches $3. Let X_n be the fortune of the player after n repetitions. Then, the stochastic process $\{X_n, n = 0, 1, \ldots\}$ is a Markov chain.

A) Find the one-step transition probability matrix of the chain.

a)
$$\begin{bmatrix} 1/6 & 2/3 & 1/6 & 0 \\ 1/6 & 2/3 & 1/6 & 0 \\ 0 & 1/6 & 2/3 & 1/6 \\ 0 & 1/6 & 2/3 & 1/6 \end{bmatrix}$$
b)
$$\begin{bmatrix} 1 & 0 & 0 & 0 \\ 1/3 & 1/3 & 1/3 & 0 \\ 0 & 1/3 & 1/3 & 1/3 \\ 0 & 0 & 0 & 1 \end{bmatrix}$$
c)
$$\begin{bmatrix} 1/3 & 1/3 & 1/3 & 0 \\ 1/3 & 1/3 & 1/3 & 0 \\ 0 & 1/3 & 1/3 & 1/3 \\ 0 & 1/3 & 1/3 & 1/3 \end{bmatrix}$$

d)
$$\begin{bmatrix} 1 & 0 & 0 & 0 \\ 1/6 & 2/3 & 1/6 & 0 \\ 0 & 1/6 & 2/3 & 1/6 \\ 0 & 0 & 0 & 1 \end{bmatrix}$$
e)
$$\begin{bmatrix} 0 & 1 & 0 & 0 \\ 1/6 & 2/3 & 1/6 & 0 \\ 0 & 1/6 & 2/3 & 1/6 \\ 0 & 0 & 1 & 0 \end{bmatrix}$$

B) Calculate $E[X_2^2]$.

a) 1/3   b) 1   c) 11/18   d) 4/3   e) 29/18

Suppose now that the man never stops playing, so that the state space of the Markov chain is the set $\{0, \pm 1, \pm 2, \ldots\}$. Suppose also that the duration T (in seconds) of a repetition of the game has an exponential distribution with mean 30. Then, the stochastic process $\{N(t), t \geq 0\}$, where $N(t)$ denotes the number of repetitions played in the interval $[0, t]$, is a Poisson process with rate $\lambda = 2$ per minute.

C) Calculate the probability that the player has completed at least three repetitions in less than two minutes.

a) $5e^{-4}$ b) $13e^{-4}$ c) $1 - 5e^{-4}$ d) $1 - 13e^{-4}$ e) $1 - \frac{47}{3}e^{-4}$

D) Calculate (approximately) the probability $P[N(25) \leq 50]$.

a) $\simeq 0$ b) $1/4$ c) $1/2$ d) $3/4$ e) $\simeq 1$

Question no. 2

Let $\{X_n, n = 1, 2, \ldots\}$ be a Bernoulli process. That is, the random variables $X_1, X_2, \ldots$ are independent and all have a Bernoulli distribution with parameter p. Calculate the autocorrelation function $R_X(n, m)$ of the process, if $p = 1/2$ and $n, m \in \{1, 2, \ldots\}$.

a) $1/4$ for all n, m b) $1/2$ for all n, m
c) $1/4$ if $n \neq m$; $1/2$ if $n = m$ d) $1/2$ if $n \neq m$; $1/4$ if $n = m$
e) none of these answers

Question no. 3

The one-step transition probability matrix $\mathbf{P}$ of a Markov chain whose state space is $\{0, 1\}$ is given by

$$\mathbf{P} = \begin{bmatrix} 1/2 & 1/2 \\ 0 & 1 \end{bmatrix}.$$

Calculate $E[X_2]$, if $P[X_0 = 0] = 1/3$.

a) $1/12$ b) $1/4$ c) $1/3$ d) $1/2$ e) $11/12$

Question no. 4

Let $N(t)$ be the number of telephone calls received at an exchange in the interval $[0, t]$. We suppose that $\{N(t), t \geq 0\}$ is a Poisson process with rate $\lambda = 10$ per hour. Calculate the probability that no calls are received during each of two consecutive 15-minute periods.

a) e^{-5} b) $e^{-2.5}$ c) $2e^{-2.5}$ d) $e^{-0.25}$ e) $2e^{-0.25}$

Question no. 5

Calculate the variance of $W(t) - 2W(\tau)$ for $t \leq \tau$, where $\{W(t), t \geq 0\}$ is a standard Brownian motion.

a) $t + 2\tau$ b) $t + 4\tau$ c) $3t + 4\tau$ d) $5t + 4\tau$ e) $9t + 2\tau$

Question no. 6

Let $\{X(t), t \geq 0\}$ be the stochastic process defined by

$$X(t) = tU + 1,$$

where U is a random variable having a U(0, 1) distribution. Find the first-order density function of the process.

a) 1 if $0 < x < 1$ b) 1 if $0 < t < 1$ c) $\frac{1}{x}$ if $0 < t < x$ d) $\frac{1}{t}$ if $0 < x < t$

e) $\frac{1}{t}$ if $1 < x < t + 1$

Question no. 7

The customers of a newspaper salesman arrive according to a Poisson process with rate $\lambda = 2$ per minute. Calculate the probability that at least one customer arrives in the interval $(t_0, t_0 + 2]$, given that there has been exactly one customer in the interval $(t_0 - 1, t_0 + 1]$.

a) $e^{-2}/2$ b) e^{-2} c) $(1 - e^{-2})/2$ d) $1 - e^{-2}$ e) $1 - (e^{-2}/2)$

Question no. 8

Let $\{W(t), t \geq 0\}$ be a standard Brownian motion. We define $X(t) = |W(t)|$ for $t \geq 0$. Is the stochastic process $\{X(t), t \geq 0\}$ Gaussian and stationary?

a) Gaussian and SSS   b) Gaussian, but not stationary
c) not Gaussian, but SSS   d) not Gaussian and not stationary
e) not Gaussian, but WSS (only)

Question no. 9

Let $\{X(t), t \geq 0\}$ be a stochastic process whose autocorrelation function is $R_X(t_1, t_2) = e^{-|t_1 - t_2|} + 1$ and autocovariance function is $C_X(t_1, t_2) = e^{-|t_1 - t_2|}$. Is the stochastic process $\{X(t), t \geq 0\}$ WSS and mean ergodic?

a) WSS and ergodic   b) not stationary, but ergodic
c) WSS, but not ergodic   d) not stationary and not ergodic
e) we cannot conclude

Question no. 10

Let $\{X(t), t \geq 0\}$ be the stochastic process defined by

$$X(t) = W(t + 1) - W(1) \quad \text{for } t \geq 0,$$

where $\{W(t), t \geq 0\}$ is a standard Brownian motion.

A) Calculate the autocovariance function of the process $\{X(t), t \geq 0\}$.

a) $\min\{t_1, t_2\} - 1$ b) $\min\{t_1, t_2\}$ c) $\min\{t_1, t_2\} + 1$ d) $2\min\{t_1, t_2\}$
e) none of these answers

B) Is the process $\{X(t), t \geq 0\}$ Gaussian and a standard Brownian motion?

a) Gaussian and a standard Brownian motion
b) Gaussian and a non-standard Brownian motion
c) Gaussian, but not a Brownian motion
d) not Gaussian and not a standard Brownian motion
e) not Gaussian and not a non-standard Brownian motion

C) Is the process $\{X(t), t \geq 0\}$ stationary and mean ergodic?

a) WSS only, and ergodic   b) SSS and ergodic
c) SSS and not ergodic   d) not stationary, but ergodic
e) not stationary and not ergodic

Question no. 11

Let $\{Y_k, k = 1, 2, \ldots\}$ be a set of independent random variables having a Bernoulli distribution with parameter $p = 1/3$. We define $X_n = \sum_{k=1}^{n} Y_k$ for $n = 1, 2, \ldots$. Then, the stochastic process $\{X_n, n = 1, 2, \ldots\}$ is a Markov chain. Calculate $p_{0,2}^3$.

a) 1/27   b) 2/27   c) 1/9   d) 2/9   e) 4/9

Question no. 12

We consider a standard Brownian motion $\{W(t), t \geq 0\}$. Calculate

$$\text{VAR}[W(4) - 2W(1)].$$

a) 0   b) 2   c) 4   d) 6   e) 8

Question no. 13

Let $\{X(t), t \geq 0\}$ be a WSS stochastic process, with zero mean and with an autocorrelation function given by $R_X(\tau) = e^{-|\tau|}$. Let

$$Y(t) := t X^2(1/t) \quad \text{for } t > 0.$$

Is the stochastic process $\{Y(t), t \geq 0\}$ WSS and mean ergodic?

a) not WSS and not ergodic   b) not WSS, but ergodic
c) WSS, but not ergodic   d) WSS and ergodic
e) not WSS and, consequently, we cannot conclude about the ergodicity

Question no. 14

Let $\{X_n, n = 0, 1, \ldots\}$ be a random walk for which

$$p_{i,i+1} = \frac{2}{3} \quad \text{and} \quad p_{i,i-1} = \frac{1}{3} \quad \text{for } i \in \{0, \pm 1, \pm 2, \ldots\}.$$

Calculate $E[X_2 \mid X_0 = 0]$.

a) 0   b) 1/9   c) 1/3   d) 4/9   e) 2/3

Question no. 15

We consider a stochastic process $\{X(t), t \geq 0\}$ for which

$$C_X(t_1, t_2) = \frac{1}{|t_2 - t_1| + 1} \quad \text{and} \quad R_X(t_1, t_2) = \frac{1}{|t_2 - t_1| + 1} + 4.$$

Is the process stationary and mean ergodic?

a) not WSS and not ergodic   b) not WSS, but ergodic
c) SSS, but not ergodic   d) WSS and ergodic
e) WSS (only), but not ergodic

Question no. 16

We define the stochastic process $\{X(t), t > 0\}$ by $X(t) = t/Y$ for $t > 0$, where Y has a uniform distribution on the interval $(0, 2)$. Find $f(x; t)$ for $x > t/2$.

a) $\dfrac{t}{x}$ b) $\dfrac{t}{2x}$ c) $\dfrac{t}{x^2}$ d) $\dfrac{t}{2x^2}$ e) $\dfrac{1}{2}$

Question no. 17

The stochastic process $\{X(t), t \geq 0\}$ is defined by

$$X(t) = N(t + 1) - N(1) \quad \text{for } t \geq 0,$$

where $\{N(t), t \geq 0\}$ is a Poisson process with rate $\lambda > 0$. Calculate $C_X(s, t)$ for $0 \leq s \leq t$.

Hint. We have: $C_N(s, t) = \lambda s$, if $0 \leq s \leq t$.

a) $\lambda(s - 1)$ b) λs c) $\lambda(t - s)$ d) λst e) λt

Question no. 18

Let $\{W(t), t \geq 0\}$ be a standard Brownian motion. We define

$$X(t) = -W(t) \quad \text{for } t \geq 0.$$

Is the stochastic process $\{X(t), t \geq 0\}$ Gaussian? Is it a Brownian motion?

a) not Gaussian and not a Brownian motion
b) not Gaussian, but a Brownian motion
c) Gaussian, but not a Brownian motion
d) Gaussian and a standard Brownian motion
e) Gaussian and a non-standard Brownian motion

Question no. 19

Let $\{X_n, n = 0, 1, 2, \dots\}$ be a Bernoulli process. That is, the random variables X_n are independent and all have a Bernoulli distribution with parameter p. Calculate the particular case $p(0, 1; n_1 = 0, n_2 = 1)$ of the second-order probability mass function of the process.

a) 0 b) p^2 c) $p(1 - p)$ d) $2p(1 - p)$ e) $(1 - p)^2$

Question no. 20

We define

$$Y(t) = \begin{cases} 0 \text{ if } N(t) \text{ is odd,} \\ 1 \text{ if } N(t) \text{ is even,} \end{cases}$$

where $\{N(t), t \geq 0\}$ is a Poisson process with rate $\lambda = 1$. We can show that $P[Y(t) = 1] = (1 + e^{-2t})/2$ for $t \geq 0$. Next, let $X_n := Y(n)$ for $n = 0, 1, \dots$. Then, $\{X_n, n = 0, 1, \dots\}$ is a Markov chain. Calculate its one-step transition probability matrix.

a) $\begin{bmatrix} 1/2 & 1/2 \\ 1/2 & 1/2 \end{bmatrix}$ b) $\begin{bmatrix} 1 & 0 \\ 0 & 1 \end{bmatrix}$ c) $\begin{bmatrix} 0 & 1 \\ 1 & 0 \end{bmatrix}$ d) $\begin{bmatrix} \frac{1+e^{-2}}{2} & \frac{1-e^{-2}}{2} \\ \frac{1-e^{-2}}{2} & \frac{1+e^{-2}}{2} \end{bmatrix}$ e) $\begin{bmatrix} \frac{1-e^{-2}}{2} & \frac{1+e^{-2}}{2} \\ \frac{1+e^{-2}}{2} & \frac{1-e^{-2}}{2} \end{bmatrix}$

Question no. 21

Let $\{X(t), t \geq 0\}$ be a Gaussian process such that $X(0) = 0$, $E[X(t)] = \mu t$ for $t > 0$, where $\mu \neq 0$, and

$$R_X(t, t + \tau) = 2t + \mu^2 t(t + \tau) \quad \text{for } t, \tau \geq 0.$$

Is the stochastic process $\{Y(t), t \geq 0\}$, where $Y(t) := X(t) - \mu t$, a Brownian motion? Is it stationary?

a) not a Brownian motion and not stationary
b) not a Brownian motion, but stationary
c) a standard Brownian motion, but not stationary
d) a non-standard Brownian motion, but not stationary
e) a non-standard Brownian motion and stationary

Question no. 22

Suppose that $\{X(t), t \geq 0\}$ is a Gaussian process, with zero mean, whose auto-covariance function is given by

$$C_X(t, t + \tau) = e^{-t} \quad \text{for } t, \tau \geq 0.$$

We define $Y(t) = X^2(t)$. Is the stochastic process $\{Y(t), t \geq 0\}$ stationary? Is it mean ergodic?

a) not stationary and not ergodic   b) not stationary, but ergodic
c) WSS only, but not ergodic   d) SSS, but not ergodic
e) WSS only, and ergodic

Question no. 23

Let $\{X_n, n = 0, 1, \ldots\}$ be a Markov chain whose state space is the set $\{0, 1\}$ and whose one-step transition probability matrix is given by

$$P = \begin{bmatrix} 0 & 1 \\ 1 & 0 \end{bmatrix}.$$

A) Calculate $C_X(t_1 = 0, t_2 = 1)$, if $P[X_0 = 0] = P[X_0 = 1] = 1/2$.
a) $-1/4$ b) $-1/8$ c) 0   d) 1/4   e) 1/2
B) Calculate $\lim_{n \to \infty} P[X_n = 0 \mid X_0 = 0]$.
a) 0   b) 1/2   c) 1   d) does not exist   e) none of these answers

Question no. 24

The failures of a certain machine happen according to a Poisson process with rate $\lambda = 1$ per week.

A) What is the probability that the machine has at least one failure during each of the first two weeks considered?
a) e^{-2} b) $1 - 2e^{-1}$ c) $(1 - e^{-1})^2$ d) $2e^{-1}$ e) $1 - e^{-2}$
B) Suppose that exactly five failures have occurred during the first four weeks considered. Let M be the number of failures during the fourth of these four weeks. Calculate $E[M \mid M > 0]$.
a) 1.24   b) 1.34   c) 1.44   d) 1.54   e) 1.64

Question no. 25

A) Let $X_1, X_2, \ldots$ be an infinite sequence of independent random variables having a Poisson distribution with parameter $\alpha = 1$. We define

$$Y_n = \sum_{k=1}^{n} X_k \quad \text{for } n = 1, 2, \ldots .$$

Then, $\{Y_n, n = 1, 2, \ldots\}$ is a Markov chain. Calculate $p_{1,3}^4$, that is, the probability of moving from state 1 to state 3 in four steps.

a) 0.1065   b) 0.1165   c) 0.1265   d) 0.1365   e) 0.1465

B) Let $\{N(t), t \geq 0\}$ be a Poisson process with rate $\lambda > 0$. We define

$$M_1(t) = N(\sqrt{t}), \quad M_2(t) = N(2t), \quad M_3(t) = N(t + 2) - N(2).$$

Which of the $\{M_k(t), t \geq 0\}$ stochastic processes is (are) also a Poisson process?

a) $M_2(t)$ only   b) $M_3(t)$ only   c) $M_2(t)$ and $M_3(t)$ only   d) none   e) all

Question no. 26

The power failures in a certain region occur according to a Poisson process with rate $\lambda_1 = 1/5$ per week. Moreover, the duration X (in hours) of a given power failure has an exponential distribution with parameter $\lambda_2 = 1/2$. Finally, we assume that the durations of the various power failures are independent random variables.

A) What is the probability that the longest, among the first three power failures observed, lasts more than four hours?

a) 0.3435   b) 0.3535   c) 0.3635   d) 0.3735   e) 0.3835

B) Suppose that there has been exactly one power failure during the first week considered. What is the probability that the failure had still not been repaired at the end of the week in question?

a) 0.0019   b) 0.0119   c) 0.0219   d) 0.0319   e) 0.0419

Question no. 27

The flow of a certain river can be in one of the following three states:

0: low flow,
1: average flow,
2: high flow.

We suppose that the stochastic process $\{X_n, n = 0, 1, \ldots\}$, where X_n represents the state of the river flow on the nth day, is a Markov chain. Furthermore, we estimate that the probability that the flow moves from state i to state j in one day is given by the formula

$$p_{i,j} = \frac{1}{2} - |i - j|\pi_i,$$

where $0 < \pi_i < 1$, for $i, j = 0, 1, 2$.

A) Calculate the probability that the river flow moves from state 0 to state 1 in one day.

a) 1/6   b) 1/4   c) 1/3   d) 5/12   e) 1/2

B) Calculate the probability that the river flow moves from state 0 to state 2 in two days.

a) 1/6   b) 1/4   c) 1/3   d) 5/12   e) 1/2

Question no. 28

A machine is made up of two components that operate independently. The lifetime T_i (in days) of component i has an exponential distribution with parameter λ_i, for $i = 1, 2$.

A) Suppose that the two components are placed in parallel, and that $\lambda_1 = \lambda_2 = \ln 2$. When the machine breaks down, the two components are replaced by new ones at the beginning of the following day. Let X_n be the number of components that function at the end of n days. Then, the stochastic process $\{X_n, n = 0, 1, \ldots\}$ is a Markov chain. Calculate its one-step transition probability matrix.

a) $\begin{bmatrix} 1/4 & 1/4 & 1/2 \\ 1/2 & 0 & 1/2 \\ 1/4 & 1/4 & 1/2 \end{bmatrix}$ b) $\begin{bmatrix} 1/4 & 1/4 & 1/2 \\ 1/2 & 1/2 & 0 \\ 1/4 & 1/4 & 1/2 \end{bmatrix}$ c) $\begin{bmatrix} 1/2 & 1/4 & 1/4 \\ 1/2 & 1/2 & 0 \\ 1/2 & 1/4 & 1/4 \end{bmatrix}$

d) $\begin{bmatrix} 1/4 & 1/4 & 1/2 \\ 0 & 1/2 & 1/2 \\ 1/4 & 1/4 & 1/2 \end{bmatrix}$ e) $\begin{bmatrix} 1/4 & 1/2 & 1/4 \\ 1/2 & 1/2 & 0 \\ 1/4 & 1/2 & 1/4 \end{bmatrix}$

B) Suppose that the two components are placed in series and that as soon as a component fails, it is replaced by a new one. Let $N(t)$ be the number of replacements in the interval $[0, t]$. We can show that the stochastic process $\{N(t), t \geq 0\}$ is a Poisson process. Give its rate λ.

Hint. If T_0 is the time between two replacements, then we can write that $T_0 = \min\{T_1, T_2\}$.

a) $\frac{1}{2}(\lambda_1 + \lambda_2)$ b) $\lambda_1 + \lambda_2$ c) $\frac{1}{2}\lambda_1\lambda_2$ d) $\lambda_1\lambda_2$ e) $\min\{\lambda_1, \lambda_2\}$

Question no. 29

Let $\{X_n, n = 0, 1, \ldots\}$ be a Markov chain whose state space is the set $\{0, 1, 2, 3, 4\}$, and whose one-step transition probability matrix is

$$P = \begin{bmatrix} 1 & 0 & 0 & 0 & 0 \\ 0.5 & 0.2 & 0.3 & 0 & 0 \\ 0 & 0 & 0 & 1 & 0 \\ 0 & 0 & 0 & 0 & 1 \\ 0 & 0 & 1 & 0 & 0 \end{bmatrix}.$$

A) Calculate the probability that the process moves from state 1 to state 2 in four steps.

a) 0.2824   b) 0.2924   c) 0.3024   d) 0.3124   e) 0.3224

B) Suppose that $X_0 = 1$. Let N be the number of times that state 1 will be visited, including the initial state. Calculate $E[N]$.

a) 1.25   b) 1.5   c) 1.75   d) 2   e) 2.5

Question no. 30

We define the stochastic process $\{X(t), 0 \le t \le 1\}$ by

$$X(t) = N(t^2) - t^2 N(1) \quad \text{for } 0 \le t \le 1,$$

where $\{N(t), t \ge 0\}$ is a Poisson process with rate $\lambda > 0$.

A) Calculate the autocorrelation function $R_X(t_1, t_2)$ of $\{X(t), 0 \le t \le 1\}$ at $t_1 = 1/4$ and $t_2 = 1/2$.

Hint. We have: $R_N(t_1, t_2) = \lambda^2 t_1 t_2 + \lambda \min\{t_1, t_2\}$.

a) $\lambda/64$ b) $\lambda/32$ c) $3\lambda/64$ d) $\lambda/16$ e) $5\lambda/64$

B) Calculate $P[X(t) > 0 \mid N(1) = 1]$ for $0 < t < 1$.

a) $1/2$ b) t c) $1 - t$ d) t^2 e) $1 - t^2$

6

Estimation and Testing

6.1 Point Estimation

Definition 6.1.1. *A* **random sample** *of size n of a random variable X is a set $X_1, \ldots, X_n$ of independent random variables whose distribution is identical to that of X. That is, X_k has the same distribution function as X, for $k = 1, \ldots, n$. The random variable X is also called the* **population** *and each X_k is an* **observation** *of X.*

Definition 6.1.2. *A* **statistic** *is a function, that does not depend on any unknown parameter, of the observations in a random sample. The distribution of a statistic is called a* **sampling distribution**.

Example 6.1.1 The *sample mean*:

$$\bar{X} := \sum_{k=1}^{n} \frac{X_k}{n} \tag{6.1}$$

is a statistic. Similarly, the *sample variance*:

$$S^2 := \sum_{k=1}^{n} \frac{(X_k - \bar{X})^2}{n - 1} \tag{6.2}$$

is a statistic. However, the quantity

$$\sum_{k=1}^{n} \frac{(X_k - \mu)^2}{n - 1}$$

is a statistic only if $\mu := E[X]$ is known.

  If the population X is Gaussian, then the sampling distribution of $\bar{X}$ is also Gaussian. Finally, if X does not have a Gaussian distribution, but n is large enough, then the sampling distribution of $\bar{X}$ is approximately Gaussian, by the central limit theorem.

Definition 6.1.3. *A* **point estimate** *of an unknown parameter, θ, of a population is a real number that corresponds to this parameter.*

Definition 6.1.4. *An* **estimator** *of an unknown parameter, θ, of a population is a statistic $T = g(X_1, \ldots, X_n)$ that corresponds to this parameter.*

Remarks. i) An estimator T is a random variable, since it is a function of a random sample.

ii) The quantity θ can, actually, be a vector: $\theta = (\theta_1, \ldots, \theta_m)$.

Example 6.1.2 The sample mean, $\bar{X}$, is an estimator of the (unknown) population mean μ. Likewise, S^2 is an estimator of $\sigma^2 := \text{VAR}[X]$. Moreover, numerical values obtained with a *particular* random sample, that is, $\bar{x}$ and s^2, are point estimates of μ and σ^2, respectively.

Properties

Definition 6.1.5. *An estimator T of an unknown parameter θ is said to be* **unbiased** *if $E[T] = \theta$. Furthermore, the* **bias** *of T is defined by*

$$\text{Bias}[T] = E[T] - \theta. \tag{6.3}$$

Remark. The estimator T $(= T(n))$ is said to be *asymptotically unbiased* if

$$\lim_{n \to \infty} E[T] = \theta.$$

Example 6.1.3 The sample mean $\bar{X}$ is an unbiased estimator of μ, because

$$E[\bar{X}] = E\left[\sum_{k=1}^{n} \frac{X_k}{n}\right] = \frac{1}{n} \sum_{k=1}^{n} E[X_k] \stackrel{\text{i.d.}}{=} \frac{1}{n} \sum_{k=1}^{n} \mu = \frac{1}{n} n\mu = \mu.$$

Likewise, S^2 is an unbiased estimator of σ^2. Indeed, we can show that

$$S^2 := \sum_{k=1}^{n} \frac{(X_k - \bar{X})^2}{n-1} = \frac{1}{n-1}\left[\sum_{k=1}^{n} X_k^2 - n\bar{X}^2\right].$$

Moreover (see Chapter 4, p. 185),

$$\text{VAR}[\bar{X}] \stackrel{\text{ind.}}{=} \sum_{k=1}^{n} \frac{\text{VAR}[X_k]}{n^2} \stackrel{\text{i.d.}}{=} \frac{1}{n^2} n\sigma^2 = \frac{\sigma^2}{n}. \tag{6.4}$$

Then, making use of the formula

$$E[X^2] = \text{VAR}[X] + (E[X])^2,$$

we find that $E[S^2] = \sigma^2$. Indeed, we have:

$$E[S^2] = \frac{1}{n-1}\left\{n(\sigma^2 + \mu^2) - n\left(\frac{\sigma^2}{n} + \mu^2\right)\right\} = \frac{1}{n-1}(n\sigma^2 - \sigma^2) = \sigma^2.$$

Definition 6.1.6. *The* **mean square error** *of the estimator T of an unknown population parameter θ is given by*

$$\text{MSE}[T] = E[(T - \theta)^2]. \tag{6.5}$$

Remarks. i) Expanding the squared expression in the above definition, we find that

$$\text{MSE}[T] = \text{VAR}[T] + (\text{Bias}[T])^2. \tag{6.6}$$

Thus, if T is an unbiased estimator of θ, then $\text{MSE}[T] = \text{VAR}[T]$.

ii) Let T_1 and T_2 be two estimators of θ. We say that T_1 is *relatively more efficient* (or simply *better*) than T_2 if

$$\text{MSE}[T_1] < \text{MSE}[T_2]. \tag{6.7}$$

Example 6.1.4 Let $X_1, \ldots, X_n$ be a random sample of a Gaussian population X with mean μ and variance σ^2, where μ and σ^2 are unknown parameters. Then, making use of Equation (6.4), we can write that

$$\text{MSE}[\bar{X}] = \text{VAR}[\bar{X}] = \frac{\sigma^2}{n}. \tag{6.8}$$

Moreover, we can show that

$$\text{MSE}[S^2] = \frac{2}{n-1}\sigma^4. \tag{6.9}$$

Actually, we find that the (biased) estimator of σ^2 obtained by dividing by n rather than by $n - 1$ in (6.2) is relatively more efficient than S^2.

Definition 6.1.7. *Let T be an estimator of θ based on a random sample of size n. We say that T is a* **consistent** *estimator of θ if T converges in probability to θ, that is, if*

$$\lim_{n\to\infty} P[|T - \theta| < \epsilon] = 1 \quad \forall \epsilon > 0. \tag{6.10}$$

Example 6.1.5 The sample mean, $\bar{X}$, is a consistent estimator of μ, because, by Bienaymé–Chebyshev's inequality, we can write that

$$P[|\bar{X} - \mu| < \epsilon] \geq 1 - \frac{\text{VAR}[\bar{X}]}{\epsilon^2} = 1 - \frac{\sigma^2}{n\epsilon^2} \uparrow 1 \quad \text{as } n \to \infty \quad \forall \epsilon > 0.$$

Note that we can also use the weak law of large numbers to show that $\bar{X}$ is a consistent estimator of μ.

The Method of Maximum Likelihood

Definition 6.1.8. *Let $X_1, \ldots, X_n$ be a random sample of a population X whose probability density (or mass) function $f_X(x; \theta)$ depends on an unknown parameter θ. The* **likelihood function** *of the sample is*

$$L(\theta) = \prod_{k=1}^{n} f_X(X_k; \theta). \tag{6.11}$$

Definition 6.1.9. *The* **maximum likelihood estimator**, θ_{ML}, *of the parameter θ is the value of θ that maximizes the likelihood function.*

Remark. We can have: $\theta = (\theta_1, \ldots, \theta_m)$.

Example 6.1.6 a) Let $X_1, \ldots, X_n$ be a random sample of $X \sim \text{Exp}(\lambda)$. Then, we have:

$$L(\lambda) = \prod_{k=1}^{n} \lambda e^{-\lambda X_k} = \lambda^n e^{-\lambda n \bar{X}} \quad \text{if } X_k \geq 0 \quad \forall k.$$

Since λ maximizes $L(\lambda)$ if and only if λ maximizes $\ln L(\lambda)$, we can consider

$$\ln L(\lambda) = n \ln \lambda - \lambda n \bar{X}.$$

Finally, we find, by differentiating $\ln L(\lambda)$ (with respect to λ) and setting this derivative equal to zero, that

$$\lambda_{ML} = 1/\bar{X}.$$

We can check that it is a maximum. Indeed, we have:

$$\frac{d^2}{d\lambda^2} \ln L(\lambda) = -\frac{n}{\lambda^2} < 0.$$

b) If $X \sim \text{Poi}(\alpha)$, we have:

$$L(\alpha) = \prod_{k=1}^{n} \frac{e^{-\alpha} \alpha^{X_k}}{X_k!} = \frac{e^{-n\alpha} \alpha^{n\bar{X}}}{\prod_{k=1}^{n} X_k!} \quad \text{if } X_k \in \{0, 1, \ldots\} \quad \forall k$$

$$\Rightarrow \quad \ln L(\alpha) = -n\alpha + n\bar{X} \ln \alpha - \sum_{k=1}^{n} \ln X_k!$$

$$\Rightarrow \quad \frac{d}{d\alpha} \ln L(\alpha) = -n + \frac{n\bar{X}}{\alpha} = 0 \Leftrightarrow \alpha = \bar{X}.$$

Since

$$\frac{d^2}{d\alpha^2} \ln L(\alpha) = -\frac{n\bar{X}}{\alpha^2} < 0$$

(at least for n large enough), we can write that

$$\alpha_{ML} = \bar{X}.$$

Remarks. i) **Method of moments.** We set, in the case when $X \sim \text{Exp}(\lambda)$,

$$E[X] = \bar{X} \Leftrightarrow \frac{1}{\lambda} = \bar{X}.$$

It follows that the estimator of the parameter λ by the method of moments, λ_M, is also given by $1/\bar{X}$. Similarly, when $X \sim \text{Poi}(\alpha)$, proceeding as above, we find at once that $\alpha_M = \bar{X} \, (= \alpha_{ML})$.

In general, we set

$$E[X^m] = \sum_{k=1}^{n} \frac{X_k^m}{n} \quad \text{for } m = 1, 2, \ldots. \tag{6.12}$$

We use j (helpful) equations to estimate the j unknown parameters of the function $f_X(x; \theta)$. For example, if $X \sim \text{U}[-\theta, \theta]$, then we use the equation obtained with $m = 2$ (because the equation for $m = 1$ does not enable us to estimate θ):

$$E[X^2] = \sum_{k=1}^{n} \frac{X_k^2}{n}.$$

That is,

$$\frac{(2\theta)^2}{12} = \sum_{k=1}^{n} \frac{X_k^2}{n} \Rightarrow \theta_M = \left[3 \sum_{k=1}^{n} \frac{X_k^2}{n} \right]^{1/2}.$$

Actually, some authors instead use the equation

$$\text{VAR}[X] = S^2 \Leftrightarrow \text{VAR}[X] = \sum_{k=1}^{n} \frac{(X_k - \bar{X})^2}{n - 1}$$

to estimate the parameter θ, which yields a different estimator:

$$\hat{\theta} = \left[3 \sum_{k=1}^{n} \frac{(X_k - \bar{X})^2}{n - 1} \right]^{1/2}.$$

However, since $E[X] = 0$, if n is large enough, then $\bar{X}$ should be approximately equal to zero, so that the two estimators will produce almost equal point estimates.

ii) Let $X \sim \text{U}[0, \theta]$. Then,

$$L(\theta) = \frac{1}{\theta^n} \quad \text{if } X_k \in [0, \theta] \, \forall k.$$

We find that $\theta_{ML} = \max\{X_1, \ldots, X_n\}$, while $\theta_M = 2\bar{X}$. For example, if we have the following particular observations: 1, 2 and 9, then $\theta_{ML} = 9$ and $\theta_M = 2 \times 4 = 8$ (which is impossible).

6.2 Estimation by Confidence Intervals

Definition 6.2.1. *An interval* $[LC, UC]$ *is called a* **(two-sided) confidence interval** *at* $100(1 - \alpha)\%$ *for* θ *if*

$$P[LC(X_1, \ldots, X_n) \leq \theta \leq UC(X_1, \ldots, X_n)] = 1 - \alpha. \qquad (6.13)$$

Remarks. i) θ is a parameter, whereas LC and UC are random variables.

ii) We call LC and UC the *lower confidence* and *upper confidence limits*, respectively. Moreover, $1 - \alpha$ is called the *confidence coefficient*.

iii) *One-sided* confidence intervals with a *lower bound* and an *upper bound* are given respectively by $[LC, \infty)$ and $(-\infty, UC]$, where now

$$P[LC \leq \theta] = P[\theta \leq UC] = 1 - \alpha. \qquad (6.14)$$

a) Confidence interval for μ; σ known

Let $X_1, \ldots, X_n$ be a random sample of a Gaussian population X whose mean μ is unknown, but whose variance σ^2 is known. Then, we can write that

$$\bar{X} \sim N(\mu, \sigma^2/n) \Rightarrow \frac{\bar{X} - \mu}{\sigma/\sqrt{n}} \sim N(0, 1). \qquad (6.15)$$

It follows that

$$P\left[-z_{\alpha/2} \leq \frac{\bar{X} - \mu}{\sigma/\sqrt{n}} \leq z_{\alpha/2} \right] = 1 - \alpha, \qquad (6.16)$$

where $z_{\alpha/2}$ is defined by (see Fig. 6.1)

$$\Phi(z_{\alpha/2}) = 1 - \frac{\alpha}{2} \quad \text{or} \quad Q(z_{\alpha/2}) = \frac{\alpha}{2}. \qquad (6.17)$$

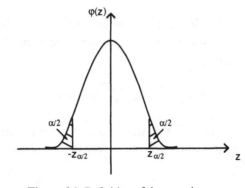

Figure 6.1. Definition of the quantity $z_{\alpha/2}$.

Table 6.1. Values of z_α for various values of α.

α	0.25	0.10	0.05	0.025	0.01	0.005	0.001	0.0005
z_α	0.674	1.282	1.645	1.960	2.326	2.576	3.090	3.291

That is, z_α is the $100(1-\alpha)$th percentile of the standard Gaussian distribution (if $100(1-\alpha)$ is an integer). We deduce from Equation (6.16) that

$$P\left[\bar{X} - z_{\alpha/2}(\sigma/\sqrt{n}) \le \mu \le \bar{X} + z_{\alpha/2}(\sigma/\sqrt{n})\right] = 1 - \alpha. \tag{6.18}$$

Therefore, the interval

$$\left[\bar{X} - z_{\alpha/2}(\sigma/\sqrt{n}), \bar{X} + z_{\alpha/2}(\sigma/\sqrt{n})\right] \tag{6.19}$$

is a $100(1-\alpha)\%$ confidence interval for μ.

Remarks. i) One-sided $100(1-\alpha)\%$ confidence intervals for μ are given by

$$\left[\bar{X} - z_\alpha(\sigma/\sqrt{n}), \infty\right) \quad \text{and} \quad \left(-\infty, \bar{X} + z_\alpha(\sigma/\sqrt{n})\right]. \tag{6.20}$$

ii) The confidence intervals given by expressions (6.19) and (6.20) are still valid (approximately), even if the population X is not Gaussian, as long as n is large enough (by the central limit theorem).

iii) The values of z_α can be obtained using a statistical software package or found in a table. The most useful values are given in Table 6.1. Furthermore, by symmetry, we have: $z_{1-\alpha} = -z_\alpha$. Finally, there are also formulas that give good approximations to z_α.

Example 6.2.1 Let $x_1, \ldots, x_{10}$ be a particular random sample of a population $X \sim N(\mu, \sigma^2 = 0.04)$. Suppose that $\bar{x} = 10.5$. Then, a confidence interval for μ with confidence coefficient equal to $1 - \alpha = 0.95$ is given by

$$10.5 \pm 1.960(0.2/\sqrt{10}) \Rightarrow [10.38, 10.62] \quad \text{(approximately)}.$$

Remark. We cannot write that $P\left[\mu \in [10.38, 10.62]\right] \simeq 0.95$, because μ is not a random variable. Once the data have been collected, we calculate a deterministic interval. Then, the parameter μ *is* or is *not* in this interval. There is no probability left.

b) Confidence interval for μ; σ unknown

If the variance σ^2 of the population X is unknown, but n is sufficiently large, then we can still use expressions (6.19) and (6.20). We only have to replace σ by its estimator S. However, if n is small (< 30), these expressions are no longer valid. In this case, X *must* have a Gaussian distribution. We consider

$$T := \frac{\bar{X} - \mu}{S/\sqrt{n}} = \frac{(\bar{X} - \mu)/(\sigma/\sqrt{n})}{\{[(n-1)(S/\sigma)^2]/(n-1)\}^{1/2}}. \tag{6.21}$$

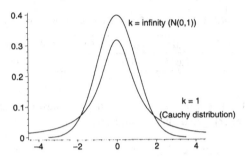

Figure 6.2. Examples of Student's t_k distributions.

We can show that $\bar{X}$ and S are independent random variables and that

$$\frac{(n-1)}{\sigma^2} S^2 \sim \chi^2_{n-1}. \tag{6.22}$$

That is, $(n-1)(S^2/\sigma^2)$ has a *chi-square* distribution with $n-1$ *degrees of freedom*. Since the numerator on the right-hand term of Equation (6.2) has a Gaussian $N(0,1)$ distribution, then, by definition, T has a *Student's t* distribution with $n-1$ *degrees of freedom*. We write: $T \sim t_{n-1}$ (see Fig. 6.2). Proceeding as in Case a) above, we find that

$$\left[\bar{X} - t_{\alpha/2,n-1}(S/\sqrt{n}), \bar{X} + t_{\alpha/2,n-1}(S/\sqrt{n})\right] \tag{6.23}$$

is a $100(1-\alpha)\%$ confidence interval for μ, where $t_{\alpha/2,n-1}$ is defined by (see Fig. 6.3)

$$P[T \leq t_{\alpha/2,n-1}] = 1 - \frac{\alpha}{2} \quad \text{if } T \sim t_{n-1}. \tag{6.24}$$

Remarks. i) The formulas for the one-sided confidence intervals are easily deduced from (6.23). We obtain:

$$\left[\bar{X} - t_{\alpha,n-1}(S/\sqrt{n}), \infty\right) \quad \text{and} \quad \left(-\infty, \bar{X} + t_{\alpha,n-1}(S/\sqrt{n})\right]. \tag{6.25}$$

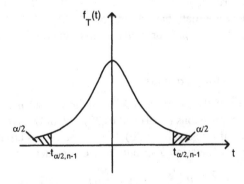

Figure 6.3. Definition of the quantity $t_{\alpha/2,n-1}$.

Table 6.2. Values of $t_{0.025,n}$ and $t_{0.05,n}$ for various values of n.

n	1	2	3	4	5	6	7	8
$t_{0.025,n}$	12.706	4.303	3.182	2.776	2.571	2.447	2.365	2.306
$t_{0.05,n}$	6.314	2.920	2.353	2.132	2.015	1.943	1.895	1.860

n	9	10	15	20	25	30	40	∞
$t_{0.025,n}$	2.262	2.228	2.131	2.086	2.060	2.042	2.021	1.960
$t_{0.05,n}$	1.833	1.812	1.753	1.725	1.708	1.697	1.684	1.645

ii) The values of $t_{\alpha,n}$ are obtained using a statistical software package or found in a table. The values of $t_{0.025,n}$ and of $t_{0.05,n}$ for many values of n are given in Table 6.2. Furthermore, by symmetry, as in the case of the standard Gaussian distribution, we have: $t_{1-\alpha,n} = -t_{\alpha,n}$.

iii) Student is the pseudonym of W. S. Gosset.[1] The probability density function of a Student distribution with k degrees of freedom is given by

$$f_T(t) = \frac{\Gamma[(k+1)/2]}{\sqrt{\pi k}\,\Gamma(k/2)} \frac{1}{[1+(t^2/k)]^{(k+1)/2}} \quad \text{for } -\infty < t < \infty. \qquad (6.26)$$

Its mean is equal to zero (for $k > 1$) and its variance to $k/(k-2)$ (for $k > 2$). Moreover, the Student distribution with 1 degree of freedom is a particular case of the *Cauchy*[2] *distribution*, whose probability density function is given by

$$f_X(x) = \frac{1}{\pi\beta\left[1+\left(\frac{x-\alpha}{\beta}\right)^2\right]} \quad \text{for } -\infty < x < \infty, \qquad (6.27)$$

where $\alpha \in \mathbb{R}$ and $\beta > 0$ are parameters. The t_1 distribution corresponds to the case when $\alpha = 0$ and $\beta = 1$. We can show that this Cauchy distribution can also be obtained from two independent $N(0, \sigma^2)$ random variables, X and Y, by defining $Z = X/Y$. Note that the mathematical expectation of this random variable does not exist, because

---

[1] William Sealy Gosset, 1876–1937, was born and died in England. He studied chemistry and mathematics at Oxford University. Next, he worked as a chemist for the Guinness brewery, in Ireland, where he invented a *statistical test* for the control of the quality of the beer. This test uses the distribution that bears his name. He published many papers on statistics. His research was motivated by practical problems.

[2] (Baron) Augustin Louis Cauchy, 1789–1857, was born and died in France. He is considered the father of mathematical analysis and the inventor of the theory of functions of a complex variable. He studied at the École Polytechnique de Paris and at the École des Ponts et Chaussées. He was professor at the École Polytechnique and at the Collège de France. His political ideas and his religious convictions caused him a lot of trouble. He wrote 789 scientific papers on mathematics, mechanics, optics, etc. His contributions to mathematical physics are very important. Many formulas or mathematical results bear his name.

$$E[X] = \int_{-\infty}^{\infty} x \frac{1}{\pi(x^2 + 1)} \, dx$$

and this improper integral does not converge. However, if we consider the *Cauchy principal value* of the integral, defined by

$$\lim_{c \to \infty} \int_{-c}^{c} x \frac{1}{\pi(x^2 + 1)} \, dx,$$

then we can write that $E[X] = 0$. Finally, given that $E[X^2] = \infty$, the variance of X is infinite (or does not exist).

Example 6.2.2 If σ is unknown in Example 6.2.1 and if the standard deviation of the *sample* is equal to 0.25, then the confidence interval becomes

$$10.5 \pm 2.262(0.25/\sqrt{10}) \Rightarrow [10.32, 10.68] \quad \text{(approximately)}.$$

Remark. Since $t_{\alpha,n} > z_\alpha$ for any $n \in \{1, 2, \dots\}$, the confidence interval obtained when the population standard deviation is unknown should generally be **wider** than the corresponding confidence interval calculated with σ known, which is logical. However, it can happen in practice that the estimator S of σ considerably underestimates the true value of the population standard deviation, so that the confidence interval in the case when σ is unknown is narrower. For instance, if we had obtained a value of s equal to 0.15 rather than 0.25 above, then the confidence interval would have been [10.39, 10.61] (approximately), which is a slightly narrower interval than the one calculated in Example 6.2.1.

6.3 (Pearson's) Chi-Square Goodness-of-Fit Test

Let X be a random variable whose probability density (or mass) function $f_X(x; \theta)$ is unknown. We want to *test* the *null hypothesis*

$$H_0: f_X = f_0 \tag{6.28}$$

against the *alternative hypothesis*

$$H_1: f_X \neq f_0, \tag{6.29}$$

where f_0 is a given function. The *procedure*, proposed by K. Pearson,[3] is the following:

---

[3] Karl Pearson, 1857–1936, was born and died in England. He studied at Cambridge University and then made his career in London. His most important works were on applications of statistics to biological problems. His contributions to statistics, in particular his goodness-of-fit test published in 1900, are contained in his papers on the *theory of evolution*. He also contributed to *regression analysis*. The term "standard deviation" is due to him. He and Fisher totally disagreed about the statistical techniques they were using. Moreover, he studied, in Germany, medieval German literature. He was even offered a post in this field at Cambridge University.

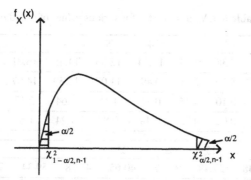

Figure 6.4. Definition of the quantities $\chi^2_{\alpha/2,n-1}$ and $\chi^2_{1-\alpha/2,n-1}$.

i) we divide the set S_X of possible values of X into k disjoint and exhaustive *classes* (or intervals);

ii) we take a random sample of size n from the population X;

iii) we calculate

$$D^2 := \sum_{j=1}^{k} \frac{(n_j - m_j)^2}{m_j}, \tag{6.30}$$

where n_j is the number of observations in the jth class (the *observed frequency*) and m_j denotes the *expected frequency under H_0*, that is, the number of observations that the jth class should count, *on average*, if the hypothesis H_0 is true;

iv) we can show that if H_0 is true and if the number n of observations is large enough, then $D^2 \approx \chi^2_{k-r-1}$. That is, D^2 has (approximately) a chi-square distribution with $k - r - 1$ degrees of freedom, where r is the number of unknown parameters of the function f_0 that we *must* estimate. We reject H_0 at the *significance level* α if and only if

$$D^2 > \chi^2_{\alpha,k-r-1}, \tag{6.31}$$

where (see Fig. 6.4)

$$P[X \le \chi^2_{\alpha,n}] = 1 - \alpha \quad \text{if } X \sim \chi^2_n. \tag{6.32}$$

Remarks. i) In general, it is preferable that $m_j \ge 5 \ \forall j$. If this is not the case, we can combine (adjacent) classes and therefore reduce k.

ii) The k intervals (or classes) are not necessarily of equal width. Likewise, it is not necessary that the corresponding probabilities be equal.

iii) Before drawing the random sample, we can state that if the null hypothesis H_0 is true, then the number N_j of observations that will fall in the jth class is a random variable having a binomial distribution with parameters n and p_j, where p_j is the

Table 6.3. Values of $\chi^2_{\alpha,n}$ for various values of α and n.

n	1	2	3	4	5	6	7	8	9
$\chi^2_{0.025,n}$	5.02	7.38	9.35	11.14	12.83	14.45	16.01	17.53	19.02
$\chi^2_{0.05,n}$	3.84	5.99	7.81	9.49	11.07	12.59	14.07	15.51	16.92
$\chi^2_{0.95,n}$	0^+	0.10	0.35	0.71	1.15	1.64	2.17	2.73	3.33
$\chi^2_{0.975,n}$	0^+	0.05	0.22	0.48	0.83	1.24	1.69	2.18	2.70

n	10	15	20	25	30	40	50	100
$\chi^2_{0.025,n}$	20.48	27.49	34.17	40.65	46.98	59.34	71.42	129.56
$\chi^2_{0.05,n}$	18.31	25.00	31.41	37.65	43.77	55.76	67.50	124.34
$\chi^2_{0.95,n}$	3.94	7.26	10.85	14.61	18.49	26.51	34.76	77.93
$\chi^2_{0.975,n}$	3.25	6.27	9.59	13.12	16.79	24.43	32.36	74.22

probability that the variable X takes a value in the jth class (if H_0 is true). We have: $m_j = np_j$. Furthermore, we know that the binomial distribution tends to a Gaussian distribution. Finally, the square of a standard Gaussian distribution is a chi-square distribution with one degree of freedom.

iv) The values of $\chi^2_{\alpha,n}$ are obtained using a statistical software package or a table. There are also formulas that give excellent approximations, in particular:

$$\chi^2_{\alpha,n} \simeq n \left[z_\alpha \sqrt{\frac{2}{9n}} + 1 - \frac{2}{9n} \right]^3 \quad \text{(Wilson–Hilferty's approximation)} \quad (6.33)$$

and

$$\chi^2_{\alpha,n} \simeq \frac{1}{2} \left[z_\alpha + \sqrt{2n-1} \right]^2 \quad \text{(Fisher's approximation).} \quad (6.34)$$

The values of $\chi^2_{\alpha,n}$ for various values of α and n are given in Table 6.3, p. 264. Contrary to z_α and $t_{\alpha,n}$, we cannot write that $\chi^2_{1-\alpha,n} = -\chi^2_{\alpha,n}$ (because the chi-square distribution is non-negative).

Example 6.3.1 We want to test the hypothesis that the lifetime X (in months) of an electronic component has an exponential distribution. We collected the following observations:

j	[0, 10)	[10, 20)	[20, 30)	[30, 40)	[40, ∞)	Σ
n_j	35	20	18	8	19	100

Moreover, the mean of the observations is equal to 25.

Thus, we want to test

$$H_0\!: X \sim \text{Exp}(\lambda)$$

against

$$H_1: X \text{ does } not \text{ have an exponential distribution.}$$

Suppose that $\alpha = 0.05$. First, we must estimate the unknown parameter λ. We already saw, in Section 6.1, p. 256, that the maximum likelihood estimator of λ is given by $1/\bar{X}$. Then, we set

$$f_X(x) = \frac{1}{25} e^{-x/25} \quad \text{for } x \geq 0.$$

Next, we find that

$$P[a \leq X < b] = e^{-a/25} - e^{-b/25} \quad \text{for } 0 \leq a < b,$$

whence we deduce the following table:

j	[0, 10)	[10, 20)	[20, 30)	[30, 40)	[40, ∞)	Σ
n_j	35	20	18	8	19	$n = 100$
$(\simeq) \, p_j$	0.33	0.22	0.15	0.10	0.20	1
$(np_j =) \, m_j$	33	22	15	10	20	100

Now, given that $m_j \geq 5 \; \forall j$, we calculate

$$D^2 = \frac{(35-33)^2}{33} + \cdots + \frac{(19-20)^2}{20} \simeq 1.35.$$

Remark. We could use d^2 to distinguish between the test statistic D^2 and the value taken by this statistic.

Since $\chi^2_{0.05,5-1-1} \simeq 7.81$, we accept the exponential model, with $\lambda = 1/25$, at significance level $\alpha = 0.05$.

Remarks. i) If we wanted to test, for instance, the hypothesis that the random variable X has a Gaussian distribution, then we would have to add the interval $(-\infty, 0)$ (in which there are of course no observations).

ii) The parameter α is actually defined by

$$\alpha = P[\text{Error of type I}] = P[\text{Reject } H_0 \mid H_0 \text{ is true}]. \tag{6.35}$$

Here, the notation for conditional probability given in Chapter 2 has been used. However, we will never have to calculate the probability of the event $\{H_0 \text{ is true}\}$. It is merely a statement that we assume to be true in the calculation of the *type I error risk*.

We also define the parameter β by the formula

$$\beta = P[\text{Error of type II}] = P[\text{Accept } H_0 \mid H_0 \text{ is false}]. \tag{6.36}$$

The quantity $1 - \beta$ is called the *power* of the test. A test generally more powerful than that of Pearson is the Kolmogorov–Smirnov test. To test normality, the Shapiro–Wilk test (for $n \leq 50$) is considered by many people to be the best.

Note that, contrary to α, the value of β is not unique in a given problem, but rather depends on the particular alternative hypothesis H_1 that we assume to be true. There are actually an infinite number of particular alternative hypotheses H_1, whereas H_0 is unique. For example, above if the hypothesis H_1 is true, then the random variable X can have any distribution, except the exponential distribution.

6.4 Tests of Hypotheses on the Parameters

I) Comparisons between a parameter and a constant

a) Test of a theoretical mean μ; σ known

Let $X_1, \ldots, X_n$ be a random sample of size n of a random variable X with unknown mean μ, but known variance σ^2. We want to test

$$H_0: \mu = \mu_0 \quad \text{against} \quad H_1: \mu \neq \mu_0. \tag{6.37}$$

Remark. A hypothesis of the form $H_0: \mu = \mu_0$ is called *simple*, whereas $H_1: \mu \neq \mu_0$ is a *multiple* or *composite* hypothesis. Moreover, the test in (6.37) is said to be *two-tailed*. A *right-tailed* or *upper-tailed* (respectively *left-tailed* or *lower-tailed*) test is given by

$$H_0: \mu = \mu_0 \quad \text{against} \quad H_1: \mu > \mu_0 \quad (\text{resp. } \mu < \mu_0). \tag{6.38}$$

In this case, we may write the null hypothesis H_0 as follows:

$$H_0: \mu \leq \mu_0 \quad (\text{resp. } \mu \geq \mu_0). \tag{6.39}$$

To perform the test in (6.37), we use the statistic

$$Z_0 := \frac{\bar{X} - \mu_0}{\sigma/\sqrt{n}}. \tag{6.40}$$

When H_0 is true, $\bar{X}$ has (at least approximately if n is large enough) a Gaussian distribution with parameters μ_0 and σ^2/n. It follows that

$$Z_0 \overset{H_0}{\sim} N(0, 1). \tag{6.41}$$

We reject H_0 at significance level α if and only if

$$|Z_0| > z_{\alpha/2}, \tag{6.42}$$

where $z_{\alpha/2}$ is such that $\Phi(z_{\alpha/2}) = 1 - \alpha/2$ (see Section 6.2, p. 258).

Remarks. i) We have, indeed:

$$P[\text{Reject } H_0 \mid H_0 \text{ is true}] = P[|Z_0| > z_{\alpha/2} \mid \mu = \mu_0] = 2(\alpha/2) = \alpha. \tag{6.43}$$

ii) In the case of the one-tailed tests, we reject H_0 if and only if

$$\begin{cases} Z_0 > z_\alpha & \text{if } H_1: \mu > \mu_0, \\ Z_0 < -z_\alpha & \text{if } H_1: \mu < \mu_0. \end{cases} \tag{6.44}$$

Example 6.4.1 Let $x_1, \ldots, x_{25}$ be a particular random sample of a Gaussian population with variance $\sigma^2 = 4$. Suppose that $\bar{x} = 2.8$. To test

$$H_0: \mu = 2 \quad \text{against} \quad H_1: \mu \neq 2,$$

we calculate

$$z_0 = \frac{2.8 - 2}{2/\sqrt{25}} = 2.$$

We then choose α, the significance level of the test. Let $\alpha = 0.02$; then $z_{\alpha/2} = z_{0.01} \simeq 2.326$. Since $|2| \leq 2.326$, we do not reject the null hypothesis. On the other hand, if we choose $\alpha = 0.05$, we obtain: $z_{\alpha/2} = z_{0.025} \simeq 1.960$. Since $|2| > 1.960$, we reject the hypothesis $H_0: \mu = 2$ at significance level $\alpha = 0.05$.

Remarks. i) Statistical software programs generally give the value of α that corresponds to the observed statistic z_0. This value is called the *P-value*. In the preceding example, we find that $z_{\alpha/2} = 2$ if and only if $\alpha \simeq 0.0455$. Thus, if we choose α smaller than 0.0455, then we cannot reject H_0.

In general, it is difficult to give an exact formula for the P-value. However, in the case of the hypothesis test on a mean with σ known, it is easy to show that

$$P = \begin{cases} 2[1 - \Phi(|z_0|)] & \text{if } H_1: \mu \neq \mu_0, \\ 1 - \Phi(z_0) & \text{if } H_1: \mu > \mu_0, \\ \Phi(z_0) & \text{if } H_1: \mu < \mu_0. \end{cases} \tag{6.45}$$

ii) In the case of the tests (especially the *one-tailed* tests) *on the parameters* of a random variable, the null hypothesis is usually the one that we think is false, or at least that we put in doubt. Consequently, we try to reject it. For this reason, we prefer to say that we *cannot* reject H_0, rather than saying that we *accept* H_0.

Type II error

If the hypothesis H_1 is true, that is, if $\mu = \mu_1 = \mu_0 + \Delta$, where $\Delta \neq 0$, then $\bar{X} \sim N(\mu_1, \sigma^2/n)$ and it follows that

$$Z_0 \sim N\left(\frac{\mu_1 - \mu_0}{\sigma/\sqrt{n}}, 1\right) \equiv N\left(\frac{\Delta\sqrt{n}}{\sigma}, 1\right). \tag{6.46}$$

Therefore,

$$\begin{aligned} \beta \; (= \beta(\Delta)) \; &= \; P[\text{Accept } H_0 \mid H_0 \text{ is false}] \\ &= \; P[-z_{\alpha/2} \leq Z_0 \leq z_{\alpha/2} \mid \mu = \mu_0 + \Delta] \end{aligned}$$

$$= P\left[-z_{\alpha/2} - \frac{\Delta\sqrt{n}}{\sigma} \leq N(0,1) \leq z_{\alpha/2} - \frac{\Delta\sqrt{n}}{\sigma} \right]$$

$$= \Phi\left(z_{\alpha/2} - \frac{\Delta\sqrt{n}}{\sigma} \right) - \Phi\left(-z_{\alpha/2} - \frac{\Delta\sqrt{n}}{\sigma} \right). \tag{6.47}$$

Remark. In the case of the one-tailed tests, we find that

$$\beta(\Delta) = \begin{cases} \Phi\left(z_\alpha - \dfrac{\Delta\sqrt{n}}{\sigma} \right) & \text{if } H_1: \mu > \mu_0, \\[2mm] \Phi\left(z_\alpha + \dfrac{\Delta\sqrt{n}}{\sigma} \right) & \text{if } H_1: \mu < \mu_0. \end{cases} \tag{6.48}$$

Example 6.4.1 (continued) If the true value of the population mean is $\mu = \mu_1 = 2.6 = 2 + 0.6$, and if $\alpha = 0.05$, we calculate

$$\beta(0.6) \simeq \Phi(1.96 - (0.6)(5/2)) - \Phi(-1.96 - (0.6)(5/2))$$

$$= \Phi(0.46) - \Phi(-3.46) \overset{\text{Tab. 3.3}}{\simeq} 0.6772 - (1 - 0.9997) = 0.6769.$$

Thus, the power of the test, when $\mu = 2.6$, is given by $1 - \beta(0.6) \simeq 0.3231$, which is weak. If we wish to increase this power, we can take more observations. Indeed, the value of β decreases as n increases (see (6.47)). For instance, if we take $n = 100$, then we obtain that $\beta(0.6) \simeq 0.1492$, so that $1 - \beta(0.6) \simeq 0.8508$. In practice, we would like $1 - \beta(\Delta)$ to be greater than or equal to 0.8 for a difference Δ that we deem *significant* between μ_0 and the true population mean.

Sample size

We can make use of Equation (6.47) to calculate the value of n required to get a certain β, given α and Δ. Indeed, if $\Delta > 0$ we have:

$$\Phi\left(-z_{\alpha/2} - \frac{\Delta\sqrt{n}}{\sigma} \right) \simeq 0, \tag{6.49}$$

so that

$$\beta \simeq \Phi\left(z_{\alpha/2} - \frac{\Delta\sqrt{n}}{\sigma} \right). \tag{6.50}$$

This last equation implies that

$$z_{1-\beta} \simeq z_{\alpha/2} - \frac{\Delta\sqrt{n}}{\sigma}. \tag{6.51}$$

Since $z_{1-\beta} = -z_\beta$, we obtain the following formula:

$$n \simeq \frac{(z_{\alpha/2} + z_\beta)^2 \sigma^2}{\Delta^2}, \tag{6.52}$$

where $\Delta = \mu - \mu_0$.

Remarks. i) If $\Delta < 0$, then

$$\Phi\left(z_{\alpha/2} - \frac{\Delta\sqrt{n}}{\sigma}\right) \simeq 1 \tag{6.53}$$

and we obtain the same formula as above for n.

ii) In the case of the one-tailed tests, we find the following *exact* result:

$$n = \frac{(z_\alpha + z_\beta)^2 \sigma^2}{\Delta^2}. \tag{6.54}$$

Example 6.4.2 Suppose that we test

$$H_0: \mu = 2 \quad \text{against} \quad H_1: \mu \neq 2$$

with $\alpha = 0.05$. Then, if $\mu = 2.5$ and $\sigma^2 = 1$, to obtain a value of β smaller than or equal to 0.10, we must take

$$n \simeq \frac{(z_{0.025} + z_{0.1})^2 \cdot 1}{(2.5 - 2)^2} \simeq \frac{(1.960 + 1.282)^2}{0.25} \simeq 42.$$

Remark. Here, we have:

$$\Phi\left(-z_{\alpha/2} - \frac{\Delta\sqrt{n}}{\sigma}\right) \simeq \Phi(-1.960 - (0.5)\sqrt{42}) \simeq \Phi(-5.20) \simeq 0.$$

b) Test of a theoretical mean μ; σ unknown

Let $X_1, \ldots, X_n$ be a random sample of a random variable X having a Gaussian distribution with mean μ and variance σ^2 unknown. To test

$$H_0: \mu = \mu_0, \tag{6.55}$$

we use the statistic

$$T_0 := \frac{\bar{X} - \mu_0}{S/\sqrt{n}}, \tag{6.56}$$

where S is the sample standard deviation. We can write (see p. 260) that

$$T_0 \overset{H_0}{\sim} t_{n-1}. \tag{6.57}$$

We reject H_0 at significance level α if and only if

$$\begin{cases} |T_0| > t_{\alpha/2, n-1} & \text{if } H_1: \mu \neq \mu_0, \\ T_0 > t_{\alpha, n-1} & \text{if } H_1: \mu > \mu_0, \\ T_0 < -t_{\alpha, n-1} & \text{if } H_1: \mu < \mu_0. \end{cases} \tag{6.58}$$

Remark. If the sample size is large enough (≥ 30), we can use the test in Case a) above; we simply replace σ by its estimator S. Moreover, in this case, the assumption of a Gaussian distribution for X is not essential.

Example 6.4.3 Let

$$5, \ 15, \ 14, \ 12, \ 4, \ 8, \ 9, \ 12, \ 10, \ 17, \ 12, \ 11$$

be a particular random sample of a Gaussian population with mean μ and variance σ^2 unknown. We choose $\alpha = 0.05$ and we want to test

$$H_0: \mu = 15 \quad \text{against} \quad H_1: \mu \neq 15.$$

We first find that $\bar{x} = 10.75$ and $s \simeq 3.84$. Next, we calculate

$$t_0 \simeq \frac{10.75 - 15}{3.84/\sqrt{12}} \simeq -3.83.$$

Finally, since $t_{0.025,11} \simeq 2.201$, we have: $|t_0| > t_{0.025,11}$ and we can reject the null hypothesis H_0 at level $\alpha = 0.05$.

Remark. Actually, we must set the hypotheses H_0 and H_1 *prior to* taking the random sample.

Type II error and sample size

By definition, in the two-tailed case,

$$\beta = P[\text{Accept } H_0 \mid H_0 \text{ is false}] = P[|T_0| \leq t_{\alpha/2,n-1} \mid \mu = \mu_0 + \Delta], \quad (6.59)$$

where $\Delta \neq 0$. Now, if $\mu = \mu_0 + \Delta$, we find that

$$T_0 := \frac{\bar{X} - \mu_0}{S/\sqrt{n}} = \frac{Y}{\sqrt{W^2/(n-1)}}, \quad (6.60)$$

where $Y \sim N(\Delta\sqrt{n}/\sigma, 1)$ and $W^2 \sim \chi^2_{n-1}$ are independent random variables. We say that T_0 has a *non-central t* distribution with $n - 1$ degrees of freedom and *non-centrality parameter* $\Delta\sqrt{n}/\sigma$. Consequently, it is not easy to calculate the *exact* value of β.

There exist approximation formulas for β. For example, we can show that

$$\beta(\delta) \simeq \Phi\left(\frac{t_{\alpha,n-1} - \delta\sqrt{n}}{\sqrt{1 + (t^2_{\alpha,n-1}/2n)}}\right) \quad \text{if } H_1: \mu > \mu_0, \quad (6.61)$$

where

$$\delta := \frac{\mu - \mu_0}{\sigma}. \quad (6.62)$$

Since σ is unknown, we must replace it by its point estimate s (if the random sample has already been taken) to calculate δ, or we can express the difference $\mu - \mu_0$ in terms of σ.

There also exist curves, called *operating characteristic curves*, that give the value of β in terms of a parameter like δ above for a fixed α and many values of n. Moreover, we can use these curves to (approximately) calculate the size n of the random sample that we must take to obtain a given value of β. Finally, we also find tables that give the value of n that corresponds to β, when α and δ are fixed.

In this book, as far as the computation of β or n is concerned, we will generally simply consider the case when the size of the random sample (taken or that must be taken) is large. The problem is then reduced to the preceding one where σ is known and we can use the formulas already obtained. We must however replace σ by s (if the data have been gathered).

c) Test of a theoretical variance σ^2

Let $X_1, \ldots, X_n$ be a random sample of a Gaussian random variable X, with mean μ and variance σ^2 unknown. To test the null hypothesis H_0: $\sigma^2 = \sigma_0^2$, we use the statistic

$$W_0^2 := \frac{(n-1)S^2}{\sigma_0^2}. \tag{6.63}$$

When H_0 is true, W_0^2 has a chi-square distribution with $n - 1$ degrees of freedom. We reject H_0 at significance level α if and only if

$$\begin{cases} W_0^2 > \chi^2_{\alpha/2,n-1} \quad \text{or} \quad W_0^2 < \chi^2_{1-(\alpha/2),n-1} & \text{if } H_1: \sigma^2 \neq \sigma_0^2, \\ \quad\quad W_0^2 > \chi^2_{\alpha,n-1} & \text{if } H_1: \sigma^2 > \sigma_0^2, \\ \quad\quad W_0^2 < \chi^2_{1-\alpha,n-1} & \text{if } H_1: \sigma^2 < \sigma_0^2. \end{cases} \tag{6.64}$$

Remark. As in the case of the test of a theoretical mean μ with σ known, we could consider the problem of testing H_0: $\sigma^2 = \sigma_0^2$ when this time (we assume that) the population mean is known. However, in practice this case is much less important.

Example 6.4.4 (See [4].) Let

$$10.2, \ 9.9, \ 9.8, \ 10.1, \ 10.1, \ 9.9, \ 10.0, \ 9.7, \ 10.1, \ 10.3$$

be a particular random sample of a random variable X that represents the weight of soap boxes. We assume that X has (approximately) a Gaussian $N(\mu, \sigma^2)$ distribution and we want to test

$$H_0: \sigma^2 = 0.04 \quad \text{against} \quad H_1: \sigma^2 > 0.04$$

at level $\alpha = 0.05$. We find that $\bar{x} = 10.01$ and $\sum_{i=1}^{10} x_i^2 = 1002.31$. It follows that

$$s^2 = \sum_{i=1}^{10} \frac{(x_i - \bar{x})^2}{10 - 1} = \frac{1}{9}\left[\sum_{i=1}^{10} x_i^2 - 10\bar{x}^2\right] \simeq \frac{0.31}{9},$$

so that

$$w_0^2 \simeq \frac{0.31}{0.04} = 7.75.$$

Finally, since $\chi_{0.05,9}^2 \simeq 16.92$, we cannot reject the null hypothesis H_0 at significance level $\alpha = 0.05$.

Remark. Here, since $s^2 \simeq 0.034$ is smaller than 0.04, we certainly could not reject H_0 with a small α.

Type II error and sample size

Suppose that the alternative hypothesis is $H_1: \sigma^2 < \sigma_0^2$ and that the true value of σ^2 is $\sigma_1^2 < \sigma_0^2$. We have:

$$\beta = P[W_0^2 \geq \chi_{1-\alpha,n-1}^2 \mid \sigma^2 = \sigma_1^2]$$

$$= P\left[\frac{\sigma_1^2}{\sigma_0^2} \frac{(n-1)S^2}{\sigma_1^2} \geq \chi_{1-\alpha,n-1}^2 \,\middle|\, \sigma^2 = \sigma_1^2\right]$$

$$= P[\chi_{n-1}^2 \geq \lambda\chi_{1-\alpha,n-1}^2], \tag{6.65}$$

where

$$\lambda := \frac{\sigma_0^2}{\sigma_1^2}. \tag{6.66}$$

Similarly, we find that

$$\beta(\lambda) = \begin{cases} P[\chi_{n-1}^2 \leq \lambda\chi_{\alpha,n-1}^2] & \text{if } H_1: \sigma^2 > \sigma_0^2, \\ P[\lambda\chi_{1-(\alpha/2),n-1}^2 \leq \chi_{n-1}^2 \leq \lambda\chi_{\alpha/2,n-1}^2] & \text{if } H_1: \sigma^2 \neq \sigma_0^2. \end{cases} \tag{6.67}$$

Now, if n is large, we may write that

$$\chi_n^2 \approx N(n, 2n). \tag{6.68}$$

Hence, we can approximately calculate $\beta(\lambda)$ with the help of the preceding formulas.

To determine the size of the random sample needed to obtain a β smaller than or equal to a given value, we can use the following formulas:

$$n \simeq \begin{cases} \dfrac{3}{2} + \dfrac{1}{2}\left[\dfrac{\sigma_0 z_\alpha + \sigma_1 z_\beta}{\sigma_1 - \sigma_0}\right]^2 & \text{(one-tailed cases)}, \\[4mm] \dfrac{3}{2} + \dfrac{1}{2}\left[\dfrac{\sigma_0 z_{\alpha/2} + \sigma_1 z_\beta}{\sigma_1 - \sigma_0}\right]^2 & \text{(two-tailed case)}, \end{cases} \tag{6.69}$$

where σ_1^2 is the true variance of the population.

Example 6.4.4 (continued) Suppose that we gathered 51 observations rather than 10 in the preceding example. If $\sigma_1^2 = 0.06$, then $\lambda = 2/3$ and we calculate

$$\beta(\lambda = 2/3) \quad = \quad P[\chi_{50}^2 \le (2/3)\chi_{0.05,50}^2]$$

$$\overset{\text{Tab. 6.3}}{\simeq} P[N(50, 100) \le 45] = \Phi(-0.5) \overset{\text{Tab. 3.3}}{\simeq} 0.31.$$

If we want to be able to detect, with a probability of at least 0.90, that σ^2 is greater than 0.04, when the population variance is in fact equal to 0.06, we must take

$$n \simeq \frac{3}{2} + \frac{1}{2}\left[\frac{0.2\,z_{0.05} + \sqrt{0.06}\,z_{0.10}}{\sqrt{0.06} - 0.2}\right]^2 \simeq 103.83.$$

Thus, we must take at least 104 observations.

Remark. We can show that if n is large enough, then we may write that

$$S \approx N\left(\sigma, \tfrac{\sigma^2}{2n}\right). \tag{6.70}$$

It follows that

$$Z_0 := \frac{S - \sigma_0}{\sigma_0/\sqrt{2n}} \overset{H_0}{\approx} N(0, 1). \tag{6.71}$$

Therefore, when n is large, we can also reject H_0 if and only if

$$\begin{cases} |Z_0| > z_{\alpha/2} & \text{if } H_1: \sigma^2 \ne \sigma_0^2, \\ Z_0 > z_\alpha & \text{if } H_1: \sigma^2 > \sigma_0^2, \\ Z_0 < -z_\alpha & \text{if } H_1: \sigma^2 < \sigma_0^2. \end{cases} \tag{6.72}$$

II) Comparisons between two parameters

a) Test of two theoretical means; variances known

Let $X_1, \ldots, X_m$ be a random sample of a random variable X with unknown mean μ_X and known variance σ_X^2, and $Y_1, \ldots, Y_n$ be a random sample of a random variable Y with unknown mean μ_Y and known variance σ_Y^2. We assume that the random variables X_i and Y_j are independent $\forall i, j$. To test

$$H_0: \mu_X - \mu_Y = \Delta, \tag{6.73}$$

we use the statistic

$$Z_0 := \frac{\bar{X} - \bar{Y} - \Delta}{\sqrt{\frac{\sigma_X^2}{m} + \frac{\sigma_Y^2}{n}}}. \tag{6.74}$$

Since $\bar{X}$ and $\bar{Y}$ are independent, we have:

$$\bar{X} - \bar{Y} \sim N\left(\mu_X - \mu_Y, \frac{\sigma_X^2}{m} + \frac{\sigma_Y^2}{n}\right) \tag{6.75}$$

(at least approximately if m and n are large enough). It follows that

$$Z_0 \stackrel{H_0}{\sim} N(0, 1). \tag{6.76}$$

We reject H_0 at significance level α if and only if

$$\begin{cases} |Z_0| > z_{\alpha/2} & \text{if } H_1 \colon \mu_X - \mu_Y \neq \Delta, \\ Z_0 > z_\alpha & \text{if } H_1 \colon \mu_X - \mu_Y > \Delta. \end{cases} \tag{6.77}$$

Remark. In the case of the tests comparing two parameters, we can limit ourselves to the two alternative hypotheses above, since we want to determine whether the two parameters differ by a constant Δ (H_0) or not (H_1), or whether the difference between *one* parameter and the *other* is smaller than or equal to Δ (H_0) or greater than Δ (H_1).

Example 6.4.5 Let $x_1, \ldots, x_{10}$ be a particular random sample of $X \sim N(\mu_X, 2)$ and $y_1, \ldots, y_{15}$ be a particular random sample of $Y \sim N(\mu_Y, 3)$. We assume that the samples are independent and we want to test

$$H_0 \colon \mu_X = \mu_Y \quad \text{against} \quad H_1 \colon \mu_Y > \mu_X.$$

We simply have to take $\Delta = 0$ and reverse the role of X and Y in the above formulas. We choose $\alpha = 0.05$. Suppose that $\bar{x} = 4$ and $\bar{y} = 5$. Then we calculate

$$z_0 = \frac{5 - 4}{\sqrt{\frac{3}{15} + \frac{2}{10}}} \simeq 1.58.$$

Since $z_{0.05} \simeq 1.645$, we cannot reject the null hypothesis $H_0 \colon \mu_X = \mu_Y$ at level $\alpha = 0.05$.

Remark. If we had obtained $z_0 = 1.68$, for instance, then we could have rejected H_0 and accepted $H_1 \colon \mu_X < \mu_Y$. However, we could not reject H_0 with $z_0 = 1.68$ if the alternative hypothesis were rather $H_1 \colon \mu_X \neq \mu_Y$, since $|1.68| = 1.68$ is not greater than $z_{0.025} \simeq 1.960$.

Type II error and sample size

Proceeding as in the case of the test of a theoretical mean with σ known, we find that

$$\beta(\delta) = \begin{cases} \Phi(z_{\alpha/2} - \delta) - \Phi(-z_{\alpha/2} - \delta) & \text{if } H_1 \colon \mu_X - \mu_Y \neq \Delta, \\ \Phi(z_\alpha - \delta) & \text{if } H_1 \colon \mu_X - \mu_Y > \Delta, \end{cases} \tag{6.78}$$

where

$$\delta := \frac{\mu_X - \mu_Y - \Delta}{\sqrt{\frac{\sigma_X^2}{m} + \frac{\sigma_Y^2}{n}}}. \tag{6.79}$$

Similarly, we find that to obtain a given β, we must draw random samples of size

$$m = n \simeq \frac{(z_{\alpha/2} + z_\beta)^2(\sigma_X^2 + \sigma_Y^2)}{(\mu_X - \mu_Y - \Delta)^2} \tag{6.80}$$

when the alternative hypothesis is $H_1: \mu_X - \mu_Y \neq \Delta$. In the case of the one-tailed test, we replace $\alpha/2$ by α in the preceding formula, and the result is now exact, rather than approximate.

Remark. In theory, it is not necessary to take $m = n$ ($= c$). We can choose m, for instance, and then the value of n that corresponds to a given β is $2c - m$. We can proceed in that manner, in particular, when it is more expensive to draw a random sample from one random variable than from the other.

b) Test of two theoretical means; variances unknown

Let $X_1, \ldots, X_m$ and $Y_1, \ldots, Y_n$ be random samples of a random variable X and of a random variable Y, respectively. We assume that all the parameters are unknown and that X and Y have independent Gaussian distributions. We want to test

$$H_0: \mu_X - \mu_Y = \Delta. \tag{6.81}$$

Case 1: $\sigma_X^2 = \sigma_Y^2$

We first assume that the variances are unknown, but equal. In practice, we perform a test of equality of variances (which will be seen further on) to make sure that this hypothesis is reasonable. We define

$$S_p^2 = \frac{(m-1)S_X^2 + (n-1)S_Y^2}{m+n-2}, \tag{6.82}$$

where S_X^2 and S_Y^2 are the sample variances, and we consider the statistic

$$T_0 := \frac{\bar{X} - \bar{Y} - \Delta}{S_p\sqrt{\frac{1}{m} + \frac{1}{n}}}. \tag{6.83}$$

We may write that

$$T_0 \overset{H_0}{\sim} t_{m+n-2}. \tag{6.84}$$

It follows that we reject H_0 at significance level α if and only if

$$\begin{cases} |T_0| > t_{\alpha/2, m+n-2} & \text{if } H_1: \mu_X - \mu_Y \neq \Delta, \\ T_0 > t_{\alpha, m+n-2} & \text{if } H_1: \mu_X - \mu_Y > \Delta. \end{cases} \tag{6.85}$$

Case 2: $\sigma_X^2 \neq \sigma_Y^2$

If we cannot assume that $\sigma_X^2 = \sigma_Y^2$, then we must use the statistic

$$T_0^* := \frac{\bar{X} - \bar{Y} - \Delta}{\sqrt{\frac{S_X^2}{m} + \frac{S_Y^2}{n}}}, \tag{6.86}$$

which, when H_0 is true, has *approximately* a t distribution with ν degrees of freedom, where

$$\nu := \frac{(a+b)^2}{\frac{a^2}{m-1} + \frac{b^2}{n-1}}, \tag{6.87}$$

with

$$a := \frac{S_X^2}{m} \quad \text{and} \quad b := \frac{S_Y^2}{n}. \tag{6.88}$$

We then replace T_0 by T_0^* and $m + n - 2$ by ν in the preceding case.

Remarks. i) We also find in several books the formula

$$\nu = \frac{(a+b)^2}{\frac{a^2}{m+1} + \frac{b^2}{n+1}} - 2. \tag{6.89}$$

ii) We can show that $\min\{m - 1, n - 1\} \leq \nu \leq m + n - 2$.

iii) When the sample sizes are large enough, we can use the standard Gaussian distribution as an approximate distribution (under H_0) for T_0 or T_0^*.

c) Test of two means with paired observations

Suppose that we gather pairs $(X_1, Y_1), \ldots, (X_n, Y_n)$ of observations of the random variables $X \sim N(\mu_X, \sigma_X^2)$ and $Y \sim N(\mu_Y, \sigma_Y^2)$, and that the conditions under which the observations are taken vary (or may vary) from one pair to the other. In this case, the observations X_k and Y_k are *not* independent. To test $H_0: \mu_X - \mu_Y = \Delta$, we define the variable $D = X - Y$ and we calculate the differences $D_k := X_k - Y_k$ for $k = 1, \ldots, n$. The problem of testing H_0 is then reduced to that of testing a theoretical mean μ_D with σ_D unknown, and we can use the formulas in Case I) b), p. 269.

Example 6.4.6 (See [11].) In a test of hardness of materials, a steel ball is pressed into each material and the diameter of the indentation is measured. This diameter is related to the hardness. We have two types of steel balls, which we use with ten different materials. We want to determine whether the two types of steel produce equivalent results. We collected the following data:

Material	1	2	3	4	5	6	7	8	9	10
Steel 1	75	46	57	43	58	32	61	56	34	65
Steel 2	52	41	43	47	32	49	52	44	57	60

Let X be the indentation diameter obtained with steel 1 and Y be the indentation diameter with steel 2. We define $D = X - Y$ and we want to test

$$H_0: \mu_D = 0 \quad \text{against} \quad H_1: \mu_D \neq 0.$$

Let $\alpha = 0.05$. We calculate the differences d_k for $k = 1, \ldots, 10$:

Material	1	2	3	4	5	6	7	8	9	10
d_k	23	5	14	−4	26	−17	9	12	−23	5

We find that $\bar{d} = 5$ and $s_D \simeq 15.85$. We calculate next

$$t_0 = \frac{\bar{d}}{s_D/\sqrt{n}} \simeq \frac{5\sqrt{10}}{15.85} \simeq 1.$$

Since $t_{0.025, 10-1} \simeq 2.262$, we do not reject H_0 at level $\alpha = 0.05$.

d) Test of the equality of two variances

Let $X \sim N(\mu_X, \sigma_X^2)$ and $Y \sim N(\mu_Y, \sigma_Y^2)$ be two independent random variables, and let S_X^2 and S_Y^2 be the variances of a random sample of size m of X and of a random sample of size n of Y, respectively. We assume that all the parameters are unknown. To test the null hypothesis

$$H_0: \sigma_X^2 = \sigma_Y^2, \tag{6.90}$$

we use the statistic

$$F_0 := \frac{S_X^2}{S_Y^2}. \tag{6.91}$$

We can show that if H_0 is true, then F_0 has a (Fisher)[4] F distribution with $m - 1$ and $n - 1$ degrees of freedom. We write: $F_0 \overset{H_0}{\sim} F_{m-1,n-1}$. We reject H_0 at significance level α if and only if

---

[4] Sir Ronald Fisher, 1890–1962, was born in England and died in Australia. He studied astronomy at Cambridge University, but he worked as a biologist in a research center, where he made contributions to genetics and statistics. He is considered a founder of the modern theory of statistics because of his numerous contributions to this field. In particular, he invented the *analysis of variance* and the concept of *maximum likelihood*, in addition to giving the exact sampling distributions of many statistics. Like Gosset's methods, the statistical methods that Fisher developed are valid for small samples, whereas Pearson worked with large samples. From 1933, he became a professor, first in London and then in Cambridge.

Table 6.4. Values of $F_{0.025,n,n}$ and $F_{0.05,n,n}$ for several values of n.

n	1	2	3	4	5	6	7	8
$F_{0.025,n,n}$	647.8	39.00	15.44	9.60	7.15	5.82	4.99	4.43
$F_{0.05,n,n}$	161.4	19.00	9.28	6.39	5.05	4.28	3.79	3.44

n	9	10	15	20	30	60	120	∞
$F_{0.025,n,n}$	4.03	3.72	2.86	2.46	2.07	1.67	1.43	1.00
$F_{0.05,n,n}$	3.18	2.98	2.40	2.12	1.84	1.53	1.35	1.00

$$\begin{cases} F_0 > F_{\alpha/2,m-1,n-1} \quad \text{or} \quad F_0 < F_{1-(\alpha/2),m-1,n-1} & \text{if } H_1: \sigma_X^2 \neq \sigma_Y^2, \\ \qquad\qquad F_0 > F_{\alpha,m-1,n-1} & \text{if } H_1: \sigma_X^2 > \sigma_Y^2. \end{cases} \tag{6.92}$$

Remarks. i) The *critical values* $F_{\alpha,m-1,n-1}$ are defined in the same manner as the other critical values (z_α, $t_{\alpha,n}$ and $\chi_{\alpha,n}^2$):

$$P[F \leq F_{\alpha,m-1,n-1}] = 1 - \alpha \quad \text{if } F \sim F_{m-1,n-1}. \tag{6.93}$$

ii) We find $F_{\alpha,m-1,n-1}$ with the help of a statistical software package or in a statistical table. Several values of $F_{0.025,m,n}$ and $F_{0.05,m,n}$ are given in Table 6.4 278, in the case when $m = n$.

The values of $F_{0.025,n_1,n_2}$ and $F_{0.05,n_1,n_2}$ for n_1 and $n_2 \in \{1, 2, \dots, 12\}$ can be found in Appendix B, p. 342. Moreover, we have the following relationship:

$$F_{1-\alpha,n_1,n_2} = \frac{1}{F_{\alpha,n_2,n_1}}. \tag{6.94}$$

Finally, for values of n_1 and n_2 large enough, we can use the approximation formula

$$F_{\alpha,n_1,n_2} \simeq \exp\left[\frac{1}{n_2} - \frac{1}{n_1} + z_\alpha \sqrt{\frac{2}{n_1} + \frac{2}{n_2}} \right]. \tag{6.95}$$

iii) As in the case of the test of a theoretical variance σ^2, an approximate test, valid when the sample sizes are large, can be constructed based on the Gaussian distribution. Indeed, if m and n are large enough, then we may write that

$$S_X \approx N\left(\sigma_X, \frac{\sigma_X^2}{2m} \right) \quad \text{and} \quad S_Y \approx N\left(\sigma_Y, \frac{\sigma_Y^2}{2n} \right), \tag{6.96}$$

so that

$$Z_0 := \frac{S_X - S_Y}{S_p \left(\frac{1}{2m} + \frac{1}{2n} \right)^{1/2}} \overset{H_0}{\approx} N(0, 1), \tag{6.97}$$

where S_p^2 is defined in (6.82). We then reject H_0: $\sigma_X^2 = \sigma_Y^2$ at significance level α if and only if

$$\begin{cases} |Z_0| > z_{\alpha/2} \text{ if } H_1: \sigma_X^2 \neq \sigma_Y^2, \\ Z_0 > z_\alpha \quad \text{ if } H_1: \sigma_X^2 > \sigma_Y^2. \end{cases} \qquad (6.98)$$

iv) The probability density function of a random variable having a Fisher distribution with m and n degrees of freedom is given by

$$f_X(x) = \frac{\Gamma[(m+n)/2]m^{m/2}n^{n/2}}{\Gamma(m/2)\Gamma(n/2)} \frac{x^{(m/2)-1}}{(mx+n)^{(m+n)/2}} \quad \text{for } x > 0. \qquad (6.99)$$

We find that

$$E[X] = \frac{n}{n-2} \quad \text{if } n > 2 \quad \text{and} \quad \text{VAR}[X] = \frac{2n^2(m+n-2)}{m(n-2)^2(n-4)} \quad \text{if } n > 4. \qquad (6.100)$$

Example 6.4.7 We consider two manufacturing processes for bottles. Let $X \sim N(\mu_X, \sigma_X^2)$ be the capacity of the bottles made with the manufacturing process currently in use, and let $Y \sim N(\mu_Y, \sigma_Y^2)$ be the capacity of the bottles made with a new manufacturing process. All the parameters are unknown, but we think that $\mu_X = \mu_Y$. We want to test at level $\alpha = 0.05$

$$H_0: \sigma_X^2 = \sigma_Y^2 \quad \text{against} \quad H_1: \sigma_X^2 > \sigma_Y^2.$$

We take at random 20 bottles made with the current process and 25 fabricated with the new process. We find that $s_X^2 \simeq 0.0144$ and $s_Y^2 \simeq 0.0064$. We calculate

$$f_0 := \frac{s_X^2}{s_Y^2} \simeq \frac{0.0144}{0.0064} = 2.25.$$

Now, we find in a statistical table that $F_{0.05,19,24} \simeq 2.04$ (2.02 according to (6.95)). Consequently, we can reject H_0 at level $\alpha = 0.05$.

6.5 Exercises, Problems, and Multiple Choice Questions, Supplementary Exercises

Solved Exercises

Exercise no. 1 (6.1)

Let

$$f_X(x; \theta) = \frac{1}{\theta^2} x e^{-x/\theta} \quad \text{for } x > 0,$$

where $\theta > 0$ is an unknown parameter. We have: $E[X] = 2\theta$ and $\text{VAR}[X] = 2\theta^2$.

a) Find the maximum likelihood estimator of θ.

b) Calculate the mean square error of the estimator θ_{ML} obtained in a).

c) What is the approximate distribution of θ_{ML}? Justify your answer.

Solution

a) We calculate

$$L(\theta) = \prod_{k=1}^{n} \frac{1}{\theta^2} X_k e^{-X_k/\theta} = \theta^{-2n} \left(\prod_{k=1}^{n} X_k \right) \exp\left\{ -\sum_{k=1}^{n} X_k/\theta \right\}$$

$$\Rightarrow \quad \ln L(\theta) = -2n \ln \theta + \sum_{k=1}^{n} \ln X_k - \sum_{k=1}^{n} \frac{X_k}{\theta}$$

$$\Rightarrow \quad \frac{d}{d\theta} \ln L(\theta) = -\frac{2n}{\theta} + \frac{1}{\theta^2} \sum_{k=1}^{n} X_k = 0 \Leftrightarrow \theta = \sum_{k=1}^{n} \frac{X_k}{2n} \Rightarrow \quad \theta_{ML} = \frac{1}{2} \bar{X}.$$

Remark. Note that $\frac{d^2}{d\theta^2} \ln L(\theta) = \frac{2n}{\theta^2} - \frac{2}{\theta^3} \sum_{k=1}^{n} X_k < 0 \Leftrightarrow \theta < \bar{X}$. We have indeed: $\theta_{ML} = \frac{1}{2} \bar{X} < \bar{X}$.

b) $E[\theta_{ML}] \overset{a)}{=} E[\bar{X}/2] = \frac{1}{2} E[X] = \frac{2\theta}{2} = \theta \Rightarrow$

$$\text{MSE}[\theta_{ML}] = \text{VAR}[\theta_{ML}] = \text{VAR}[\bar{X}/2] = \frac{\text{VAR}[X]}{4n} = \frac{2\theta^2}{4n} = \frac{\theta^2}{2n}.$$

c) By the central limit theorem, we can write that $\bar{X}$ has approximately a Gaussian $N(E[X], \frac{1}{n} \text{VAR}[X])$ distribution. Then, $\theta_{ML} = \frac{1}{2} \bar{X} \overset{b)}{\approx} N(\theta, \frac{\theta^2}{2n})$.

Exercise no. 2 (6.3)

An engineer responsible for the control of the quality in a company that manufactures modems picks, at the end of each workday, two modems at random (and without replacement) among those manufactured on that day. Let X be the number of defective modems among the two examined. We suppose that $X \sim B(n = 2, p)$.

a) We can show that the maximum likelihood estimator of p, based on m observations, is given by $p_{ML} = \bar{X}/2$, where $\bar{X}$ is the mean of a random sample (of size m) of X. Calculate the mean square error of p_{ML}.

b) We gathered the following data over a 30-day period:

j	0	1	2
n_j	20	8	2

where j is the number of defective modems in the sample of size 2 and n_j is the number of days on which we observed j defectives in the sample. Test the hypothesis that $X \sim B(n = 2, p = 1/5)$. Use $\alpha = 0.05$.

Hint. We have: $\chi^2_{0.05,1} \simeq 3.84$, $\chi^2_{0.05,2} \simeq 5.99$ and $\chi^2_{0.05,3} \simeq 7.81$.

Solution

a) We first calculate

$$E[p_{ML}] \equiv E[\bar{X}/2] = \frac{1}{2}E[\bar{X}] = \frac{1}{2}E[X] = \frac{2p}{2} = p.$$

It follows that $\text{Bias}[p_{ML}] = 0$, and then

$$\text{MSE}[p_{ML}] = \text{VAR}[p_{ML}] \equiv \text{VAR}[\bar{X}/2] = \frac{1}{4}\frac{\text{VAR}[X]}{m} = \frac{p(1-p)}{2m}.$$

b) Since $p_0 = (4/5)^2 = 16/25$ and $p_2 = (1/5)^2 = 1/25$, we can construct the following table:

j	0	1	2	Σ
n_j	20	8	2	30
p_j	16/25	8/25	1/25	1
$30p_j = m_j$	19.2	9.6	1.2	30

Given that $m_2 = 1.2 < 5$, we combine the last two classes. We obtain:

j	0	1 or 2	Σ
n_j	20	10	30
p_j	16/25	9/25	1
m_j	19.2	10.8	30

Next, we calculate

$$D^2 = \frac{(20 - 19.2)^2}{19.2} + \frac{(10 - 10.8)^2}{10.8} \simeq 0.0926.$$

Since $D^2 \simeq 0.0926$ is smaller than $\chi^2_{0.05, 2-0-1} \simeq 3.84$, we can accept the model $X \sim B(n = 2, p = 1/5)$, with $\alpha = 0.05$.

Exercise no. 3 (6.3)

Let X be the number of flaws in the paint of new cars. We suppose that $X \sim \text{Poi}(\lambda)$. We constructed the following table with the help of a random sample of 100 cars:

Number of flaws	0	1	2	3
Number of cars	40	36	20	4

a) Calculate the maximum likelihood estimator of the parameter λ, based on the above data.

b) Test the fit of the model $X \sim \text{Poi}(\lambda = 1)$ to the data. Use $\alpha = 0.05$.

Hint. We have: $\chi^2_{0.05, 2} \simeq 5.99$ and $\chi^2_{0.05, 3} \simeq 7.81$.

Solution

a) We have (see Section 6.1, p. 256):

$$L(\lambda) = \prod_{k=1}^{100} \frac{e^{-\lambda}\lambda^{X_k}}{(X_k)!} = \frac{e^{-100\lambda}\lambda^{\sum_{k=1}^{100} X_k}}{\prod_{k=1}^{100} (X_k)!}$$

$$\Rightarrow \quad \ln L(\lambda) = -100\lambda + \left(\sum_{k=1}^{100} X_k\right)\ln\lambda - \sum_{k=1}^{100}\ln X_k!$$

$$\Rightarrow \quad \frac{d}{d\lambda}\ln L(\lambda) = -100 + \frac{1}{\lambda}\sum_{k=1}^{100} X_k = 0 \Rightarrow \quad \lambda_{ML} = \sum_{k=1}^{100}\frac{X_k}{100}.$$

Next, we calculate

$$\bar{x} = (0 \times 40 + 1 \times 36 + 2 \times 20 + 3 \times 4)\frac{1}{100} = 0.88.$$

Thus, $\lambda_{ML} = 0.88$.

b) We have: $X \overset{H_0}{\sim} \text{Poi}(\lambda = 1) \Rightarrow p_k := p_X(k) = e^{-1}\frac{1^k}{k!}$ $\forall k \geq 0$. We then complete the table as follows:

k	0	1	2	3+	Σ
n_k	40	36	20	4	$100 = n$
$(\simeq) p_k$	0.368	0.368	0.184	0.080	1
$(100 p_k =) m_k$	36.8	36.8	18.4	8	100

Since $m_k \geq 5$ $\forall k$, next we calculate

$$D^2 \simeq \frac{(40 - 36.8)^2}{36.8} + \frac{(36 - 36.8)^2}{36.8} + \frac{(20 - 18.4)^2}{18.4} + \frac{(4 - 8)^2}{8} \simeq 2.44.$$

Given that $D^2 < \chi^2_{0.05,4-0-1} \simeq 7.81$, we accept the proposed model at significance level $\alpha = 0.05$.

Exercise no. 4 (6.4)

We believe that the courier service of company A is faster than that of company B. Independent observations of the delivery times (in days) of the two courier services yielded:

$$n_A = 5, \quad \bar{x}_A = 1.2; \quad n_B = 10, \quad \bar{x}_B = 1.5.$$

Moreover, we assume that the standard deviations of the delivery times are $\sigma_A = \sigma_B = 0.2$.

a) What additional assumptions must we make to be able to perform the comparison? Set also the hypotheses H_0 and H_1.

b) What is the smallest value of α for which we can reject H_0 if $z_0 = -2.74$?

Solution

a) Let X_A (respectively X_B) be the delivery time with company A (resp. B). We assume that X_A and X_B have a Gaussian distribution, so that

$$X_A \sim N(\mu_A, \sigma_A^2 = 0.04) \quad \text{and} \quad X_B \sim N(\mu_B, \sigma_B^2 = 0.04).$$

Furthermore, we assume that X_A and X_B are independent random variables. Finally, we set:

$$H_0: \mu_A = \mu_B \quad \text{and} \quad H_1: \mu_A < \mu_B.$$

b) We reject H_0 if and only if

$$z_0 < -z_\alpha \quad \Leftrightarrow \quad -2.74 < -z_\alpha.$$

We have:

$$\Phi(2.74) \stackrel{\text{Tab. 3.3}}{\simeq} 0.9969.$$

Therefore, we can write that

$$\alpha_{min} \simeq (1 - 0.9969)^+ = 0.0031^+.$$

Exercise no. 5 (6.4)

A manufacturer of synthetic rubber asserts that the average hardness of its rubber is equal to 64.3 degrees Shore. Previous experiments show that the standard deviation of the hardness is equal to two degrees Shore. We believe that the manufacturer's assertion either underestimates or overestimates the average rubber hardness. Therefore, we perform a test of hypotheses. If the true mean (μ) is indeed 64.3 degrees, the probability of reaching this conclusion must be equal to 0.95. Moreover, if the difference between μ and 64.3 is ± 1 degree, the test performed should conclude, with a probability of at least 0.9, that μ is not equal to 64.3. We assume that the rubber hardness has (approximately) a Gaussian distribution.

a) How many observations must we take?

b) If $\bar{x} = 65$ and n is the value computed in a), what is the conclusion of the test?

c) Calculate the probability of concluding that the average rubber hardness is equal to 64.3, if the true average hardness is 65 degrees Shore and if we use the sample size computed in a).

d) How many observations must we draw if we want to reject H_0 with a type I error risk of 0.05 when $\bar{x} = 65$?

Solution

a) We have: $\alpha = 0.05$ and $\beta\,(|\mu - 64.3| = 1) = 1 - 0.9 = 0.1$. Therefore, we must take

$$n \simeq \frac{(z_{0.025} + z_{0.10})^2\,(2)^2}{(\mu - 64.3)^2} \overset{\text{Tab. 6.1}}{\simeq} (1.960 + 1.282)^2\,\frac{4}{1} \simeq 42.04.$$

Thus, we must draw $n = 43$ (or 42) observations.

b) We want to test

$$H_0\colon \mu = 64.3 \quad \text{against} \quad H_1\colon \mu \neq 64.3$$

at significance level $\alpha = 0.05$. We reject H_0 if and only if

$$\bar{x} < C_1 := 64.3 - z_{0.025}\frac{2}{\sqrt{43}} \simeq 64.3 - 0.6 = 63.7$$

or

$$\bar{x} > C_2 := 64.3 + 1.960\frac{2}{\sqrt{43}} \simeq 64.3 + 0.6 = 64.9.$$

Since $\bar{x} = 65 > 64.9$, we reject the null hypothesis H_0 at level $\alpha = 0.05$.

Remark. If $\bar{x} = 65$ with $n = 42$ (rather than 43), then $C_2 \simeq 64.9$ still and the conclusion is the same.

c) We want to calculate $\beta\,(\mu = 65)$. We can write that

$$\beta\,(\mu = 65) \overset{\text{b)}}{\simeq} P\left[63.7 \leq \bar{X} \leq 64.9 \mid \mu = 65\right]$$

$$= P\left[63.7 \leq N\left(65, \tfrac{2}{\sqrt{43}}\right) \leq 64.9\right]$$

$$\simeq P\left[-4.26 \leq N(0,1) \leq -0.33\right]$$

$$\simeq P\left[N(0,1) \leq -0.33\right] \overset{\text{Tab. 3.3}}{\simeq} 0.37.$$

Remark. We could have used Equation (6.47) (p. 268) instead.

d) We seek n (minimal) such that

$$65 > 64.3 + z_{0.025}\frac{2}{\sqrt{n}} \quad \Leftrightarrow \quad \sqrt{n} > 5.6 \quad \Leftrightarrow \quad n > 31.3.$$

Thus, we must take at least 32 observations.

Exercise no. 6 (6.4)

We study the thermal efficiency X (in %) of Diesel engines made by a major car maker. We assume that X has (approximately) a Gaussian distribution with standard deviation equal to 2. Tests performed on 25 engines yielded the following results: $\bar{x} = 31.4$ and $s = 1.6$.

a) Test the hypothesis $H_0: \mu = 32.3$ against $H_1: \mu \neq 32.3$. Use $\alpha = 0.05$.

b) Compute the probability of making a type II error in a) if the average thermal efficiency is in fact 31.3%.

c) Based on the data collected, can we conclude that the standard deviation of X is actually smaller than 2%? Perform a test of hypotheses at significance level $\alpha = 0.01$.

Hint. We have: $\chi^2_{0.99,24} \simeq 10.86$.

Solution

a) We want to test

$$H_0: \mu = 32.3 \quad \text{against} \quad H_1: \mu \neq 32.3$$

at significance level $\alpha = 0.05$. We reject H_0 if and only if

$$\bar{x} < C_1 := 32.3 - z_{0.025}\frac{2}{\sqrt{25}} \simeq 31.516$$

or

$$\bar{x} > C_2 := 32.3 + 1.960\frac{2}{\sqrt{25}} \simeq 33.084.$$

Since $\bar{x} = 31.4 < 31.516$, we reject H_0 at level $\alpha = 0.05$.

b) We seek

$$\beta\,(\mu = 31.3) \simeq P\left[31.516 \leq \bar{X} \leq 33.084 \mid \mu = 31.3\right]$$

$$\simeq P\left[31.516 \leq N\left(31.3, \tfrac{4}{25}\right) \leq 33.084\right] = P\,[0.54 \leq N(0, 1) \leq 4.46]$$

$$\overset{\text{Tab. 3.3}}{\simeq} 1 - 0.7054 = 0.2946.$$

c) We want to test

$$H_0: \sigma^2 = 4 \quad \text{against} \quad H_1: \sigma^2 < 4$$

at level $\alpha = 0.01$. We reject H_0 if and only if

$$\frac{(25-1)}{4}s^2 < \chi^2_{1-0.01,25-1} \simeq 10.86.$$

Since

$$s^2 = (1.6)^2 = 2.56 \quad \Rightarrow \quad 6s^2 = 15.36 \geq 10.86,$$

we cannot reject H_0 at level $\alpha = 0.01$.

Unsolved Problems

Problem no. 1

Let $X_1, \ldots, X_n$ be a random sample of the random variable X whose probability density function is

$$f_X(x) = \frac{x}{2\theta^2} \quad \text{if } 0 \le x \le 2\theta \quad (= 0 \text{ elsewhere}).$$

We have: $\text{VAR}[X] = 2\theta^2/9$.

a) We propose the following estimator of the parameter θ: $T = \bar{X}$. Calculate the mean square error of T.

b) Calculate the maximum likelihood estimator of θ.

Problem no. 2

Let $1.1, 1.3, 1.9, 2.1, 2.5$ and 3.4 be a particular random sample of $X \sim \text{N}(\mu, \sigma^2)$. Calculate

a) a one-sided 95% confidence interval with an upper bound for μ;

b) a one-sided 90% confidence interval with a lower bound for μ, if we suppose that $\sigma^2 = 0.5$.

Problem no. 3

We collected the following observations of a random variable X:

j	1	2	3	4
n_j	24	16	6	4

Test the hypothesis H_0: $X \sim \text{Geom}(p = 1/2)$. Use $\alpha = 0.05$.

Problem no. 4

Let $X_1, \ldots, X_n$ be a random sample of size n of a discrete random variable X whose probability mass function is given by

$$p_X(k; \theta) = \theta(1-\theta)^{k-1} \quad \text{for } k = 1, 2, \ldots,$$

where $\theta \in (0, 1)$ is an unknown parameter.

a) Find the maximum likelihood estimator of the parameter θ.

b) We collected the following observations of the random variable X:

k	1	2	3
n_k	20	15	15

Test the hypothesis

$$H_0: p_X(k) = (1/2)^k \quad \text{for } k = 1, 2, \ldots \quad \text{against} \quad H_1: p_X(k) \ne (1/2)^k.$$

Use $\alpha = 0.10$.

Hint. We have: $\chi^2_{0.10,1} \simeq 2.706$, $\chi^2_{0.10,2} \simeq 4.605$ and $\chi^2_{0.10,3} \simeq 6.251$.

Problem no. 5

Let $X_1, \ldots, X_n$ be a random sample of size n of a random variable X having a Gaussian $N(\theta, 4)$ distribution, where θ is an unknown parameter.

a) Suppose that $n = 25$ and $\sum_{k=1}^{25} X_k = 2.5$. Find a constant c such that the interval $(-\infty, c]$ is a 95% confidence interval for θ.

b) We propose the following two estimators of the unknown parameter θ:

$$T_1 = \sum_{k=1}^{n} \frac{X_k}{n} \quad \text{and} \quad T_2 = \sum_{k=1}^{n} d X_k,$$

where d is a positive constant.

  i) Calculate the mean square error of T_2.

  ii) Find for what values of d the estimator T_2 is relatively more efficient than T_1, if $n = 4$ and $\theta = 1$.

Problem no. 6

Let

$$f_X(x; \theta) = \begin{cases} \frac{1}{\theta} e^{-(x-1)/\theta} & \text{if } x > 1, \\ 0 & \text{elsewhere,} \end{cases}$$

where $\theta > 0$.

a) Find θ_{ML}, that is, the maximum likelihood estimator of the parameter θ.

b) Calculate the mean square error of θ_{ML}.

Hint. We have: $E[X] = \theta + 1$ and $VAR[X] = \theta^2$.

c) Let $T := c\bar{X}$ be an estimator of θ, where $\bar{X}$ is the mean of a random sample of size n of X. Find the value of the constant c that minimizes the mean square error of T.

Problem no. 7

Let $X_1, \ldots, X_n$ be a random sample of size n of a random variable X whose probability density function is

$$f_X(x; \theta) = \frac{1}{\theta} e^{-x/\theta} \quad \text{for } x \geq 0.$$

We can show that the maximum likelihood estimator of the unknown parameter θ (> 0) is given by $\theta_{ML} = \bar{X}$.

a) Calculate the mean square error of θ_{ML}.

b) A particular random sample of size $n = 80$ of X has been drawn. The mean and the variance of the sample are equal to 2 and 5, respectively. Calculate an approximate 95% confidence interval for the parameter θ.

c) Give an approximate 95% confidence interval for θ, if in b) the sample variance is unknown. Justify your answer.

Problem no. 8

Let X be a random variable of continuous type whose probability density function is given by

$$f_X(x; \beta) = \begin{cases} \frac{1}{3!\beta^4} x^3 e^{-x/\beta} & \text{if } x > 0, \\ 0 & \text{if } x \le 0, \end{cases}$$

where $\beta > 0$ is a parameter.

a) Given a random sample of size n, $X_1, \ldots, X_n$, find the maximum likelihood estimator of β.

b) Consider a random sample of size four ($n = 4$): X_1, X_2, X_3, X_4, and the following two estimators for β: i) the estimator obtained in a) (with $n = 4$); ii) $\hat{\beta} := \frac{1}{40}(X_1 + 2X_2 + 3X_3 + 4X_4)$. Compare these two estimators with regard to their bias and their mean square error, and determine which one is preferable.

Hint. For the random variable X considered, we have: $E[X] = 4\beta$ and $\text{VAR}[X] = 4\beta^2$.

Problem no. 9

Let

$$p_X(x; \theta) = \begin{cases} (1 - \theta)/3 & \text{if } x = 0, \\ 1/3 & \text{if } x = 1, \\ (1 + \theta)/3 & \text{if } x = 2, \end{cases}$$

where $-1 < \theta < 1$ is an unknown parameter. A random sample of size $n = 30$ of X has enabled us to construct the following table:

i	0	1	2
n_i	10	12	8

a) We propose the following estimator of the parameter θ: $\hat{\theta} := \frac{3}{2}(\bar{X} - 1)$. Calculate the mean square error of $\hat{\theta}$.

Hint. We have: $E[X] = 1 + \frac{2}{3}\theta$ and $\text{VAR}[X] = \frac{2}{3} - \frac{4}{9}\theta^2$.

b) Test the hypothesis

$$H_0: p_X(x; \theta) = p_X(x; \theta = 0) = \frac{1}{3} \quad \text{for } x = 0, 1, 2.$$

Use $\alpha = 0.05$.

Problem no. 10

A particular random sample of nine cigarettes of a certain brand contains an average of 4.2 mg of nicotine per cigarette. The cigarette maker claims that the content X of nicotine in its cigarettes is not, on average, greater than 3.5 mg. We assume that $X \approx N(\mu; 1.96)$.

a) Can we put the cigarette maker's assertion in doubt? Use $\alpha = 0.05$.

b) What is the probability of concluding, in a), that the cigarette maker is wrong if μ = 3.8?

c) What is the smallest number of cigarettes that we should examine if we want the probability computed in b) to be greater than or equal to 90%?

d) To better protect the consumer, is it preferable to perform the test in a) with $\alpha = 0.05$ or $\alpha = 0.01$? Justify your answer.

Problem no. 11

We want to compare the compressive forces (in kilograms per square centimeter) of two types of concrete. We want to show that the second type of concrete has a greater compressive force, on average, than the first one. Measurements made with two specimens are given below:

Concrete 1	295	319	304	302
Concrete 2	318	316	312	318

a) Define the variables and specify the necessary basic assumptions.

b) Compare, at significance level $\alpha = 0.05$, the average compressive forces, assuming that $\sigma_1^2 \neq \sigma_2^2$.

c) Show, using a test of hypotheses, that the variances are actually not significantly different at level $\alpha = 0.05$.

d) Perform again the test of comparison of the population means in b) with $\sigma_1^2 = \sigma_2^2$.

Problem no. 12

Let $X \sim N(\mu, \sigma^2)$. A particular random sample of size $n = 25$ of X yielded $\bar{x} = 0.2$ and $s = 0.5$.

a) We want to test $H_0: \mu = 0$ against $H_1: \mu \neq 0$ at significance level $\alpha = 0.05$. Calculate the value of the statistic used to perform the test.

b) We also want to test $H_0: \sigma^2 = 0.2$ against $H_1: \sigma^2 > 0.2$ at significance level $\alpha = 0.05$. What is the value of the percentile used to make the decision (that is, the critical value)?

Problem no. 13

We are interested in the lifetime X (in tens of thousands of kilometers) of tires of a certain brand. We wish to test the hypothesis $H_0: \mu_X = 50$ against $H_1: \mu_X < 50$ at significance level $\alpha = 0.05$. We assume that $X \sim N(\mu_X, 25)$ (approximately).

a) What is the value of β if $n = 9$ and $\mu_X = 45$?

b) What is the smallest value of n for which $\beta \leq 0.10$ if $\mu_X = 45$?

Problem no. 14

Let $X_1 \sim N(\mu_1, \sigma_1^2)$ and $X_2 \sim N(\mu_2, \sigma_2^2)$ be two independent random variables. Particular random samples of X_1 and X_2 yielded:

$$n_1 = 9, \quad \bar{x}_1 = 5, \quad s_1 = 2; \quad n_2 = 10, \quad \bar{x}_2 = 3, \quad s_2 = 1.$$

a) Is the variance σ_1^2 significantly greater than σ_2^2? Use $\alpha = 0.05$.

b) If we assume that the theoretical variances are equal, can we then state that the theoretical means are significantly different? Use $\alpha = 0.05$.

Problem no. 15

We assume that all the students taking a certain course are equally gifted. We also assume that the mark obtained by a student on the final exam has (approximately) a Gaussian distribution. The results obtained on this exam, by the students of two groups with different teachers, are the following:

Group	n	$\bar{x}$	s
1	49	10.5	4.1
2	35	9.3	3.8

Let X_i be the mark that a student taking the course with teacher i will have on the final exam, for $i = 1, 2$. We assume that $X_1 \approx N(\mu_1, \sigma_1^2)$ and $X_2 \approx N(\mu_2, \sigma_2^2)$ are independent random variables.

a) Can we conclude that the difference between the variations in the marks is significant if we use a significance level $\alpha = 0.10$? Specify the null hypothesis H_0 and the alternative hypothesis H_1.

Hint. Use the fact that $F_{0.05, 48, 34} > F_{0.05, 50, 50}$.

b) Can we postulate that the teacher has a significant effect on the results of his students? Use $\alpha = 0.05$ and specify H_0 and H_1.

c) What is the smallest number of observations that we need to be able to detect, with a probability of at least 90%, an actual one-mark difference between the means μ_1 and μ_2?

d) What is, approximately, the value of β for the test performed in b) if $\mu_1 - \mu_2 = 2$?

Problem no. 16

The ohmic resistance of a certain electronic component must be equal, on average, to 400 ohms. A random sample of 16 components, drawn from a large batch of components, yielded the following observations:

392 396 386 389 388 387 403 397 401 391 400 402 394 406 406 400

We assume that the ohmic resistance has approximately a Gaussian distribution.

a) Can we affirm, at significance level $\alpha = 0.05$, that the batch meets the norm of 400 ohms?

b) Calculate the probability of making a type II error with the test performed in a) (that is, with the rejection constants for $\bar{x}$ computed in a)) if the ohmic resistance has actually a Gaussian $N(405, 49)$ distribution.

c) Under the same assumptions as in b), how many observations must we take if the probability of making a type II error must be smaller than or equal to 0.05?

Problem no. 17

We use two identical machines to manufacture a certain part. We wish to determine whether the two machines have the same variability with respect to an important characteristic of this part. Particular random samples, drawn from the production of each machine, yielded the following results:

Machine A	140	135	140	138	135	138	140
Machine B	135	138	136	140	138	135	139

We have: $\bar{x}_A = 138$, $s_A \simeq 2.24$, $\bar{x}_B = 137.29$ and $s_B \simeq 1.98$.

a) Specify the assumptions that we must make to be able to perform the test.

b) Perform the appropriate statistical test, using a significance level equal to 0.10. What is the conclusion?

c) Test the hypothesis that the 14 observations collected come from a uniform distribution over the interval [135, 140]. Use the significance level $\alpha = 0.05$ and the following intervals: [135, 137.5) and [137.5, 140].

Problem no. 18

We are interested in the breaking strength of metal rods made by two companies. We have a batch of rods manufactured by each company. We draw at random ten rods from each batch and we measure the strength needed to break each one of them. We obtain the following results:

Batch 1	53.2	53.9	53.1	50.9	42.2	52.9	55.8	41.9	50.0	49.0
Batch 2	55.0	64.5	53.0	57.8	56.1	58.0	55.8	50.8	54.0	52.2

We assume that the breaking strength of the metal rods has approximately a Gaussian distribution.

a) Test the hypothesis that the average breaking strength is the same in the two batches, at significance level $\alpha = 0.05$.

Hint. We have: $t_{0.025,18} \simeq 2.101$.

b) Test, assuming that the 20 rods actually come from the same company, the hypothesis that i) the average breaking strength of a metal rod is equal to 55 and ii) the variance of the breaking strength of a metal rod is 20. Use $\alpha = 0.05$.

Hint. We have: $t_{0.025,19} \simeq 2.093$, $\chi^2_{0.025,19} \simeq 32.85$ and $\chi^2_{0.975,19} \simeq 8.91$.

Problem no. 19

Measurements of percentages of elongation have been made on ten steel parts. Five of these parts were treated with method A (aluminum only) and the other five with method B (aluminum plus calcium), yielding the following results:

Method A (%)	28	29	31	33	30
Method B (%)	34	27	30	36	33

a) Test the equality of variances. Use $\alpha = 0.10$.

b) Can we conclude that the two methods give, on average, the same results? Use $\alpha = 0.05$.

Multiple Choice Questions

Question no. 1

Let $X_1, \ldots, X_n$ be a random sample of a random variable X having a Bernoulli distribution with parameter p.

A) Find the maximum likelihood estimator p_{ML} of the parameter p.

a) $1/(n\bar{X})$ b) $n\bar{X}$ c) $1/\bar{X}$ d) $\bar{X}/n$ e) $\bar{X}$

B) Give the exact and the approximate distribution of $n\bar{X}$.

a) $B(1, p)$ and $Poi(np)$ b) $B(1, p)$ and $N(np, npq)$ c) $B(1, p)$ and $N(0, 1)$
d) $B(n, p)$ and $N(p, pq/n)$ e) $B(n, p)$ and $N(np, npq)$

C) Give a formula for an approximate $100(1 - \alpha)\%$ confidence interval for p.

a) $p \pm z_{\alpha/2}\left[\frac{p(1-p)}{n}\right]^{1/2}$ b) $\bar{X} \pm z_{\alpha/2}\left[\frac{p(1-p)}{n}\right]^{1/2}$ c) $\bar{X} \pm z_{\alpha/2}\left[\frac{\bar{X}(1-\bar{X})}{n}\right]^{1/2}$

d) $p \pm z_{\alpha/2}\frac{p(1-p)}{n}$ e) $\bar{X} \pm z_{\alpha/2}\frac{\bar{X}(1-\bar{X})}{n}$

Question no. 2

Let $X_1, \ldots, X_n$ be a random sample of a random variable X having a uniform distribution on the interval $[0, \theta]$. We propose the following estimator of the unknown parameter θ: $\hat{\theta} = 2\bar{X}$. Calculate the mean square error of $\hat{\theta}$.

Hint. We have: $VAR[X] = \theta^2/12$.

a) 0 b) $\frac{\theta^2}{12}$ c) $\frac{\theta^2}{3}$ d) $\frac{\theta^2}{12n}$ e) $\frac{\theta^2}{3n}$

Question no. 3

Let X be a random variable having a Gaussian distribution with unknown parameters. A particular random sample $x_1, \ldots, x_{25}$ of X has yielded the following results:

$$\sum_{k=1}^{25} x_k = 175 \quad \text{and} \quad \sum_{k=1}^{25} x_k^2 = 1550.$$

Calculate a two-sided 95% confidence interval for $E[X]$.

a) $7 \pm 0.7360\, t_{0.025,24}$ b) $7 \pm 0.7360\, t_{0.025,25}$ c) $7 \pm 0.7360\, z_{0.25}$
d) $7 \pm 0.5417\, t_{0.025,24}$ e) $7 \pm 0.5417\, z_{0.025}$

Question no. 4

We can show that a theoretical confidence interval at approximately 95% for the parameter α of a random variable having a Poisson distribution is given by $\bar{X} \pm 1.96\, STD[\bar{X}]$, where $\bar{X}$ is the mean of a random sample of size n of X. A particular random sample of size $n = 1000$ has yielded a sample mean of 0.4. Calculate (approximately) the confidence interval for α, based on this particular sample.

a) 0.4 ± 0.02 b) 0.4 ± 0.03 c) 0.4 ± 0.04 d) 0.4 ± 0.05
e) we cannot calculate it

Question no. 5

We want to test the hypothesis that a random variable X has a Gaussian $N(0, 1)$ distribution. A (particular) random sample of size $n = 100$ of X has enabled us to construct the following table:

j	$(-\infty, -0.674)$	$[-0.674, 0)$	$[0, 0.674)$	$[0.674, \infty)$
n_j	20	25	25	30

Calculate the statistic D^2 used to carry out the test.

Hint. We have: $Q(0.674) \simeq 0.25$.

a) 0   b) 4/25   c) 1   d) 2   e) 4

Question no. 6

Let

$$f_X(x; \theta) = \frac{1}{2\theta} \quad \text{if } -\theta < x < \theta,$$

where $\theta > 0$ is an unknown parameter. We have: $E[X^2] = \theta^2/3$ and $E[X^4] = \theta^4/5$.

A) To estimate the unknown parameter $\beta := \theta^2$, we propose the following estimator:

$$\hat{\beta} := \frac{3}{n} \sum_{k=1}^{n} X_k^2,$$

where $X_1, \ldots, X_n$ is a random sample of size n of X. Calculate the mean square error of $\hat{\beta}$.

a) $\frac{4}{5}\frac{\beta}{n}$ b) $\frac{4}{45}\frac{\beta^2}{n^2}$ c) $\frac{4}{5}\frac{\beta^2}{n^2}$ d) $\frac{4}{45}\frac{\beta^2}{n}$ e) $\frac{4}{5}\frac{\beta^2}{n}$

B) We collected the following observations of the random variable X:

j	$(-2, -1)$	$[-1, 0)$	$[0, 1)$	$[1, 2)$
n_j	27	28	32	33

We want to test the hypothesis

$$H_0: f_X(x; \theta) = f_X(x; \theta = 2) = \frac{1}{4} \quad \text{if } -2 < x < 2.$$

Calculate the statistic D^2 used to carry out the test and give the number d of degrees of freedom of D^2 (under H_0).

a) $D^2 \simeq 0.8667; d = 2$ b) $D^2 \simeq 0.8667; d = 3$ c) $D^2 \simeq 0.8739; d = 2$
d) $D^2 \simeq 0.8739; d = 3$ e) none of these answers

Question no. 7

Let $X_1, \ldots, X_n$ be a random sample of a random variable X having an exponential distribution with parameter λ. We define $\theta = \sqrt{\lambda}$. Find the maximum likelihood estimator of the parameter θ.

a) $1/\sqrt{\bar{X}}$ b) $1/\bar{X}$ c) $1/\bar{X}^2$ d) $\sqrt{\bar{X}}$ e) $\bar{X}^2$

Question no. 8

The following data have been collected by tossing a die 90 times:

j	1	2	3	4	5	6
n_j	13	12	16	18	15	16

Let X be the number obtained by tossing the die. We want to test the hypothesis

$$H_0: P[X \text{ is even}] = P[X \text{ is odd}].$$

Calculate the statistic D^2 used to perform the test and give the number d of degrees of freedom of D^2 under the null hypothesis H_0.

a) $D^2 = 0; d = 0$ b) $D^2 = 0.0\bar{4}; d = 1$ c) $D^2 = 0.0\bar{4}; d = 2$
d) $D^2 = 1.6; d = 4$ e) $D^2 = 1.6; d = 5$

Question no. 9

A particular 95% confidence interval for the mean μ of a population $X \sim N(\mu, 9)$, calculated from a random sample of size $n = 99$, is given by $10 \pm z_{0.025} \frac{3}{\sqrt{99}}$. A 100th observation, x_{100}, is taken. Calculate the new confidence interval, if $x_{100} = 11$.

a) $10 \pm z_{0.025}(0.3)$ b) $10.01 \pm z_{0.025}(0.3)$ c) $10 \pm z_{0.025}(0.299)$
d) $10.01 \pm z_{0.025}(0.299)$ e) none of these answers

Question no. 10

We want to test the hypothesis that the following data come from a $B(3, \frac{1}{2})$ distribution:

j	0	1	2
n_j	6	18	16

Calculate the statistic D^2 used to make the decision and give the number d of degrees of freedom of D^2 under the null hypothesis.

a) $D^2 \simeq 1.87; d = 2$ b) $D^2 \simeq 1.87; d = 3$ c) $D^2 \simeq 3.87; d = 3$
d) $D^2 \simeq 5.87; d = 2$ e) none of these answers

Question no. 11

Suppose that X has a uniform distribution on the interval $[0, \theta]$. To estimate the unknown parameter θ, we propose the estimator $\hat{\theta} = \bar{X}$, where $\bar{X}$ is the mean of a random sample of size n of X. Calculate the mean square error of $\hat{\theta}$.

a) $\frac{\theta^2}{12n}$ b) $\frac{\theta^2}{12n} + \frac{\theta^2}{4}$ c) $\frac{\theta^2}{3n}$ d) $\frac{\theta^2}{12}$ e) $\frac{\theta^2}{12} + \frac{\theta^2}{4n}$

Question no. 12

Let X be a random variable having a Poisson distribution with unknown parameter $\theta > 0$. We consider a random sample of size n (> 30) of X. Use the central limit theorem to obtain an approximate $100(1 - \alpha)\%$ confidence interval for θ.

a) $\bar{X} \pm z_{\alpha/2}\theta^{1/2}$ b) $\bar{X} \pm z_{\alpha/2}(\theta/n)^{1/2}$ c) $\bar{X} \pm z_{\alpha/2}(\bar{X}/n)$
d) $\bar{X} \pm z_{\alpha/2}(\bar{X}/n^{1/2})$ e) $\bar{X} \pm z_{\alpha/2}(\bar{X}/n)^{1/2}$

Question no. 13

Suppose that $X \sim N(0, \theta)$, where θ is an unknown parameter. Find the estimator of θ by the method of maximum likelihood.

a) $\bar{X}^2$ b) $\frac{1}{n^2}\sum_{k=1}^{n} X_k^2$ c) $\frac{1}{n}\sum_{k=1}^{n} X_k^2$ d) $\frac{1}{n}\sum_{k=1}^{n}(X_k - \bar{X})^2$

e) $\frac{1}{n-1}\sum_{k=1}^{n}(X_k - \bar{X})^2$

Question no. 14

Let $X \sim N(\mu, \mu^2)$. Obtain a formula for an approximate $100(1-\alpha)\%$ confidence interval for μ.

a) $\bar{X} \pm z_{\alpha/2}(\mu)$ b) $\bar{X} \pm z_{\alpha/2}(\mu/n^{1/2})$ c) $\bar{X} \pm z_{\alpha/2}(|\bar{X}|/n)$

d) $\bar{X}^2 \pm z_{\alpha/2}(|\bar{X}|/n^{1/2})$ e) $\bar{X} \pm z_{\alpha/2}(|\bar{X}|/n^{1/2})$

Question no. 15

Let

$$f_X(x; \theta) = \frac{1}{2\theta}e^{-|x|/\theta} \quad \text{for } x \in \mathbb{R},$$

where θ is a positive parameter. We can show that $\text{VAR}[X] = 2\theta^2$.

A) Find the maximum likelihood estimator of θ.

a) 0   b) $\frac{1}{n}\sum_{k=1}^{n} X_k$ c) $\frac{1}{n}\sum_{k=1}^{n} |X_k|$ d) $\frac{1}{n}(\sum_{k=1}^{n} X_k^2)^{1/2}$ e) $\frac{1}{\sqrt{n}}(\sum_{k=1}^{n} X_k^2)^{1/2}$

B) We consider the following estimator of the parameter $\beta := \theta^2$:

$$\hat{\beta} := \frac{1}{2}\sum_{k=1}^{n} \frac{X_k^2}{n-1}.$$

Calculate the bias of $\hat{\beta}$.

a) 0   b) β c) $\frac{n}{n-1}\beta$ d) $\frac{1}{n-1}\beta$ e) $\frac{1}{n}\beta$

Question no. 16

Two makers of batteries for laptop computers claim that with their batteries the computers can be used for at least three hours before the batteries have to be recharged. Batteries from these two makers have been tested with ten different computers. The data are the following:

Computer	1	2	3	4	5	6	7	8	9	10
T_1	2.9	2.8	2.9	3.2	3.0	3.1	2.7	2.9	2.7	2.9
T_2	3.1	3.2	3.3	3.0	2.9	2.9	3.1	3.2	2.8	3.2

where T_i is the functioning time (in hours) obtained with a brand i battery, for $i = 1, 2$. We assume that T_i has (approximately) a Gaussian distribution.

A) We believe that the standard deviation of the random variable T_1 is greater than 0.2. Calculate the statistic w_0^2 needed to perform the test.

a) 0.636   b) 1.145   c) 5.725   d) 10.305   e) 11.450

B) Next, we want to test the hypothesis that there is not a significant difference between the mean of T_1 and that of T_2. Calculate the appropriate statistic.

a) -2.21 b) -2.06 c) -1.96 d) 2.21   e) 7.00

C) We find that the standard deviation of the random sample of T_2 is (about) 0.1636. How many observations should we take (approximately) in order that $\beta \simeq 0.10$, if $\mu_{T_2} = 2.9$, when we perform the test of H_0: $\mu_{T_2} = 3$ against H_1: $\mu_{T_2} < 3$ at significance level $\alpha = 0.05$?

a) 23   b) 25   c) 28   d) 31   e) 34

Question no. 17

A company has invented a device that is supposed to reduce fuel consumption by at least 10%.

A) Let R be the reduction in fuel consumption obtained with the device in question. We suppose that R has a gamma distribution with parameters α and λ.

i) If $\alpha = 1$, we find that the maximum likelihood estimator of the parameter $\theta := 1/\lambda$ is $\theta_{ML} = \bar{R}$, where $\bar{R}$ is the mean of a random sample of size n of R. Calculate the mean square error of θ_{ML}.

Hint. We have: $E[R] = \alpha/\lambda$ and $VAR[R] = \alpha/\lambda^2$.

a) $\frac{1}{n\theta^2}$ b) $\frac{1}{n\theta}$ c) $\frac{n}{\theta^2}$ d) $\frac{\theta}{n}$ e) $\frac{\theta^2}{n}$

ii) Estimate the parameter α by the method of moments, if $\alpha = \lambda$.

a) $\bar{R}$ b) $1/\bar{R}$ c) $\left[\left(\frac{1}{n}\sum_{k=1}^{n} R_k^2\right) - 1\right]^{-1}$ d) $\left[\frac{1}{n}\sum_{k=1}^{n} R_k^2\right]^{-1}$ e) $\frac{1}{n}\sum_{k=1}^{n} R_k^2$

B) The device has been installed on ten cars of the same make and their fuel consumption has been measured over a 200-km distance. The results are given in the table below:

Car	1	2	3	4	5	6	7	8	9	10
X	17.5	18.3	19.1	16.4	18.9	17.8	20.2	19.4	17.6	18.4

where X is the fuel consumption with the device installed. We assume that X has (approximately) a Gaussian distribution. Knowing that the average fuel consumption of this make of car, without the device, is equal to 10 liters/100 km, we wish to test at significance level $\alpha = 0.025$

$$H_0: \mu_X = 18 \quad \text{against} \quad H_1: \mu_X > 18.$$

i) Give the value of the statistic and of the percentile used to make the decision.

a) 1.04; 2.262   b) 1.14; 2.262   c) 3.29; 2.262   d) −0.95; 1.960
e) 0.95; 1.960

ii) Use a linear interpolation, based on the fact that $\Phi(1) \simeq 0.8413$ and $\Phi(1.5) \simeq 0.9332$, to compute approximately the value of β if the true value of the mean of X is $\mu_X = 19$ and if $\sigma_X = 1$.

a) 0.092   b) 0.102   c) 0.112   d) 0.122   e) 0.132

iii) What is the smallest number of observations that must be taken to be able to detect, with a probability of at least 0.95, that μ_X is greater than 18 if we suppose that the true value of μ_X is 19 and that $\sigma_X = 1$?

a) 12   b) 13   c) 14   d) 15   e) 16

Question no. 18

Let $X_1, \ldots, X_n$ be a random sample of a random variable X whose probability density function is defined by

$$f_X(x; \theta) = \theta x^{\theta - 1} \quad \text{if } 0 < x < 1,$$

where $\theta > 0$ is an unknown parameter.

A) Find the estimator of θ by the method of moments.

a) $\bar{X}$ b) $1 - \bar{X}$ c) $\bar{X}(1 - \bar{X})$ d) $(1 - \bar{X})/\bar{X}$ e) $\bar{X}/(1 - \bar{X})$

B) In order to check whether the model $f_X(x; \theta)$ proposed above is adequate for a certain random variable X, we perform a (Pearson's) goodness-of-fit test. We gather 162 observations of X and we group them into three intervals. We obtain the following table:

Interval	$(0, \frac{1}{3})$	$[\frac{1}{3}, \frac{2}{3})$	$[\frac{2}{3}, 1)$
Number of observations	14	62	86

Moreover, we find that the value of the maximum likelihood estimator of the parameter θ calculated from these 162 observations is $\theta_{ML} = 2$.

Calculate the statistic D^2 used to carry out the test and give the number d of degrees of freedom associated with D^2.

a) $D^2 \simeq 2.25$; $d = 1$ b) $D^2 \simeq 3.25$; $d = 1$ c) $D^2 \simeq 4.25$; $d = 1$

d) $D^2 \simeq 2.25$; $d = 2$ e) $D^2 \simeq 3.25$; $d = 2$

Question no. 19

A maker of electronic chips asserts that the average size μ (in micrometers) of certain components of its chips does not exceed 2. In order to check this assertion, we decide to perform a test of hypotheses. The test used should be such that there is a 0.05 probability of rejecting the maker's assertion if μ is equal to 2. This probability should be 0.9 if μ is equal to 2.5.

We assume that the size of the components in question has (approximately) a Gaussian $N(\mu, 0.16)$ distribution.

A) Give the hypotheses that we want to test.

a) $H_0: \mu = 2$; $H_1: \mu = 2.5$ b) $H_0: \mu = 2$; $H_1: \mu \neq 2$

c) $H_0: \mu = 2$; $H_1: \mu < 2$ d) $H_0: \mu = 2$; $H_1: \mu > 2$

e) $H_0: \mu = 2.5$; $H_1: \mu > 2.5$

B) What is the smallest size of the random sample that must be drawn?

a) 2   b) 6   c) 15   d) 18   e) 35

C) If $\mu = 2.4$ and $n = 10$, what is the probability of rejecting the maker's assertion?

a) $\Phi(1.1)$ b) $\Phi(1.2)$ c) $\Phi(1.3)$ d) $\Phi(1.4)$ e) $\Phi(1.5)$

Question no. 20

The lifetime X of the transistors produced by a certain company has a Gaussian distribution with a standard deviation of 30 days. The company aims at an average lifetime of 210 days. In order to estimate the mean μ of X, the engineer responsible for the control of the quality examines 36 transistors taken at random and obtains a sample mean of 225 days. A 95% confidence interval for μ is then given by [215.2, 234.8].

A) We want to reduce by half the width of the confidence interval for μ by increasing the size of the sample. What value of n must be used?

a) 54   b) 72   c) 90   d) 108   e) 144

B) We now want to reduce by half the width of the confidence interval for μ, while keeping the sample size equal to $n = 36$. What value of α must then be chosen?

a) $\frac{1}{2}Q(0.98)$ b) $Q(0.98)$ c) $2Q(0.98)$ d) $\frac{1}{2}Q(1.96)$ e) $2Q(1.96)$

C) Next, the engineer wants to test, at level 0.025, whether the data indicate that the average lifetime is greater than 210 days.

i) Give the hypotheses that we want to test.

a) $H_0: \mu = 210; H_1: \mu < 210$ b) $H_0: \mu = 210; H_1: \mu \neq 210$
c) $H_0: \mu = 210; H_1: \mu > 210$ d) $H_0: \mu = 210; H_1: \mu > 225$
e) $H_0: \mu = 225; H_1: \mu < 210$

ii) With the test used in i), what is the probability of concluding that the average lifetime is not greater than 210 days, when it is actually 220 days?

a) $Q(0.04)$ b) $Q(-0.04)$ c) $Q(0.14)$ d) $Q(-0.14)$ e) $Q(0.24)$

D) Suppose that the population variance is unknown. What is the value of the statistic used to test the hypothesis that the mean of X is greater than 210 days, if the sum of the squares of the 36 observations is equal to 1,875,000?

a) 2.02   b) 2.12   c) 2.22   d) 2.32   e) 2.42

Question no. 21

Let X be a random variable having the following probability density function:

$$f_X(x; \theta) = \begin{cases} 2\theta x e^{-\theta x^2} & \text{if } x \geq 0, \\ 0 & \text{if } x < 0, \end{cases}$$

where $\theta > 0$ is an unknown parameter.

A) Find the maximum likelihood estimator of the unknown parameter θ.

a) $\bar{X}$ b) $(\bar{X})^2$ c) $(\bar{X})^{-2}$ d) $\frac{\sum_{k=1}^{n} X_k^2}{n}$ e) $\frac{n}{\sum_{k=1}^{n} X_k^2}$

B) We collected 30 observations of the random variable X, with which the following frequency table was constructed:

$[0, \frac{\sqrt{2}}{2})$	$[\frac{\sqrt{2}}{2}, 1)$	$[1, \infty)$
12	8	10

We want to test, at significance level $\alpha = 0.05$, the hypothesis

$$H_0: f_X(x; \theta) = f_X(x; \theta = 1) = 2xe^{-x^2} \quad \text{against} \quad H_1: f_X(x) \neq 2xe^{-x^2}$$

for $x \geq 0$. Give the value of the statistic used to perform the test, as well as the number d of degrees of freedom associated with this statistic (under H_0).

a) 0.20; $d = 1$ b) 0.20; $d = 2$ c) 0.40; $d = 1$ d) 0.40; $d = 2$
e) 0.60; $d = 1$

Question no. 22

A tire manufacturer claims that the average braking distance for a car equipped with its brand Z tires is not greater than the average braking distance obtained with the more expensive brand W tires. We believe that this statement is false and we decide to test it by measuring the braking distance (in meters) from 100 km/h to a complete stop. The data are the following:

Brand Z tires	40	41	39	44	45
Brand W tires	38	40	38	41	45

Let Z be the braking distance obtained with the brand Z tires, and W the braking distance obtained with those of brand W. We assume that $Z \sim N(\mu_Z, \sigma_Z^2)$ and $W \sim N(\mu_W, \sigma_W^2)$ (approximately).

A) Suppose that the measurements have been made with five different makes of cars, the first time using the brand Z tires, and the second time using the brand W tires. Give the value of the statistic used to test, at level $\alpha = 0.05$, the hypothesis H_0: $\mu_Z = \mu_W$. Give also the value of the percentile c to which the statistic in question is compared.

a) 0.81; $c \simeq 1.860$ b) 0.81; $c \simeq 2.306$ c) 2.75; $c \simeq 1.860$
d) 2.75; $c \simeq 2.132$ e) 2.75; $c \simeq 2.776$

B) Suppose that the ten braking distance measurements have been made with the same car, on ten different days. Calculate, assuming that $\sigma_Z = \sigma_W = 2.5$, the value of the appropriate statistic to test at level $\alpha = 0.05$ the hypothesis $H_0 : \mu_Z = \mu_W$, and give the value of the percentile c to which this statistic is compared.

a) 0.81; $c \simeq 1.645$ b) 0.81; $c \simeq 1.860$ c) 0.89; $c \simeq 1.645$
d) 0.89; $c \simeq 1.860$ e) 0.89; $c \simeq 2.306$

C) Under the same assumptions as in B), what is the probability of detecting that the average braking distance μ_Z is greater than the mean μ_W when, in fact, $\mu_Z - \mu_W = 1.25$?

a) $Q(0.791)$ b) $Q(0.854)$ c) $Q(2.436)$ d) $\Phi(0.791)$ e) $\Phi(0.854)$

Question no. 23

Suppose that $X \sim N(-\theta, \theta)$, where $\theta > 0$ is an unknown parameter.

A) We propose the following estimator of the parameter θ: $\hat{\theta} := -\bar{X}$. Calculate the mean square error of $\hat{\theta}$.

a) θ^2 b) $\theta^2 + \frac{\theta}{n}$ c) $\frac{\theta}{n}$ d) $\frac{\theta^2}{n}$ e) $\theta^2 - \frac{\theta}{n}$

B) Use the estimator $\hat{\theta} = -\bar{X}$ to obtain a formula giving an approximate $100(1-\alpha)\%$ confidence interval for θ, based on a random sample of size $n = 100$, if we assume that $\bar{X} < 0$.

a) $-\bar{X} \pm z_\alpha \frac{(-\bar{X})^{1/2}}{10}$ b) $-\bar{X} \pm z_{\alpha/2} \frac{(-\bar{X})}{10}$ c) $-\bar{X} \pm z_{\alpha/2} \frac{(-\bar{X})^{1/2}}{10}$

d) $-\bar{X} \pm z_{\alpha/2} \frac{\theta^{1/2}}{10}$ e) $-\bar{X} \pm z_\alpha \frac{\theta^{1/2}}{10}$

Question no. 24

The speed X of the microchips of a certain company is supposed to be 2 GHz. We assume that X has (approximately) a Gaussian $N(\mu, \sigma^2)$ distribution.

A) We take a random sample of size $n = 9$ of X and we compute the average speed $\bar{x}$ of the microchips. If the sample standard deviation s is equal to 0.2, for what values of $\bar{x}$ can we conclude that the average speed of the microchips is smaller than 2 GHz? Use $\alpha = 0.025$.

a) $\bar{x} < 1.85$ b) $\bar{x} > 1.85$ c) $\bar{x} < 1.95$ d) $\bar{x} > 1.95$ e) $|\bar{x} - 2| < 0.15$

B) Suppose that the average speed of the microchips is actually equal to 2.1 GHz, and that $\sigma = 0.25$. What is the probability of rejecting the hypothesis H_0: $\mu = 2$ (accepting H_1: $\mu < 2$ instead) at level $\alpha = 0.05$, if we draw a random sample of size $n = 16$?

a) $\Phi(2.2)$ b) $\Phi(3.2)$ c) $Q(2.2)$ d) $Q(3.2)$ e) $2\Phi(2.2) - 1$

C) What is the value of the statistic used to test the hypothesis H_0: $\sigma^2 = 0.02$ against H_1: $\sigma^2 > 0.02$ if a random sample of size $n = 10$ yielded

$$\sum_{k=1}^{10} x_k = 20.17 \quad \text{and} \quad \sum_{k=1}^{10} x_k^2 = 40.775?$$

What is the conclusion of the test if $\alpha = 0.05$?

a) 4.6; we reject H_0 b) 4.6; we accept H_0 c) 9.2; we reject H_0
d) 9.2; we accept H_0 e) 18.4; we reject H_0

Question no. 25

Suppose that X is a random variable having a Weibull distribution with parameters $\beta > 0, \gamma \in \mathbb{R}$ and $\delta = 1$. That is,

$$f_X(x; \beta, \gamma) = \beta(x - \gamma)^{\beta-1} \exp\{-(x - \gamma)^\beta\} \quad \text{for } x \geq \gamma.$$

A) Let $\gamma = 0$. Give the value of the estimator of the unknown parameter β by the method of moments, if a random sample of X yielded $\bar{x} = 6$.

Hint. Set $y = x^\beta$ in the calculation of the mathematical expectation of X and use the fact that

$$\int_0^\infty x^{\alpha-1} e^{-x} \, dx = \Gamma(\alpha).$$

a) 1/5   b) 1/4   c) 1/3   d) 1/2   e) 1

B) Let $\beta = 2$. Estimate the parameter γ by the method of maximum likelihood with the help of a single observation, X_1, of X.

a) $X_1 - \frac{1}{2}$ b) $X_1 - \frac{\sqrt{2}}{2}$ c) X_1 d) $X_1 + \frac{\sqrt{2}}{2}$ e) $X_1 + \frac{1}{2}$

C) Test, at significance level $\alpha = 0.05$, the hypothesis H_0 that

$$f_X(x; \beta, \gamma) = f_X(x; \beta = 1, \gamma = 0) = e^{-x} \quad \text{for } x \geq 0,$$

if we collected the following data:

j	[0, 1)	[1, 2)	[2, 3)	[3, ∞)
n_j	80	40	20	10

Give the value of the statistic D^2 and the conclusion of the test.

a) 5.94; we accept H_0 b) 6.94; we accept H_0 c) 6.94; we reject H_0
d) 7.94; we accept H_0 e) 7.94; we reject H_0

Question no. 26
   A car maker, in its publicity, claims that the power X (in horsepower) of the engines of its cars of brand A is 200 hp. Tests made with 16 independent vehicles gave an average power of 196 hp and a sample standard deviation s of 5 hp. We assume that X has approximately a Gaussian $N(\mu, \sigma^2)$ distribution.

A) We can state, with a risk α of type I error equal to 0.05, that the average power of the engines is in fact smaller than 200 hp. Give the values of the statistic and the percentile that were both used to make the decision.

a) -3.2; -2.131 b) -3.2; -1.753 c) -3.2; -1.645 d) 3.2; 1.753
e) 3.2; 2.131

B) Based on the data collected, can we conclude, if $\alpha = 0.05$, that $\sigma_X > 4$? Give the value of the statistic used and the conclusion of the test.

a) 4.69; we accept H_0 b) 18.75; we accept H_0 c) 18.75; we reject H_0
d) 23.44; we accept H_0 e) 23.44; we reject H_0

C) What is the smallest number of observations that should be taken in B) to be able to state, with a probability of at least 0.9, that $\sigma_X > 4$ if, in fact, $\sigma_X = 6$?

a) 23   b) 24   c) 25   d) 26   e) 27

Question no. 27
   The random variable X has the probability density function

$$f_X(x; \theta) = \frac{3}{2\theta^3} x^2 \quad \text{if } -\theta \leq x \leq \theta,$$

where θ (> 0) is an unknown parameter.

A) Find the estimator of the parameter θ by the method of moments.

a) 0   b) $\bar{X}$ c) $\left(\frac{5}{3}\bar{X}\right)^{1/2}$ d) $\frac{5}{3n}\sum_{k=1}^{n} X_k^2$ e) $\left(\frac{5}{3n}\sum_{k=1}^{n} X_k^2\right)^{1/2}$

B) We constructed the frequency table below from a random sample of size 80 of X:

j	[-2, -1)	[-1, 0)	[0, 1)	[1, 2]
n_j	32	4	5	39

Give the value of the statistic D^2 used to test

$$H_0: f_X(x; \theta) = f_X(x; \theta = 2) = \frac{3}{16}x^2 \quad \text{if } -2 \leq x \leq 2.$$

How many degrees of freedom does the statistic D^2 have (under H_0)?

a) $D^2 = 27/35$; $d = 2$ b) $D^2 = 32/35$; $d = 2$ c) $D^2 = 32/35$; $d = 3$
d) $D^2 = 37/35$; $d = 3$ e) $D^2 = 37/35$; $d = 4$

Question no. 28

We wish to compare the fuel consumption (in liters per 100 km) of two makes of cars, A and B. We believe that the cars of brand B use less fuel, on average, than those of brand A. Let X_A (respectively X_B) be the fuel consumption of the cars of brand A (resp. B). We assume that $X_A \approx N(\mu_A, \sigma_A^2)$ and $X_B \approx N(\mu_B, \sigma_B^2)$ are independent random variables. Random samples of X_A and X_B have been taken. The data are the following:

X_A	10.5	10.9	9.8	10.2	11.9	9.9	10.7	10.1	10.6
X_B	9.7	10.0	10.3	9.4	9.6	9.8	10.2	10.0	

A) We must first test the equality of variances. We find that if $\alpha = 0.05$, then we conclude that $\sigma_A^2 = \sigma_B^2$. What is the value of the statistic used to test, at significance level $\alpha = 0.05$, the hypothesis that the average fuel consumption of the brand B cars is smaller than that of the brand A cars? What is the alternative hypothesis H_1?

a) 2.46; $H_1: \mu_A > \mu_B$ b) 2.56; $H_1: \mu_A > \mu_B$ c) 2.46; $H_1: \mu_A < \mu_B$
d) 2.56; $H_1: \mu_A < \mu_B$ e) 2.66; $H_1: \mu_A > \mu_B$

B) If we choose $\alpha = 0.10$ instead, then we must conclude that $\sigma_A^2 \neq \sigma_B^2$. What is, in this case, the value of the statistic used to test, at level $\alpha = 0.10$, the hypothesis in A)? What is the conclusion of the test?

a) 2.56; we accept H_0 b) 2.66; we accept H_0 c) 2.56; we reject H_0
d) 2.66; we reject H_0 e) 2.56; we cannot conclude

C) Suppose that we exclude the observation 11.9 from the random sample of X_A, because this observation is doubtful. What is then the value of the statistic that we use to test the hypothesis $H_0: \sigma_A^2 = \sigma_B^2$ against $H_1: \sigma_A^2 \neq \sigma_B^2$? What is the conclusion of the test if $\alpha = 0.05$?

a) 1.68; we accept H_0 b) 1.78; we accept H_0 c) 2.68; we accept H_0
d) 2.78; we reject H_0 e) 3.68; we reject H_0

Supplementary Exercises

Question no. 1

Obtain the maximum likelihood estimator of the parameter θ in the probability density function

$$f_X(x; \theta) = \frac{\theta}{2}e^{-\theta|x|} \quad \text{for } x \in \mathbb{R}.$$

Question no. 2

Suppose that the random variable X has the following probability density function:

$$f_X(x; \alpha, \beta) = \frac{1}{\alpha} e^{-x/\beta} \quad \text{for } x \geq 0,$$

where $\alpha > 0$ and $\beta > 0$ are unknown constants. Find the estimator of the parameter α by the method of maximum likelihood.

Question no. 3

Suppose that $X_1 \sim N(\mu_{X_1}, 1)$ and $X_2 \sim N(\mu_{X_2}, \sigma_{X_2}^2)$ are independent random variables. Random samples from X_1 and X_2 yielded the following table:

	n	$\bar{x}$	s
X_1	10	0.1	0.9
X_2	100	-0.2	1.1

a) Test the hypothesis $H_0: \mu_{X_1} = \mu_{X_2}$ against $H_1: \mu_{X_1} > \mu_{X_2}$ at significance level $\alpha = 0.05$.

b) What is (approximately) the probability of making a type II error in a) if, in fact, $\mu_{X_1} - \mu_{X_2} = 0.5$?

Question no. 4

A random variable having a *generalized Pareto distribution* is such that

$$f_X(x; \alpha, \beta) = \alpha(\beta - 2)(1 + \alpha x)^{-\beta+1} \quad \text{for } x > 0,$$

where $\alpha > 0$ and $\beta > 2$ are unknown parameters. Estimate α and β by the method of moments, if we assume that $\beta > 4$.

Question no. 5

The tensile strength X (in kN/m^2) of a certain synthetic fiber has a Gaussian $N(\mu, 100)$ distribution. The company that makes this fiber claims that the average tensile strength is greater than or equal to 240 kN/m^2. In order to check the assertion, we perform a statistical test. A random sample of size $n = 18$ yielded $\bar{x} = 237$ kN/m^2.

a) What is the alternative hypothesis H_1 for this test?

b) What is the value of the statistic Z_0 used to perform the test?

c) Calculate the value of β if $\mu = 235$, $n = 18$ and $\alpha = 0.05$.

d) What is the smallest number of observations that must be taken if the probability of rejecting the company's assertion must be of at least 0.975 when $\mu = 236$ and $\alpha = 0.05$?

Question no. 6

In order to compare heat loss from steel pipes and from glass pipes, we consider seven pairs of pipes (made up of a steel pipe and a glass pipe). The pipes in a given pair have the same diameter, but diameters differ from one pair to another. Water is passed through the pipes at the same initial temperature and the heat loss (in °C) is measured over a 50-meter distance. The results obtained are the following:

Pair number	1	2	3	4	5	6	7
Steel pipe	4.6	3.7	4.2	1.9	4.8	6.1	4.7
Glass pipe	2.5	4.2	2.0	1.8	2.7	3.2	3.0

We assume that the heat loss has approximately a Gaussian distribution.

a) What is the value of the statistic used to test the hypothesis that the heat loss is, on average, greater in steel pipes than in glass pipes?

b) What is the number of degrees of freedom associated with the test statistic in a)?

Question no. 7

Let $f_X(x; \theta) = 1/\theta$ for $0 < x < \theta$. We want to test $H_0: \theta = 1$ against $H_1: \theta > 1$. To do so, we take a single observation, X_1, of X and we reject H_0 if and only if $X_1 > 0.9$.

a) What is the type I error risk, α, of the test?

b) What is the value of β if $\theta = 1.5$?

Question no. 8

The precision of the manufacturing process of bolts is under statistical control as long as the standard deviation of the diameter of the bolts does not exceed 1.5 mm. We want to test, at significance level $\alpha = 0.05$, the hypothesis that the manufacturing process is under control. We assume that the diameter X of the bolts has approximately a Gaussian $N(\mu, \sigma^2)$ distribution.

a) What is the conclusion of the test if the sum of 20 measurements of the diameter is equal to 109.8 and the sum of the squares of the measurements is 655.6?

Hint. We have: $\chi^2_{0.05, 19} \simeq 30.144$.

b) What is the type II error risk of the test performed in a) if the standard deviation of X is actually equal to 2.2 mm?

c) Determine the (smallest) size of the random sample needed to obtain a type II error risk of 0.05 when the standard deviation of X is equal to 2.2 mm.

Question no. 9

The strength level of material A (respectively B) has approximately a Gaussian distribution: $X_A \sim N(\mu_A, \sigma_A^2)$ (resp. $X_B \sim N(\mu_B, \sigma_B^2)$). To test the hypothesis $H_0: \mu_A = \mu_B$ against $H_1: \mu_A \neq \mu_B$, we took two independent random samples:

X_A	2.08	3.77	2.46	2.20	2.58	3.35	3.52
X_B	3.13	3.07	3.66	3.51	3.22		

We compute

$$\bar{x}_A = 2.85, \quad s_A^2 \simeq 0.464; \quad \bar{x}_B = 3.32, \quad s_B^2 \simeq 0.065.$$

a) Perform the test with a type I error risk α of 0.10.

b) Perform the test at significance level $\alpha = 0.05$.

Remark. In a) and b), we must first test the equality of variances with the same type I error risk (10% and 5%, respectively).

Question no. 10 (See [11].)

The average percentage of waste produced by a certain manufacturing operation is supposed to be smaller than or equal to 7%. We took several days at random and we calculated the percentages of waste:

$$6.5151, \quad 7.4970, \quad 7.4601, \quad 6.3723, \quad 8.3257, \quad 9.8199, \quad 9.5618.$$

We assume that the percentage X of waste has (approximately) a Gaussian $N(\mu, \sigma^2)$ distribution.

a) Based on the data, can we conclude that μ is significantly greater than 7%? Use $\alpha = 0.01$.

Hint. We have: $t_{0.01,6} \simeq 3.143$.

b) If it is important to detect a ratio $(\mu - 0.07)/\sigma$ of 0.5 with a probability of at least 95%, what is the smallest number of observations that must be taken?

c) For a ratio $(\mu - 0.07)/\sigma$ of 2, what is the power of the test performed in a)?

Question no. 11 (See [11].)

The diameter of a steel ball has been measured by 12 individuals, using two types of pairs of calipers. The following data, where for instance 5 represents 0.265 and 4 represents 0.264, have been collected:

Individual	1	2	3	4	5	6	7	8	9	10	11	12
Calipers 1	5	5	6	7	7	5	7	7	5	8	8	5
Calipers 2	4	5	4	6	7	8	4	5	5	7	8	9

a) Is there a significant difference between the means of the populations from which the two samples come? Use $\alpha = 0.05$ and specify the basic assumptions that we must make to perform the test.

Hint. We have: $t_{0.025,11} \simeq 2.201$.

b) Is the standard deviation of the first population significantly greater than 0.001? Use $\alpha = 0.10$.

Hint. We have: $\chi^2_{0.10,11} \simeq 17.28$.

7

Simple Linear Regression

7.1 Introduction: The Model

We saw in Chapter 6 how to test the hypothesis that the data we collected come from a given distribution, such as a Gaussian distribution. When the random variable of interest is a function of a deterministic variable (that we can control), *regression* can help us find the form of the relationship between the random and deterministic variables. In this chapter, we will treat in detail the basic problem, namely that of finding the best *linear* relationship between the random variable and the deterministic variable. We will also briefly consider the case when the relationship between the two variables is assumed to be non-linear.

Let Y be a random variable and x a deterministic variable (that is, non-random). We have a random sample $(x_1, Y_1), \ldots, (x_n, Y_n)$ and we want to find a mathematical relationship that expresses Y in terms of x. The variable x is called the **independent** variable and Y is called the **dependent** or **response** variable.

In the case of **simple linear regression**, the model that we propose is of the form

$$Y = \beta_0 + \beta_1 x + \epsilon, \tag{7.1}$$

where ϵ is an error term. We assume that each observation Y_i of Y satisfies the equation

$$Y_i = \beta_0 + \beta_1 x_i + \epsilon_i, \tag{7.2}$$

where $\epsilon_i \sim N(0, \sigma^2)$ for $i = 1, \ldots, n$, and that the random variables ϵ_i are *independent*. Note that we take for granted that the variance of ϵ_i is the same for all values of i.

We can thus write that

$$Y_i \sim N(\beta_0 + \beta_1 x_i, \sigma^2) \quad \text{for } i = 1, \ldots, n, \tag{7.3}$$

because β_0 and β_1 are parameters (that is, constants).

Remark. The model is known as the *simple linear regression model*, because there is a single independent variable, and the model is linear with respect to the *parameters*. The model

$$Y_i = \beta_0 + \beta_1 x_i + \beta_2 x_i^2 + \epsilon_i \tag{7.4}$$

is linear with respect to the parameters and is a polynomial of degree two, while

$$Y_i = \beta_0 e^{\beta_1 x_i} + \epsilon_i \tag{7.5}$$

is a non-linear model with respect to the parameters.

The best estimators of the parameters β_0 and β_1, that is, the *minimum variance unbiased estimators* of β_0 and β_1, are obtained using the **method of least squares**. We define the sum

$$SS = \sum_{i=1}^{n} \epsilon_i^2 = \sum_{i=1}^{n} (Y_i - \beta_0 - \beta_1 x_i)^2. \tag{7.6}$$

The estimators $\hat{\beta}_0$ and $\hat{\beta}_1$ of β_0 and β_1 by the method of least squares are the values of β_0 and β_1 that minimize the sum SS. We set

$$\frac{\partial SS}{\partial \beta_0} = -2 \sum_{i=1}^{n} (Y_i - \beta_0 - \beta_1 x_i) = 0 \tag{7.7}$$

and

$$\frac{\partial SS}{\partial \beta_1} = -2 \sum_{i=1}^{n} x_i (Y_i - \beta_0 - \beta_1 x_i) = 0. \tag{7.8}$$

The solution of these two equations, called the **normal equations**, is

$$\hat{\beta}_1 = \frac{\sum_{i=1}^{n}(x_i - \bar{x})(Y_i - \bar{Y})}{\sum_{i=1}^{n}(x_i - \bar{x})^2} = \frac{\sum_{i=1}^{n} x_i Y_i - n\bar{x}\bar{Y}}{\sum_{i=1}^{n} x_i^2 - n\bar{x}^2} \quad \text{and} \quad \hat{\beta}_0 = \bar{Y} - \hat{\beta}_1 \bar{x}. \tag{7.9}$$

To predict the value of Y when $x = x_i$, we then use the equation

$$\hat{Y}_i = \hat{\beta}_0 + \hat{\beta}_1 x_i. \tag{7.10}$$

Remarks. i) In the case when we assume that the errors ϵ_i are independent and all have a Gaussian $N(0, \sigma^2)$ distribution, as above, we find that the estimators given by the method of least squares and those obtained with the method of maximum likelihood are the same.

ii) The ϵ_i's are called the (theoretical) **residuals**. The quantity ϵ_i represents the difference between the observation Y_i and the value calculated from the proposed model $Y_i = \beta_0 + \beta_1 x_i$.

iii) The prediction equation should (theoretically) be used only for values of x in the interval $[x_{(1)}, x_{(n)}]$, where

$$x_{(1)} := \min\{x_1, \ldots, x_n\} \quad \text{and} \quad x_{(n)} := \max\{x_1, \ldots, x_n\}. \tag{7.11}$$

iv) Even if we know that $Y = 0$ when $x = 0$, we do not set $\beta_0 = 0$. We generally obtain better results with β_0 in the model. However, if we indeed wish to propose the **regression model through the origin**:

$$Y = \beta x + \epsilon, \tag{7.12}$$

where $\epsilon \sim N(0, \sigma^2)$, we easily find that the estimator of the parameter β obtained from the method of least squares, based on a random sample $(x_1, Y_1), \ldots, (x_n, Y_n)$, is

$$\hat{\beta} = \frac{\sum_{i=1}^{n} x_i Y_i}{\sum_{i=1}^{n} x_i^2}. \tag{7.13}$$

Example 7.1.1 (See [17].) We want to determine how the tensile strength of a certain alloy depends on the percentage of zinc it contains. We have the following data:

% of zinc	4.7	4.8	4.9	5.0	5.1
Tensile strength	1.2	1.4	1.5	1.5	1.7

We obtain the graph in Fig. 7.1.

We consider the simple linear regression model:

$$Y = \beta_0 + \beta_1 x + \epsilon,$$

where x is the percentage of zinc and Y is the tensile strength.

Remark. It is important to correctly identify the random variable and the deterministic variable in the problem.

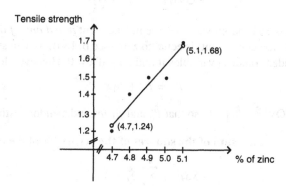

Figure 7.1. Graph in Example 7.1.1.

We find that

$$\bar{x} = 4.9, \quad \bar{y} = 1.46, \quad \sum_{i=1}^{5} x_i^2 = 120.15 \quad \text{and} \quad \sum_{i=1}^{5} x_i y_i = 35.88.$$

Then, we have:

$$\hat{\beta}_1 = \frac{\sum_{i=1}^{5} x_i y_i - 5\bar{x}\bar{y}}{\sum_{i=1}^{5} x_i^2 - 5\bar{x}^2} = \frac{35.88 - 5(4.9)(1.46)}{120.15 - 5(4.9)^2} = 1.1$$

and

$$\hat{\beta}_0 = \bar{y} - \hat{\beta}_1\bar{x} = 1.46 - (1.1)(4.9) = -3.93.$$

Thus, the prediction equation is given by

$$\hat{y} = -3.93 + 1.1x.$$

7.2 Tests of Hypotheses

Since the estimators $\hat{\beta}_0$ and $\hat{\beta}_1$ are linear combinations of independent Gaussian random variables, they also have a Gaussian distribution. Moreover, we can show that $\hat{\beta}_0$ and $\hat{\beta}_1$ are *unbiased* estimators of β_0 and β_1, and that

$$\text{VAR}[\hat{\beta}_0] = \sigma^2 \left(\frac{1}{n} + \frac{\bar{x}^2}{SS_x} \right) \quad \text{and} \quad \text{VAR}[\hat{\beta}_1] = \frac{\sigma^2}{SS_x}, \tag{7.14}$$

where

$$SS_x := \sum_{i=1}^{n} (x_i - \bar{x})^2 = \sum_{i=1}^{n} x_i^2 - n\bar{x}^2 = \sum_{i=1}^{n} x_i^2 - \frac{\left(\sum_{i=1}^{n} x_i \right)^2}{n}. \tag{7.15}$$

Remark. The estimators $\hat{\beta}_0$ and $\hat{\beta}_1$ are generally *not* independent. In fact, we can show that

$$\text{COV}[\hat{\beta}_0, \hat{\beta}_1] = -\frac{\sigma^2 \bar{x}}{SS_x}. \tag{7.16}$$

Since two Gaussian random variables are independent *if and only if* their covariance (or their correlation coefficient) is equal to zero (see p. 180), we can state that $\hat{\beta}_0$ and $\hat{\beta}_1$ are independent random variables if and only if $\bar{x} = 0$. However, let

$$\hat{\beta}_0' = \hat{\beta}_0 + \hat{\beta}_1\bar{x}. \tag{7.17}$$

We find that $\text{COV}[\hat{\beta}_0', \hat{\beta}_1] = 0$, so that $\hat{\beta}_0'$ and $\hat{\beta}_1$ *are* independent estimators.

Next, we define the **sum of the squares of the errors** (or of the residuals)

$$SS_E = \sum_{i=1}^{n} (Y_i - \hat{Y}_i)^2. \tag{7.18}$$

We can show that

$$E[SS_E] = (n - 2)\sigma^2. \tag{7.19}$$

It follows that an unbiased estimator of the variance σ^2 is given by (the **mean square of the errors**)

$$\hat{\sigma}^2 = MS_E := \frac{SS_E}{n - 2}. \tag{7.20}$$

a) To test the null hypothesis

$$H_0: \beta_0 = \beta_{00}, \tag{7.21}$$

we use the statistic

$$T_0 := \frac{\hat{\beta}_0 - \beta_{00}}{\sqrt{MS_E \left(\frac{1}{n} + \frac{\bar{x}^2}{SS_x} \right)}} \overset{H_0}{\sim} t_{n-2}. \tag{7.22}$$

We then reject H_0 at significance level α if and only if

$$\begin{cases} |T_0| > t_{\alpha/2,n-2} & \text{if } H_1: \beta_0 \neq \beta_{00}, \\ T_0 > t_{\alpha,n-2} & \text{if } H_1: \beta_0 > \beta_{00}, \\ T_0 < -t_{\alpha,n-2} & \text{if } H_1: \beta_0 < \beta_{00}. \end{cases} \tag{7.23}$$

b) Similarly, to test

$$H_0: \beta_1 = \beta_{10}, \tag{7.24}$$

we use the statistic

$$T_0 := \frac{\hat{\beta}_1 - \beta_{10}}{\sqrt{MS_E/SS_x}} \overset{H_0}{\sim} t_{n-2} \tag{7.25}$$

and we reject H_0 at significance level α if and only if

$$\begin{cases} |T_0| > t_{\alpha/2,n-2} & \text{if } H_1: \beta_1 \neq \beta_{10}, \\ T_0 > t_{\alpha,n-2} & \text{if } H_1: \beta_1 > \beta_{10}, \\ T_0 < -t_{\alpha,n-2} & \text{if } H_1: \beta_1 < \beta_{10}. \end{cases} \tag{7.26}$$

Particular Case: Test of the (Global) Significance of Regression

The test of

$$H_0: \beta_1 = 0 \quad \text{against} \quad H_1: \beta_1 \neq 0 \tag{7.27}$$

is very important, because if we cannot reject the null hypothesis H_0, then we must conclude that there is no *linear* relationship between x and Y (or, at least, we cannot conclude that there is a *significant* linear relationship between x and Y). To perform the test, we can proceed as above. In practice, we perform instead, equivalently, an **analysis of the variance**. Note first that when we choose $\beta_{10} = 0$, we have:

$$T_0^2 = \frac{\hat{\beta}_1^2}{MS_E/SS_x}. \tag{7.28}$$

We can show that

$$\hat{\beta}_1^2 SS_x = SS_R := \sum_{i=1}^{n} (\hat{Y}_i - \bar{Y})^2. \tag{7.29}$$

We say that the sum of squares SS_R is the **sum of squares due to regression**. It can be shown that the sums SS_R and SS_E are *independent* random variables having chi-square distributions with 1 and $n-2$ degrees of freedom, respectively. It follows that

$$F_0 := T_0^2 = \frac{SS_R}{MS_E} \overset{H_0}{\sim} F_{1,n-2}. \tag{7.30}$$

Remarks. i) Actually, by definition, the square of a random variable having a Student distribution with n degrees of freedom is a random variable having a Fisher distribution with 1 and n degrees of freedom. Therefore, we could have written directly that $T_0^2 \sim F_{1,n-2}$, when the null hypothesis H_0 is true.

ii) Since SS_R has a single degree of freedom, the mean square MS_R and the sum of squares SS_R are equal.

Finally, we can show that

$$E[SS_R] = \sigma^2 + \beta_1^2 SS_x \overset{H_0}{=} \sigma^2. \tag{7.31}$$

Since $E[MS_E] = \sigma^2$, the larger the value of the statistic F_0, the smaller the probability that the null hypothesis H_0 is true. We reject H_0 at significance level α if and only if

$$F_0 > F_{\alpha,1,n-2}. \tag{7.32}$$

The procedure to test the global significance of regression can be summarized using an analysis of variance table (see Table 7.1, p. 313), in which the **total sum of squares** SS_T is defined by

$$SS_T = \sum_{i=1}^{n} (Y_i - \bar{Y})^2 = \sum_{i=1}^{n} Y_i^2 - n\bar{Y}^2. \tag{7.33}$$

In practice, we first calculate the sum of squares SS_T. Next, we have: $SS_R \equiv \hat{\beta}_1^2 SS_x = \hat{\beta}_1 SS_{xY}$, where

$$SS_{xY} := \sum_{i=1}^{n} (x_i - \bar{x})(Y_i - \bar{Y}) = \sum_{i=1}^{n} x_i Y_i - n\bar{x}\bar{Y}. \tag{7.34}$$

We can then complete the analysis of variance table.

Table 7.1. Analysis of variance.

Source of variation	Sum of squares	Degrees of freedom	Mean squares	F_0
Regression	SS_R	1	MS_R	$\dfrac{MS_R}{MS_E}$
Error	SS_E	$n-2$	MS_E	
Total	SS_T	$n-1$		

Example 7.2.1 We found in Example 7.1.1 that

$$\hat{y} = -3.93 + 1.1x.$$

We now want to test

$$H_0: \beta_1 = 0 \quad \text{against} \quad H_1: \beta_1 \neq 0.$$

We have:

$$SS_T = \sum_{i=1}^{5} y_i^2 - 5\bar{y}^2 = 10.79 - 5(1.46)^2 = 0.132$$

and

$$SS_R = \hat{\beta}_1 SS_{xY} = (1.1)\left(\sum_{i=1}^{5} x_i y_i - 5\bar{x}\bar{y}\right) = (1.1)(35.88 - 35.77) = 0.121.$$

Thus, $SS_E = SS_T - SS_R = 0.011$ and

$$f_0 = \frac{MS_R}{MS_E} = \frac{0.121}{0.011/3} = 33.$$

We find in the table giving the values of $F_{0.05,n_1,n_2}$, Appendix A, p. 343, that $F_{0.05,1,3} \simeq 10.1$. Since $33 > 10.1$, we can reject H_0 at significance level $\alpha = 0.05$. Note that in fact we have: $F_{0.05,1,3} = t_{0.025,3}^2$. Therefore, we can also use Table 6.2, p. 261, to obtain the critical values $F_{0.05,1,n-2}$.

7.3 Confidence Intervals and Ellipses

Using the following results:

$$\frac{\hat{\beta}_1 - \beta_1}{\hat{\sigma}/\sqrt{SS_x}} \sim t_{n-2} \quad \text{and} \quad \frac{\hat{\beta}_0 - \beta_0}{\hat{\sigma}\sqrt{\dfrac{1}{n} + \dfrac{\bar{x}^2}{SS_x}}} \sim t_{n-2}, \tag{7.35}$$

where $\hat{\sigma} = \sqrt{MS_E} = \sqrt{SS_E/(n-2)}$, we can calculate confidence intervals for the parameters β_0 and β_1. We find that the two-sided confidence intervals at $100(1-\alpha)\%$

are given (in compact form) respectively by

$$\hat{\beta}_0 \pm t_{\alpha/2,n-2}\hat{\sigma}\sqrt{\frac{1}{n} + \frac{\bar{x}^2}{SS_x}} \tag{7.36}$$

and

$$\hat{\beta}_1 \pm t_{\alpha/2,n-2}\hat{\sigma}\sqrt{\frac{1}{SS_x}}. \tag{7.37}$$

Similarly, we can show that a $100(1-\alpha)\%$ confidence interval for the average value of the random variable Y when $x = \xi \in [x_{(1)}, x_{(n)}]$, that is, for $\beta_0 + \beta_1\xi$, is

$$\hat{\beta}_0 + \hat{\beta}_1\xi \pm t_{\alpha/2,n-2}\hat{\sigma}\sqrt{\frac{1}{n} + \frac{(\xi - \bar{x})^2}{SS_x}}. \tag{7.38}$$

We see that the width of the interval is a minimum when $\xi = \bar{x}$.

Now, the preceding formula is valid when we wish to compute a confidence interval for $\beta_0 + \beta_1\xi$, based on the n observations collected. In the case when we wish to compute a confidence interval for a *new* observation (that is, for a *prediction*) of Y at $x = \xi$ instead, the term σ^2 is added to the variance and the formula becomes

$$\hat{\beta}_0 + \hat{\beta}_1\xi \pm t_{\alpha/2,n-2}\hat{\sigma}\sqrt{1 + \frac{1}{n} + \frac{(\xi - \bar{x})^2}{SS_x}}. \tag{7.39}$$

Remark. If we want a confidence interval for the *mean* of m new observations of Y at $x = \xi$, we only have to replace the term 1 (added in front of $\frac{1}{n}$) by $\frac{1}{m}$ in the preceding formula.

Finally, we can show that the interior of the *ellipse* defined by

$$n(\beta_0 - \hat{\beta}_0)^2 + 2n\bar{x}(\beta_0 - \hat{\beta}_0)(\beta_1 - \hat{\beta}_1) + (\beta_1 - \hat{\beta}_1)^2 \sum_{i=1}^{n} x_i^2 = 2\hat{\sigma}^2 F_{\alpha,2,n-2} \tag{7.40}$$

is a $100(1-\alpha)\%$ confidence region for the pair (β_0, β_1).

Example 7.3.1 A two-sided 95% confidence interval for a new observation of Y when $x = \bar{x} = 4.9$, in Example 7.1.1, is given by

$$-3.93 + 1.1(4.9) \pm t_{0.025,3}\hat{\sigma}\sqrt{1 + \frac{1}{5} + 0} \simeq 1.46 \pm 0.21,$$

because we have: $\hat{\sigma} = \sqrt{MS_E} = \sqrt{0.011/3}$ (see Example 7.2.1, p. 313) and $t_{0.025,3} \overset{\text{Tab. 6.2}}{\simeq} 3.182$.

7.4 The Coefficient of Determination

Definition 7.4.1. *The **coefficient of determination**, R^2, is obtained by dividing the sum of squares SS_R by the sum of squares SS_T:*

$$R^2 = \frac{SS_R}{SS_T}. \tag{7.41}$$

The coefficient R^2, also called the **squared correlation coefficient**, enables us to measure the fit of the model to the data. If all the data points fall on the regression line, then we have: $R^2 = 1$. In general, R^2 takes on a value in the interval $[0, 1]$ and gives us the *percentage of the total variation SS_T* that is explained by the regression model. It follows that

$$1 - R^2 = \frac{SS_E}{SS_T} \tag{7.42}$$

is the proportion of SS_T that is *not* explained by the model.

Remarks. i) The expression *correlation coefficient* is in fact not correct, because x is not a random variable. Indeed, a correlation coefficient is only defined for two random variables, X and Y.

ii) The quantity $R = \sqrt{R^2}$ is called the *fit index*. It is often used to measure the quality of the regression model. However, it is important to know that if the value of R is large, this does not *necessarily* mean that the (simple) linear regression model is the right one. Similarly, if the value of R is small, we should not conclude that the model must be rejected. The index fit is actually a measure of the improvement to the fit obtained by including the term $\beta_1 x$ in the model, in comparison to the regression model $Y = \beta_0 + \epsilon$.

Example 7.4.1 Using the results of Example 7.2.1, we find that

$$R^2 = \frac{0.121}{0.132} \simeq 0.917$$

with the data in Example 7.1.1.

7.5 The Analysis of Residuals

We already called the quantities $\epsilon_i = Y_i - \beta_0 - \beta_1 x_i$ the (theoretical) *residuals*. We define the **sample residuals** by

$$e_i = y_i - \hat{y}_i. \tag{7.43}$$

The analysis of the residuals, which is generally carried out with the help of a statistical software program, enables us to check, in particular, the validity of the assumption made at the beginning of this chapter, namely that the error terms ϵ_i all have a

Gaussian distribution (with zero mean and common variance σ^2). If this assumption is true, then the quantities

$$z_i := \frac{e_i}{\hat{\sigma}}, \tag{7.44}$$

called the **standardized residuals**, should be particular observations of a random variable Z having approximately a standard Gaussian distribution. Since

$$P[-2 < Z < 2] \simeq 0.95 \quad \text{and} \quad P[-3 < Z < 3] \simeq 0.997, \tag{7.45}$$

there should not be more than (about) 5% of the z_i's such that $|z_i| \geq 2$, and almost no z_i's such that $|z_i| \geq 3$.

Remark. If the size of the random sample is small, then we must use the Student distribution. That is, we must compare the z_i's to the values taken by a random variable having a t_{n-2} distribution.

Example 7.5.1 The regression equation obtained with the data provided in Example 7.1.1 is

$$\hat{y} = -3.93 + 1.1x,$$

and we have: $\hat{\sigma} = \sqrt{0.011/3}$. Making use of these results, we can build the following table:

x_i	y_i	$\hat{y}_i$	e_i	z_i
4.7	1.2	1.24	−0.04	−0.67
4.8	1.4	1.35	0.05	0.83
4.9	1.5	1.46	0.04	0.67
5.0	1.5	1.57	−0.07	−1.17
5.1	1.7	1.68	0.02	0.33

We see that all the standardized residuals are small. Therefore, the normality assumption of the random variables ϵ_i seems reasonable.

Remark. Since $n = 5$, we should compare the residuals to the values taken by a t_3 distribution. Now, we have:

$$P[-3.182 \leq T \leq 3.182] \simeq 0.05$$

if $T \sim t_3$. Therefore, the residuals could have been located in the interval $[-3.182, 3.182]$. However, the size of the particular random sample is a little small here to carry out an analysis of residuals.

The analysis of the residuals also enables us to check the other basic assumptions made in simple linear regression:

a) the model is of the form $Y = \beta_0 + \beta_1 x$;

b) the errors ϵ_i are independent and all have the same variance σ^2.

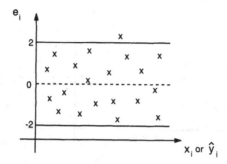

Figure 7.2. Residuals forming a uniform band.

To do so, we plot the graphs of the residuals e_i against the x_i's and the $\hat{y}_i$'s. In all cases, the points in these graphs should form a uniform band, as in Fig. 7.2.

Figure 7.3, p. 318, shows cases when we can conclude that

i) the model is not linear in x (graphs (a) and (b));
ii) the variance of the errors is not constant (graphs (c) and (d));
iii) both at the same time (graphs (e) and (f)).

Remark. If there exists at least one value of the independent variable x for which we have at least two observations of the dependent variable Y (and if there are at least three different values of x in all), then we can perform the following test of the fit of the model to the data: let $Y_{i,1}, \ldots, Y_{i,n_i}$ be n_i observations of the random variable Y when $x = x_i$, for $i = 1, \ldots, k$. We define the **sum of the squares due to pure error**:

$$SS_{PE} = \sum_{i=1}^{k} \sum_{j=1}^{n_i} (Y_{i,j} - \bar{Y}_i)^2, \tag{7.46}$$

where

$$\bar{Y}_i := \frac{1}{n_i} \sum_{j=1}^{n_i} Y_{i,j}, \tag{7.47}$$

and we set

$$SS_E = SS_{PE} + SS_L, \tag{7.48}$$

where SS_L is the **sum of squares due to the lack of fit of the model**. Next, we consider the statistic

$$F_0 := \frac{SS_L/(k-2)}{SS_{PE}/(n-k)}, \tag{7.49}$$

where $n := \sum_{i=1}^{k} n_i$ is the total number of observations in the random sample. We reject, at significance level α, the hypothesis that the model is adequate for the data when $F_0 > F_{\alpha,k-2,n-k}$.

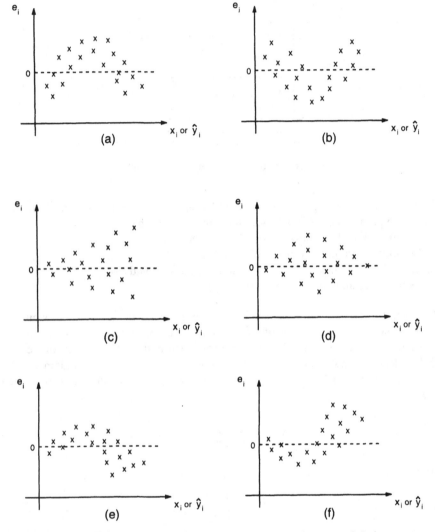

Figure 7.3. Residuals showing at least one assumption not satisfied.

7.6 Curvilinear Regression

Sometimes, a non-linear regression model can be reduced to a linear regression model by transforming the Y variable or the x variable (or both at the same time). For example, if we propose the model (without an error term)

$$Y = \beta_0 e^{\beta_1 x}, \tag{7.50}$$

then it suffices to take the (natural) logarithm on each side of the equation, and next to set $Y' = \ln Y$ and $\beta_0' = \ln \beta_0$. We thus obtain the model

$$Y' = \beta'_0 + \beta_1 x. \tag{7.51}$$

Similarly, the model

$$Y = \frac{1}{\exp(\beta_0 + \beta_1 x)} \tag{7.52}$$

becomes

$$Y' = \beta_0 + \beta_1 x \tag{7.53}$$

if we define $Y' = -\ln Y$.

Remarks. i) If we assume in (7.51) that

$$Y'_i = \beta'_0 + \beta_1 x_i + \epsilon_i, \tag{7.54}$$

where $\epsilon_i \sim N(0, \sigma^2)$ for $i = 1, \ldots, n$, then this implies that the original model is

$$Y_i = \beta_0 e^{\beta_1 x_i + \epsilon_i}. \tag{7.55}$$

An error term of the form e^{ϵ_i} may not be realistic in many situations. On the other hand, assuming that the error term has a *lognormal* distribution can be interesting in some cases.

ii) The non-linear models that can be linearized are called *intrinsically* linear.

iii) We realize that the relationship between x and Y is not linear by looking at the graph of the y_i's against the x_i's, or at the graph of the e_i's against the $\hat{y}_i$'s.

Example 7.6.1 We have the following data:

x	1	2	3	7
Y	13.0	21.9	29.8	30.4
	11.8	24.7	24.1	35.7

where x is the time (in days) elapsed after the setting of a cement and Y is the tensile strength (in kilograms per square centimeter) of the cement. We propose the curvilinear regression model

$$Y = \beta_0 e^{-\beta_1 x}.$$

a) Transform the variables to obtain a linear model.

b) Estimate the parameters β_0 and β_1 by the method of least squares.

c) Calculate a 95% prediction interval for Y when $x = 5$ and 10.

Remark. When there is more than one value of Y for a given value of x, we should specify in the problem whether all the observations are independent (actually, we should always specify in the problem that the observations are independent). Here, we assume that we indeed have *eight* values taken by independent observations.

Solution. a) We have:

$$Y = \beta_0 e^{-\beta_1 x} \quad \Leftrightarrow \quad \ln Y = \ln \beta_0 - \beta_1 x.$$

Therefore, we simply have to set $Y' = \ln Y$, $\beta_0' = \ln \beta_0$ and $\beta_1' = -\beta_1$ to obtain

$$Y' = \beta_0' + \beta_1' x.$$

Remark. We could also have set $x' = -x$, rather than $\beta_1' = -\beta_1$.

b) We must first transform the data. We obtain:

x	1	2	3	7
$\simeq Y'$	2.565	3.086	3.395	3.414
	2.468	3.207	3.182	3.575

Next, we calculate $\hat{\beta}_0'$ and $\hat{\beta}_1'$ using the formulas in Section 7.1. We find that

$$\hat{\beta}_0' \simeq 2.69 \quad \text{and} \quad \hat{\beta}_1' \simeq 0.13.$$

It follows that

$$\hat{\beta}_0 = e^{\hat{\beta}_0'} \simeq 14.74 \quad \text{and} \quad \hat{\beta}_1 = -\hat{\beta}_1' \simeq -0.13.$$

c) The 95% prediction interval for Y', when $x = \xi$, is given by

$$\hat{\beta}_0' + \hat{\beta}_1' \xi \pm \underbrace{t_{0.025,6}}_{2.447} \hat{\sigma} \sqrt{1 + \frac{1}{8} + \frac{(\xi - \bar{x})^2}{SS_x}}.$$

We find that $\bar{x} = 3.25$, $SS_x = 41.5$ and (with the transformed data) $SS_T \simeq 1.114$. It follows that $SS_E = SS_T - (\hat{\beta}_1')^2 SS_x \simeq 0.413$, and then

$$\hat{\sigma} \simeq \left(\frac{0.413}{6} \right)^{1/2} \simeq 0.262.$$

Thus, we obtain

$$\xi = 5 \Rightarrow 3.34 \pm 0.70 \quad \text{and} \quad \xi = 10 \Rightarrow 3.99 \pm 0.96.$$

Finally, in terms of the variable Y, the prediction intervals become

$$Y \mid \{x = 5\} \in [14.0, 56.8] \quad \text{and} \quad Y \mid \{x = 10\} \in [20.7, 141.2].$$

Remarks. i) Looking at the data, we see that Y increases when x increases. Therefore, the fact that $\hat{\beta}_1$ is negative is logical.

ii) Since β_1 is a *real* parameter, the model considered is equivalent to the model $Y = \beta_0 e^{\beta_1 x}$.

iii) If we had proposed the model $Y = \beta_0 + \beta_1 \ln x$, for instance, then it would have been sufficient to set $x' = \ln x$ to obtain the simple linear regression model. Moreover, if we assume that

$$Y_i = \beta_0 + \beta_1 x_i' + \epsilon_i,$$

where $\epsilon_i \sim N(0, \sigma^2)$ for $i = 1, \ldots, n$, then we have:

$$Y_i = \beta_0 + \beta_1 \ln x_i + \epsilon_i.$$

That is, if we only transform the independent variable x, then the error term ϵ_i in the model is unchanged.

7.7 Correlation

When x is not a deterministic variable, but rather a *random* variable X, we assume that $(X_1, Y_1), \ldots, (X_n, Y_n)$ is a random sample of size n of the random vector (X, Y). If we assume further that X has a Gaussian $N(\mu_X, \sigma_X^2)$ distribution and $Y \sim N(\mu_Y, \sigma_Y^2)$, then we can show that the expected value and the variance of Y, given that $X = x$, are given by (see p. 179)

$$E[Y \mid X = x] = \mu_Y + \rho_{X,Y} \frac{\sigma_Y}{\sigma_X}(x - \mu_X) \tag{7.56}$$

and

$$\mathrm{VAR}[Y \mid X = x] = \sigma_Y^2(1 - \rho_{X,Y}^2). \tag{7.57}$$

It follows that we can write that

$$E[Y \mid X = x] = \beta_0 + \beta_1 x, \tag{7.58}$$

where

$$\beta_0 := \mu_Y - \rho_{X,Y} \frac{\sigma_Y}{\sigma_X} \mu_X \quad \text{and} \quad \beta_1 := \rho_{X,Y} \frac{\sigma_Y}{\sigma_X}. \tag{7.59}$$

We can show that the maximum likelihood estimators of the parameters β_0 and β_1 are given by

$$\hat{\beta}_1 = \frac{\sum_{i=1}^n (X_i - \bar{X})(Y_i - \bar{Y})}{\sum_{i=1}^n (X_i - \bar{X})^2} = \frac{\sum_{i=1}^n Y_i(X_i - \bar{X})}{\sum_{i=1}^n (X_i - \bar{X})^2} \tag{7.60}$$

and

$$\hat{\beta}_0 = \bar{Y} - \hat{\beta}_1 \bar{X}. \tag{7.61}$$

Note that these estimators are of the same form as those obtained by the method of least squares in Section 7.1.

Now, the estimator $\hat{\rho}_{X,Y}$ of the theoretical correlation coefficient $\rho_{X,Y}$ is the **sample correlation coefficient** $R_{X,Y}$, defined by

$$R_{X,Y} = \frac{\sum_{i=1}^{n}(X_i - \bar{X})(Y_i - \bar{Y})}{\left[\sum_{i=1}^{n}(X_i - \bar{X})^2 \sum_{i=1}^{n}(Y_i - \bar{Y})^2\right]^{1/2}} := \frac{SS_{XY}}{\sqrt{SS_X SS_Y}}. \tag{7.62}$$

Remark. We find that the square of the sample correlation coefficient, $r_{X,Y}$, of the observations in a particular random sample of a pair (X, Y) is equal to the coefficient of determination: $r_{X,Y}^2 = R^2$. However, the coefficient of determination was defined for a deterministic variable x (and a random variable Y). As we already mentioned, it is not rigorous to state that the coefficient of determination is the square of the sample correlation coefficient of the variables x and Y.

Test of Hypotheses

To test the hypothesis

$$H_0: \rho_{X,Y} = 0 \quad \text{against} \quad H_1: \rho_{X,Y} \neq 0, \tag{7.63}$$

we use the statistic

$$T_0 := \sqrt{n-2}\,\frac{R_{X,Y}}{\sqrt{1 - R_{X,Y}^2}} \overset{H_0}{\sim} t_{n-2}. \tag{7.64}$$

We reject H_0 at significance level α if and only if $|T_0| > t_{\alpha/2,n-2}$.

Remarks. i) Since $SS_T = SS_Y$ and $\hat{\beta}_1 = SS_{XY}/SS_X$, we can write that

$$R_{X,Y} = \hat{\beta}_1 \left(\frac{SS_X}{SS_T}\right)^{1/2}. \tag{7.65}$$

Therefore, testing $H_0: \rho_{X,Y} = 0$ is equivalent, from a mathematical point of view, to testing $H_0: \beta_1 = 0$.

ii) To test the more general hypothesis $H_0: \rho_{X,Y} = \rho_0$, we can use the statistic

$$Z_0 := \frac{\sqrt{n-3}}{2}\left[\ln\left(\frac{1 + R_{X,Y}}{1 - R_{X,Y}}\right) - \ln\left(\frac{1 + \rho_0}{1 - \rho_0}\right)\right], \tag{7.66}$$

which has approximately a standard Gaussian distribution if the hypothesis H_0 is true and if the size n of the random sample is large enough. We reject H_0 at significance level α if and only if $|Z_0| > z_{\alpha/2}$.

Example 7.7.1 The following data are the maximum and minimum temperatures (in degrees Fahrenheit) registered during a winter week in an American city:

	Sun.	Mon.	Tue.	Wed.	Thu.	Fri.	Sat.
Maximum	6	11	14	12	5	−2	−9
Minimum	−22	−17	−15	−9	−24	−29	−35

Let Y be the maximum temperature and X be the minimum temperature. We compute $\bar{x} \simeq -21.57$ and $\bar{y} \simeq 5.29$. Moreover,

$$SS_X = \sum_{i=1}^{7} x_i^2 - 7\bar{x}^2 \simeq 463.71;$$

$$SS_Y = \sum_{i=1}^{7} y_i^2 - 7\bar{y}^2 \simeq 411.43;$$

$$SS_{X,Y} = \sum_{i=1}^{7} x_i y_i - 7\bar{x}\bar{y} \simeq 414.14.$$

It follows that

$$\hat{\beta}_1 = \frac{SS_{X,Y}}{SS_X} \simeq 0.89 \quad \text{and} \quad \hat{\beta}_0 = \bar{y} - \hat{\beta}_1\bar{x} \simeq 24.55,$$

so that we estimate $\mu_{Y|x} := E[Y \mid X = x]$ by

$$\hat{\mu}_{Y|x} \simeq 24.55 + 0.89x.$$

Remark. We could also compute

$$\hat{\mu}_{X|y} \simeq -26.89 + 1.01y.$$

The sample correlation coefficient $r_{X,Y}$ is given by

$$r_{X,Y} = \frac{SS_{X,Y}}{\sqrt{SS_X SS_Y}} \simeq 0.9482.$$

Since $R^2 = r_{X,Y}^2 \simeq 0.8990$, we say that the model explains about 89.9% of the variation in the data.

Finally, to test

$$H_0: \rho_{X,Y} = 0 \quad \text{against} \quad H_1: \rho_{X,Y} \neq 0,$$

we calculate the statistic

$$t_0 := \sqrt{n-2}\,\frac{r_{X,Y}}{\sqrt{1 - r_{X,Y}^2}} \simeq 6.67.$$

Given that $t_{0.025,5} \overset{\text{Tab. 6.2}}{\simeq} 2.571$, we can reject H_0 at significance level $\alpha = 0.05$ and conclude that the correlation coefficient $\rho_{X,Y}$ is not equal to zero.

7.8 Exercises, Problems, and Multiple Choice Questions, Supplementary Exercises

Solved Exercises

Exercise no. 1 (7.2)

Researchers think that there is a relationship between the strength Y (in tens of kilograms) of a certain metal part and the time x (in minutes) allowed for its cooling after its fabrication. They measured the strength of ten parts cooled over different time periods and obtained the results that are summarized below:

$$\bar{x} = 40.9, \quad \bar{y} = 74.4, \quad \sum_{i=1}^{10} x_i^2 = 17,077, \quad \sum_{i=1}^{10} y_i^2 = 55,504, \quad \sum_{i=1}^{10} x_i y_i = 30,436.$$

We propose the simple linear regression model:

$$Y = \beta_0 + \beta_1 x + \varepsilon, \quad \text{where } \varepsilon \sim N\left(0, \sigma^2\right).$$

a) Estimate the parameters β_0 and β_1 by the method of least squares.

b) Estimate the parameter σ.

c) Test the hypothesis $H_0: \beta_1 = 0$ against $H_1: \beta_1 \neq 0$. Use $\alpha = 0.05$.

Solution

a) We have:

$$\hat{\beta}_1 = \frac{\sum_{i=1}^{10} x_i y_i - 10\bar{x}\bar{y}}{\sum_{i=1}^{10} x_i^2 - 10\bar{x}^2} = \frac{30,436 - (10)(40.9)(74.4)}{17,077 - (10)(40.9)^2} = \frac{6.4}{348.9} \simeq 0.0183$$

and

$$\hat{\beta}_0 = \bar{y} - \hat{\beta}_1 \bar{x} \simeq 74.4 - (0.0183)(40.9) \simeq 73.6498.$$

b) We calculate

$$SS_T = \sum_{i=1}^{10} y_i^2 - 10\bar{y}^2 = 55,504 - (10)(74.4)^2 = 150.4$$

and

$$SS_R = \hat{\beta}_1^2 \left(\sum_{i=1}^{10} x_i^2 - 10\bar{x}^2\right) \stackrel{a)}{=} \hat{\beta}_1^2 (348.9) \stackrel{a)}{\simeq} 0.117.$$

Then, we can write that

$$\hat{\sigma} = \left(\frac{SS_T - SS_R}{10 - 2}\right)^{1/2} \simeq 4.33.$$

c) We reject H_0: $\beta_1 = 0$ if and only if

$$f_0 := \frac{SS_R}{\hat{\sigma}^2} \overset{\text{b)}}{\simeq} \frac{0.117}{(4.33)^2} \simeq 0.0062 > F_{0.05,1,10-2}.$$

Since $F_{0.05,1,8} \overset{\text{p. 343}}{\simeq} 5.32$, we cannot reject H_0 if $\alpha = 0.05$.

Exercise no. 2 (7.4)

We study the tensile strength Y of a certain alloy as a function of the percentages x_1 and x_2 of two of its components. We took ten specimens of this alloy, that were produced under different conditions, and we obtained the following data:

Y	213	220	216	225	235	218	239	243	233	240
x_1	13	15	14	18	19	20	22	17	16	18
x_2	2.1	2.3	2.2	2.5	3.2	2.4	3.4	4.1	4.0	4.3

Remark. We have: $\sum_{i=1}^{10} x_{1i} = 172$, $\sum_{i=1}^{10} x_{1i}^2 = 3028$, $\sum_{i=1}^{10} x_{2i} = 30.5$ and $\sum_{i=1}^{10} x_{2i}^2 = 99.65$.

a) We first propose the simple linear regression models:

$$Y = \beta_0 + \beta_i x_i + \varepsilon, \quad \text{where } \varepsilon \sim N\left(0, \sigma^2\right),$$

for $i = 1, 2$. Test, at significance level $\alpha = 0.05$, the null hypothesis H_0: $\beta_i = 0$ against H_1: $\beta_i \neq 0$, for $i = 1, 2$.

b) Based on the quantities R^2 and $\hat{\sigma}^2$, which variable, x_1 or x_2, seems to better explain the tensile strength of the alloy?

Remark. The smaller the value of $\hat{\sigma}^2$ is, the more adequate the model seems to be for the data.

c) We also consider the model

$$Y = \exp\left(\beta_1 x_1 + \varepsilon\right), \quad \text{where } \varepsilon \sim N\left(0, \sigma^2\right).$$

Use the data collected to estimate the parameter β_1 by the method of least squares.

Solution

a) We first calculate $SS_T = 1105.6$. In the case of the simple linear regression model with the variable x_1, we have:

$$SS_R = \hat{\beta}_1^2 \left(\sum_{i=1}^{10} x_{1i}^2 - 10\bar{x}_1^2\right) = 69.6\,\hat{\beta}_1^2.$$

We find that $\hat{\beta}_1 \simeq 2.221$. It follows that $SS_R \simeq 343.33$ and

$$SS_E \simeq 762.27 \Rightarrow \hat{\sigma}^2 \simeq \frac{762.27}{8} \simeq 95.3 \Rightarrow f_0 \simeq \frac{343.33}{95.3} \simeq 3.6.$$

p. 343

We reject H_0: $\beta_1 = 0$ at significance level $\alpha = 0.05$ if and only if $f_0 > F_{0.05,1,10-2} \simeq$ 5.32. Thus, we cannot reject H_0.

When we consider the model with the variable x_2, we obtain that $SS_R \simeq 942.11$, so that

$$SS_E \simeq 163.49 \Rightarrow \hat{\sigma}^2 \simeq 20.4 \Rightarrow f_0 \simeq \frac{942.11}{20.4} \simeq 46.1.$$

Therefore, here we can reject H_0: $\beta_2 = 0$ at significance level $\alpha = 0.05$, since 46.1 > 5.32.

b) In the case of the variable x_1, we have: $\hat{\sigma}^2 \overset{a)}{\simeq} 95.3$ and

$$R^2 = \frac{SS_R}{SS_T} \overset{a)}{\simeq} 0.31,$$

while for the variable x_2, we obtain that $\hat{\sigma}^2 \overset{a)}{\simeq} 20.4$ and

$$R^2 \overset{a)}{\simeq} \frac{942.11}{1105.6} \simeq 0.85.$$

Given that R^2 is larger and that the value of $\hat{\sigma}^2$ is smaller with the variable x_2, it is x_2 that better explains the tensile strength.

c) We have:

$$\ln Y = \beta_1 x_1 + \varepsilon.$$

We set

$$SS(\beta_1) = \sum_{i=1}^{10} (\ln y_i - \beta_1 x_{1i})^2.$$

Then, we have:

$$\frac{d}{d\beta_1} SS(\beta_1) = 0 \Leftrightarrow 2\sum_{i=1}^{10} (\ln y_i - \beta_1 x_{1i})(-x_{1i}) = 0 \Rightarrow \hat{\beta}_1 = \frac{\sum_{i=1}^{10} x_{1i} \ln y_i}{\sum_{i=1}^{10} x_{1i}^2}.$$

We find that

$$\sum_{i=1}^{10} x_{1i} \ln y_i \simeq 934.5 \quad \text{and} \quad \sum_{i=1}^{10} x_{1i}^2 = 3028.$$

It follows that

$$\hat{\beta}_1 \simeq \frac{934.5}{3028} \simeq 0.31.$$

Unsolved Problems

Problem no. 1

The following (independent) data have been collected:

	1	2	3	4	5	6	7	8	9	10
x	12	12	10	10	10	3	2	20	20	30
Y	20	40	30	80	50	50	90	30	40	40

We find that

$$\sum_{i=1}^{10} x_i = 129, \quad \sum_{i=1}^{10} y_i = 470, \quad \sum_{i=1}^{10} x_i^2 = 2301, \quad \sum_{i=1}^{10} y_i^2 = 26,500,$$

$$\sum_{i=1}^{10} x_i y_i = 5250, \quad SS_T = 4410.$$

We consider the model

$$Y = \beta_0 + \beta_1 x + \epsilon, \quad \text{where } \epsilon \sim N\left(0, \sigma^2\right).$$

a) What percentage of the total variation in the values of Y can be explained by a linear relationship between the variables Y and x?

b) Can we reject the hypothesis H_0: $\beta_1 = 0$ (against H_1: $\beta_1 \neq 0$) if we choose $\alpha = 0.05$?

Problem no. 2

On a Monday, a pisciculturist scatters a large quantity of baby fish into a fish pond. Each subsequent Monday, he catches at random 500 fish, marks them with a number (different every week), and then puts them back into the water. After a few hours, he throws a net, counts the total number of fish that he caught and the number of those marked with the number of the current week, before putting the fish back into the water. He obtains the following chronological table giving the number of marked fish and the total number of fish caught:

Week	1	2	3	4	5	6	7	8
No. marked	17	19	18	23	24	26	27	29
No. caught	1004	1011	1008	1015	1003	1017	1003	1013

Remark. The precision of the calculations is important in this problem.

a) Let P be the proportion of marked fish in the fish pond. The breeder considers the following model, from the fourth week:

$$P = \alpha e^{\beta t} \quad \text{for } t \in [4, 8].$$

Estimate the parameters α and β by the method of least squares.

b) By extrapolating the proposed model, what is the expected number of fish in the fish pond after 40 weeks?

c) Assuming that the fish remain edible and that they gain 7% in weight every week, from the fourth week until maturity after 50 weeks, should the pisciculturist harvest the fish after eight weeks or wait for maturity?

Remark. The value of the fish is proportional to the weight times the number of fish.

Problem no. 3 (See [22].)

An industrial engineer proposed the model

$$T = \gamma \beta^{-k}$$

for the time T (in seconds) needed to accomplish a simple manual task, as a function of the number k of times the task has been practiced, where β and γ are parameters that depend on the task and on the individual. We have the following data:

T	22.4	21.3	19.7	15.6	15.2	13.9	13.7
k	0	1	2	3	4	5	6

a) Transform the proposed model into a linear model and estimate the parameters β and γ by the method of least squares.

b) We suppose that

$$T = \gamma \beta^{-k} e^{\epsilon}, \quad \text{where } \epsilon \sim N\left(0, \sigma^2\right).$$

i) Estimate the parameter σ.

ii) Test the hypothesis $H_0: \beta = 1$ against $H_1: \beta \neq 1$ with a type I error risk α equal to 0.05.

iii) Calculate an approximate 95% confidence interval for γ.

Problem no. 4

We study the production Y (in cubic meters per second) of a manufacturing process, as a function of the temperature x (in degrees Celsius). For temperatures increasing by 100 degrees from 100 to 600 °C, the production increased from 49 to 68. We have:

$$\sum_{i=1}^{6} y_i = 369, \quad \sum_{i=1}^{6} y_i^2 = 22,947 \quad \text{and} \quad \sum_{i=1}^{6} x_i y_i = 135,400.$$

a) We propose the simple linear regression model:

$$Y = \beta_0 + \beta_1 x + \varepsilon, \quad \text{where } \varepsilon \sim N\left(0, \sigma^2\right).$$

i) Estimate the parameters β_0 and β_1 by the method of least squares.

ii) Calculate the percentage of variation explained by the model.

iii) Test, at significance level $\alpha = 0.01$, the hypothesis H_0: $\beta_1 = 0$ against H_1: $\beta_1 \neq 0$.

Hint. We have: $t_{0.005,4} \simeq 4.604$.

b) We also propose the following curvilinear regression model:

$$Y = \alpha_0 + \alpha_1\sqrt{x} + \varepsilon, \quad \text{where } \varepsilon \sim N\left(0, \sigma^2\right).$$

We then find that $\sum_{i=1}^{6} \sqrt{x_i}\, y_i \simeq 6847.934$.

i) Calculate the point estimates of the parameters α_0 and α_1 by the method of least squares.

ii) Calculate the percentage of variation explained by the model.

Problem no. 5

A trucking company wishes to determine the relationship between the age of a truck and the number of days per year it spends being repaired. To do so, it took six trucks at random and obtained the following data:

x	8	1	6	3	5	2
Y	9	16	1	4	0	10

where x is the age of the truck (in years) and Y is the number of days it spent being repaired over a one-year period.

a) Without doing any calculations, explain why, among the three models that follow, model (2) seems the most appropriate:

$$\begin{align}
(1) \qquad & Y = \beta_0 + \beta_1 x + \varepsilon, \\
(2) \qquad & Y = \beta_0 + \beta_1 x^2 + \varepsilon, \\
(3) \qquad & Y = \beta_0 + \beta_1 e^{-x} + \varepsilon,
\end{align}$$

where $\varepsilon \sim N\left(0, \sigma^2\right)$.

b) Based only on the coefficient of determination R^2, which model, between models (1) and (3) above, is the better one?

Multiple Choice Questions

Question no. 1

We study the relationship between the number x of years elapsed since receiving their first degree and the annual salary Y (in thousands of dollars) of engineers. The values taken by the observations in a random sample of size $n = 10$ yielded the following table:

x	1	2	3	4	5
Y	28	30	40	55	45
	31	35	36	40	60

We propose the model $Y = \beta_0 + \beta_1 x^2 + \varepsilon$, where $\varepsilon \sim N(0, \sigma^2)$. We then have: $SS_T = 1016$ and $\hat{\sigma}^2 \simeq 33.5$.

A) Estimate the parameter β_0 by the method of least squares.

a) 0   b) 1   c) 6.1   d) 21.7   e) 29

B) Calculate the percentage of variation explained by the proposed model.

a) 54.2%   b) 73.2%   c) 73.6%   d) 85.6%   e) 85.8%

C) Estimate the annual salary of an engineer having ten years experience, according to the proposed model.

a) 100,000   b) 104,000   c) 105,000   d) 121,700   e) 129,000

Question no. 2

The real speed of three makes of cars, when their speedometers indicate 100 km/h, has been measured for four cars of each make:

Make 1	105	103	108	101
Make 2	101	97	99	104
Make 3	102	96	98	103

We propose the model $V = \beta_0 + \beta_1 x + \epsilon$, where V is the real speed, x is the speed indicated by the speedometer and $\epsilon \sim N(0, \sigma^2)$. Estimate the parameter β_1 by the method of least squares.

a) 0   b) 1   c) 1.1   d) ∞ e) indeterminate

Question no. 3

We are interested in the relationship between two random variables, X and Y. We collected five pairs (x_i, y_i) of particular observations of (X, Y). The results are summarized below:

$$\sum_{i=1}^{5} x_i = 5, \quad \sum_{i=1}^{5} y_i = 15, \quad \sum_{i=1}^{5} x_i^2 = 55, \quad \sum_{i=1}^{5} y_i^2 = 51, \quad \sum_{i=1}^{5} x_i y_i = 51.$$

A) Calculate the sample correlation coefficient, $r_{X,Y}$.

a) 0   b) 0.194   c) 0.4   d) 0.775   e) 0.8

B) Suppose that $X \sim N(\mu_X, \sigma_X^2)$ and $Y \sim N(\mu_Y, \sigma_Y^2)$. Estimate the mean of Y, given that $X = 0$.

a) -3 b) 0   c) 1.2   d) 3   e) 4.8

C) If, with other data, we obtained $R^2 = 0.04$ and $\hat{\beta}_1 > 0$, what is then the value of the statistic used to test the hypothesis $H_0: \rho = 0$ against $H_1: \rho \neq 0$?

a) -0.577 b) -0.069 c) 0.069   d) 0.354   e) 0.577

Question no. 4

We consider the following data:

x	1	1.5	2	2.5	3
Y	13	16	18	18	19

We propose the curvilinear regression model

$$Y = \beta_0 + \beta_1 \ln x + \epsilon, \quad \text{where } \epsilon \sim N(0, \sigma^2).$$

We calculate

$$\sum_{i=1}^{5} \ln x_i \simeq 3.1135, \quad \sum_{i=1}^{5} y_i = 84, \quad \sum_{i=1}^{5} y_i \ln x_i \simeq 56.3309,$$

$$\sum_{i=1}^{5} \ln^2 x_i \simeq 2.6914, \quad \sum_{i=1}^{5} y_i^2 = 1434.$$

A) Estimate the parameter β_1 by the method of least squares.

a) 2.80 b) 5.35 c) 11.20 d) 13.47 e) 17.12

B) Calculate the value of the statistic F_0 used to test the null hypothesis $H_0: \beta_1 = 0$.

a) 4.11 b) 16.92 c) 50.77 d) 72.30 e) 75.20

C) What is the percentage of variation explained by the model?

a) 5.58% b) 12.62% c) 80.0% d) 88.22% e) 94.42%

Supplementary Exercises

Question no. 1

Preliminary calculations made with a series of particular observations (x_i, y_i), $i = 1, \ldots, 100$, of a random variable Y and a deterministic variable x yielded

$$\sum_{i=1}^{100} x_i = 500, \qquad \sum_{i=1}^{100} y_i = 1000, \qquad \sum_{i=1}^{100} x_i^2 = 24,775,$$

$$\sum_{i=1}^{100} (y_i - \bar{y})^2 = 39,600, \quad r_{x,Y} = 0.9,$$

where $r_{x,Y}$ is the sample "correlation coefficient" defined by

$$r_{x,Y} = \frac{\sum_{i=1}^{100} (x_i - \bar{x})(y_i - \bar{y})}{\left[\sum_{i=1}^{100} (x_i - \bar{x})^2 \sum_{i=1}^{100} (y_i - \bar{y})^2\right]^{1/2}}.$$

A first look at the graph of these data points suggest a simple linear regression model:

$$Y = \beta_0 + \beta_1 x + \epsilon, \quad \text{where } \epsilon \sim N(0, \sigma^2).$$

a) Calculate $\sum_{i=1}^{100} (x_i - \bar{x}) (y_i - \bar{y})$.

b) Estimate all the parameters in the model.

c) Perform an analysis of the variance and test the null hypothesis $H_0: \beta_1 = 0$ at significance level $\alpha = 0.05$.

Hint. We have: $F_{0.05,1,98} \simeq 3.94$.

Question no. 2

We have the following table showing the progression of an epidemic over time:

t	10	12	15	20	23	25	27	30
N	40	35	30	25	20	20	15	10

where t denotes the number of weeks elapsed since the outbreak of the epidemic and N is the number (in tens) of persons infected. We wish to forecast the value of N after 40 weeks. We propose the following model:

$$N = \beta_0 + \beta_1 t + \varepsilon, \quad \text{where } \varepsilon \sim N\left(0, \sigma^2\right).$$

We find that

$$n = 8, \qquad \sum_{i=1}^{8} t_i = 162, \qquad \sum_{i=1}^{8} t_i^2 = 3652,$$

$$\sum_{i=1}^{8} n_i = 195, \quad \sum_{i=1}^{8} n_i^2 = 5475, \quad \sum_{i=1}^{8} t_i n_i = 3435.$$

a) Estimate all the parameters in the model and calculate the table of analysis of variance.

b) Give the forecasted value of N after 40 weeks.

c) *Reminder.* We define a $100(1 - \alpha) \%$ one-sided confidence interval with an upper bound for a parameter θ by the determination of a statistic UC such that

$$P[\theta < UC] = 1 - \alpha.$$

Calculate a 99% one-sided confidence interval with an upper bound for the value of N after 40 weeks.

Hint. We have: $t_{0.01,6} \simeq 3.143$.

Question no. 3

We are interested in the relationship between the speed x (in kilometers per hour) reached in 10 seconds by 15 luxury cars and the braking distance Y (in meters) until a complete stop. The following data have been collected:

x	104.2	106.1	105.6	106.3	101.7	104.4	102.0	103.8
Y	39.8	40.4	39.9	40.8	33.7	39.5	33.0	37.0

x	104.0	101.5	101.9	100.6	104.9	106.2	103.1
Y	37.0	33.2	33.9	29.9	39.5	40.6	35.1

a) Estimate the parameters in the model

$$Y = \beta_0 + \beta_1 x + \epsilon, \quad \text{where } \epsilon \sim N(0, \sigma^2).$$

b) Calculate the table of analysis of variance and test the null hypothesis $H_0: \beta_1 = 0$ at significance level $\alpha = 0.05$.

c) Calculate 95% confidence intervals for the coefficients β_0 and β_1.

d) Perform an analysis of the residuals to check whether the basic assumptions are satisfied.

e) Calculate a 95% prediction interval for Y when $x = 100$, 103 and 106.

Hint. We have: $t_{0.025,13} \simeq 2.160$.

Question no. 4

The data that follow show the effect of time on the content of hydrogen for two (independent) steel specimens stored at 20 °C:

t	1	2	6	17	30
H	7.7	7.5	6.1	5.7	4.2
	8.4	8.1	6.8	5.3	4.5

where t denotes the time (in hours) and H is the content of hydrogen (in parts per million). We propose the model

$$H = \beta_0 + \beta_1 \ln t + \varepsilon, \quad \text{where } \varepsilon \sim N(0, \sigma^2).$$

a) Estimate all the parameters in the model.

b) Calculate the table of analysis of variance and test the null hypothesis $H_0: \beta_1 = 0$ at significance level $\alpha = 0.05$.

Question no. 5

We seek to establish a relationship between the diameter x (in centimeters) of the filament of an electric light bulb and its lifetime Y (in hours). We have the following data:

x	0.15	0.20	0.25	0.30	0.40	0.50	0.60	0.60
Y	120	165	204	238	296	373	410	403

x	0.60	0.70	0.70	0.80	0.80	0.80	0.90	1.00
Y	420	462	455	520	518	525	580	600

a) Estimate the parameters of the simple linear regression model.

b) Calculate the analysis of variance table and test the null hypothesis $H_0: \beta_1 = 0$ at significance level $\alpha = 0.05$.

c) Perform an analysis of the residuals to check whether the basic assumptions are satisfied.

Question no. 6

The law of perfect gases relating the pressure P (in kilograms per square centimeter) and the volume v (in cubic centimeters):

$$Pv^\gamma = c$$

can be written, after transformation, in the form

$$Y = \alpha - \gamma x,$$

where $Y := \ln P$, $x := \ln v$ and $\alpha := \ln c$. Particular observations of v and P yielded

$$(v, P) = (50, 64.7), (60, 51.3), (70, 40.5), (90, 25.9), (100, 7.8).$$

Remark. Useful calculations:

$$\sum_{i=1}^{100} x_i \simeq 21.35984, \quad \sum_{i=1}^{100} x_i^2 \simeq 91.57317, \quad \sum_{i=1}^{100} x_i y_i \simeq 72.26249,$$

$$\sum_{i=1}^{100} y_i \simeq 17.11712, \quad \sum_{i=1}^{100} y_i^2 \simeq 61.40147.$$

a) Estimate the parameters α and γ by the method of least squares.

b) Calculate the analysis of variance table of the model $Y = \alpha - \gamma x$, as well as the percentage of variation explained.

c) Calculate a confidence interval at 95% for the average value of P when $v = 80$.

Question no. 7

The regression model through the origin is given by

$$Y = \beta x + \epsilon, \quad \text{where } \epsilon \sim N(0, \sigma^2).$$

a) Find the estimator of the parameter β by the method of least squares, if we have n observations of the pair (x, Y).

b) We suppose that a recently built structure sinks into the ground according to the model

$$Z = 10 - 10e^{-\beta x},$$

where x is the age (in months) of the structure and Z is the sinking (in centimeters). Find a transformation that reduces the above model to the regression model through the origin.

Question no. 8

To evaluate a measurement process, an operator who knows the measuring instrument well obtains two measurements, X and Y, of 15 parts in a batch. A measurement process is said to have a good *repeatability* if the correlation coefficient, $r_{X,Y}$, of the random sample of the pair (X, Y) is greater than 0.80. The data are the following:

X	34	56	6	50	33	43	49	17	54	24	35	46	10	51	25
Y	45	44	19	55	17	32	37	5	54	18	26	43	16	55	11

Remark. Useful calculations: $\bar{x} = 35.53$, $\bar{y} = 31.8$, $s_X \simeq 16.20$ and $s_Y \simeq 17.01$.

a) Does the measurement process have a good repeatability?

b) Calculate the covariance $s_{X,Y} := \sum_{i=1}^{15} (x_i - \bar{x})(y_i - \bar{y})$ of the particular observations of X and Y.

c) We consider the pairs

$$(X_1, Y_1), \ldots, (X_{15}, Y_{15}), (Y_1, X_1), \ldots, (Y_{15}, X_{15}).$$

They constitute a random sample of size $n = 30$ of a pair of random variables that we denote by (U, V).

i) Explain why the standard deviation S_U of the observations of U and that of the observations of V are equal.

ii) We assume that U and V have a Gaussian distribution, so that

$$E[V \mid U = u] = \beta_0 + \beta_1 u.$$

Estimate, using the 30 particular observations of (U, V), the parameters β_0 and β_1 by the method of maximum likelihood.

Question no. 9

During the 1840s, the physicist J. Forbes measured in 17 places, in the Alps and in Scotland, the barometric pressure Y (in inches of mercury) and the boiling temperature x (in degrees Fahrenheit) of water. At the time, barometers were fragile and it was much easier to determine the boiling temperature of water in mountainous regions than to use a barometer. From this temperature, the barometric pressure could be determined. The data collected by Forbes are presented below:

x	194.5	194.3	197.0	198.4	199.4	199.9
Y	20.79	20.79	22.40	22.67	23.15	23.35

x	200.9	201.1	201.4	201.3	203.6	204.6
Y	23.89	23.99	24.02	24.01	25.14	26.57

x	209.5	208.6	210.7	211.9	212.2
Y	28.49	27.76	29.04	29.88	30.06

a) Draw the graph of the data points (x_i, y_i) and that of the data points $(x_i, \ln y_i)$. Which graph looks more like a straight line?

b) Perform a regression analysis according to the simple linear regression model. That is, estimate the parameters in the model, calculate the analysis of variance table and perform an analysis of the residuals.

c) Perform a regression analysis of the model

$$\ln Y = \gamma_0 + \gamma_1 x + \epsilon, \quad \text{where } \epsilon \sim N(0, \sigma^2).$$

d) Which model is superior, based on the percentage of variation explained by the model and on the behavior of the residuals?

Question no. 10

Let $(x_1, y_1), \ldots, (x_{10}, y_{10})$ be ten particular observations of a pair of random variables (X, Y). We have: $\sum_{i=1}^{10} x_i = \sum_{i=1}^{10} y_i = 55, s_X = s_Y = 3.03$ and $\sum_{i=1}^{10} x_i y_i = 380$.

a) Calculate the sample correlation coefficient, $r_{X,Y}$.

b) Calculate $\sum_{i=1}^{10} x_i^2 - 10\bar{x}^2$.

Question no. 11

An experiment similar to that of Forbes, described in no. 9, was realized by J. Hooker in the Himalayas. His data are the following:

x	Y	x	Y	x	Y
210.8	29.211	193.6	21.212	185.6	17.062
210.2	28.559	193.4	20.480	184.6	16.881
208.4	27.972	191.4	19.758	184.1	16.959
202.5	24.697	191.1	19.490	184.1	16.817
200.6	23.726	190.6	19.386	183.2	16.385
200.1	23.369	189.5	18.869	182.4	16.235
199.5	23.030	188.8	18.356	181.9	16.106
197.0	21.892	188.5	18.507	181.9	15.928
196.4	21.928	186.0	17.221	181.0	15.919
196.3	21.654	185.7	17.267	180.6	15.376
195.6	21.605				

a) Perform a regression analysis of the model

$$\ln Y = \gamma_0 + \gamma_1 x + \epsilon, \quad \text{where } \epsilon \sim N(0, \sigma^2).$$

b) Compare the results to those in part c) of exercise no. 9.

c) Perform a regression analysis of the model in a), using the 48 data of Hooker and Forbes together.

Question no. 12

In a simple linear regression problem, we find that the parameter β_1 is not significantly different from zero.

a) We then propose the model $Y = \beta_0 + \varepsilon$, where $\varepsilon \sim N(0, \sigma^2)$. Estimate β_0 by the method of least squares.

b) We also propose the model $Y = \beta_0 + \beta_1 x^2 + \varepsilon$. Find the value $\hat{\beta}_0$ of β_0 that minimizes the function $SS(\beta_0, \hat{\beta}_1) := \sum_{i=1}^{n}(Y_i - \beta_0 - \hat{\beta}_1 x_i^2)^2$.

Question no. 13

The voltage U (in volts) of a capacitor, with initial charge u_0, is, after t seconds, given by the equation

$$U = u_0 e^{-\beta t}.$$

An experiment yielded the following results:

t	0	1	2	3	4	5	6	7	8	9	10
U	100	75	55	40	30	20	15	10	10	5	5

a) Make a transformation to obtain a linear model with respect to the parameters.

b) Fit the new model by the method of least squares.

c) Calculate 95% confidence intervals for u_0 and β.

Question no. 14

We propose the model $Y = \beta_0 + \beta_1 x + \varepsilon$, where $\varepsilon \sim N(0, \sigma^2)$, relating the variables x and Y. We have the following results: $\sum_{i=1}^{5} y_i = 10$, $\sum_{i=1}^{5} y_i^2 = 50$ and the sum of squares SS_R is equal to 25.

a) Calculate the coefficient of determination, R^2.

b) Calculate $\hat{\sigma}^2$.

Question no. 15

Boyle's law relates the pressure P of a gas to the volume v it occupies, according to the equation

$$P v^\alpha = \beta,$$

where α and β are two constants. Fit the model using the following data:

v	0.2	1.0	0.8	1.0	1.6	0.4
P	0.6	1.0	1.5	2.0	2.5	3.5

Question no. 16

We consider the model $Y = \beta_0 x^{\beta_0 \beta_1}$.

a) Make a transformation to linearize the model. Express the new variables Y' and x', as well as the new parameters β_0' and β_1', in terms of the old ones.

b) An analysis of variance table yielded a statistic f_0 equal to 12. Can we then reject, at significance level $\alpha = 0.01$, the hypothesis $H_0: \beta_1' = 0$, if we have ten particular observations of (x, Y)? Give the value of the percentile used to perform the test.

Hint. We have: $t_{0.005,8} \simeq 3.355$.

Question no. 17 (See [2].)

A screw manufacturer wishes to inform its customers about the relationship between the nominal length x and the actual length Y of its screws (in inches). The following values, taken by independent observations of Y, have been collected:

x	1/4	1/4	1/4	1/2	1/2	1/2
Y	0.262	0.262	0.245	0.496	0.512	0.490

x	3/4	3/4	3/4	1	1	1
Y	0.743	0.744	0.751	0.976	1.010	1.004

x	5/4	5/4	5/4	3/2	3/2	3/2
Y	1.265	1.254	1.252	1.498	1.518	1.504

x	7/4	7/4	7/4	2	2	2
Y	1.738	1.759	1.750	2.005	1.992	1.992

We propose the simple linear regression model

$$Y = \beta_0 + \beta_1 x + \epsilon, \quad \text{where } \epsilon \sim N(0, \sigma^2).$$

a) Estimate all the parameters in the model.

b) Obtain a 95% confidence interval for the average value of Y when $x = 1$, using only the three particular observations of Y for this value of x.

c) Obtain a 95% confidence interval for the average value of Y when $x = 1$, using all the data available.

d) Calculate the "correlation coefficient" (see no. 1 above) of the pairs of data (x, y).

Appendix A: Mathematical Formulas

Logarithms

$$\ln ab = \ln a + \ln b; \quad \ln a/b = \ln a - \ln b; \quad \ln a^b = b \ln a$$

Geometric Series

If $|r| < 1$, we find that

$$\sum_{k=0}^{\infty} ar^k = \frac{a}{1-r}; \qquad \sum_{k=0}^{\infty} ak\, r^k = \frac{ar}{(1-r)^2};$$

$$\sum_{k=1}^{\infty} ar^k = \frac{ar}{1-r}; \qquad \sum_{k=0}^{n} ar^k = \frac{a(1-r^{n+1})}{1-r}.$$

Limits

L'Hospital's rule. Suppose that $\lim_{x \to x_0} f(x) = \lim_{x \to x_0} g(x) = 0$, or that $\lim_{x \to x_0} f(x) = \lim_{x \to x_0} g(x) = \pm\infty$. Then, under some conditions, we can write that

$$\lim_{x \to x_0} \frac{f(x)}{g(x)} = \lim_{x \to x_0} \frac{f'(x)}{g'(x)}.$$

If the functions $f'(x)$ and $g'(x)$ satisfy the same conditions as $f(x)$ and $g(x)$, we can repeat the process. Moreover, the constant x_0 may be equal to $\pm\infty$.

Derivatives

Derivative of a quotient:

$$\frac{d}{dx} \frac{f(x)}{g(x)} = \frac{g(x)f'(x) - f(x)g'(x)}{g^2(x)} \quad \text{if } g(x) \neq 0.$$

Remark. This formula can also be obtained by differentiating the product $f(x)h(x)$, where $h(x) := 1/g(x)$.

Chain rule. If $u = g(x)$, then we have:

$$\frac{d}{dx} f(u) = \frac{d}{du} f(u) \cdot \frac{du}{dx} = f'(u)g'(x) = f'[g(x)]g'(x).$$

For example, if $u = x^2$, we calculate

$$\frac{d}{dx} \exp(u^2) = \exp(u^2)2u \cdot 2x = 4x^3 \exp(x^4).$$

Integrals

Integration by parts:

$$\int u\, dv = uv - \int v\, du.$$

Integration by substitution. If the inverse function g^{-1} exists, then we can write that

$$\int_a^b f(x)\, dx = \int_c^d f[g(y)]g'(y)\, dy,$$

where $a = g(c) \Leftrightarrow c = g^{-1}(a)$ and $b = g(d) \Leftrightarrow d = g^{-1}(b)$.
   For example,

$$\int_0^4 e^{x^{1/2}}\, dx \overset{y=x^{1/2}}{=} \int_0^2 2y\, e^y\, dy = 2ye^y\big|_0^2 - 2\int_0^2 e^y\, dy = 2e^2 + 2.$$

In this example, $y = g^{-1}(x) = x^{1/2}$ and $x = g(y) = y^2$.

Appendix B: Quantiles of the Sampling Distributions

Gaussian distribution

α	0.25	0.10	0.05	0.025	0.01	0.005	0.001	0.0005
z_α	0.674	1.282	1.645	1.960	2.326	2.576	3.090	3.291

Student distribution

n	1	2	3	4	5	6	7	8
$t_{0.025,n}$	12.706	4.303	3.182	2.776	2.571	2.447	2.365	2.306
$t_{0.05,n}$	6.314	2.920	2.353	2.132	2.015	1.943	1.895	1.860

n	9	10	15	20	25	30	40	∞
$t_{0.025,n}$	2.262	2.228	2.131	2.086	2.060	2.042	2.021	1.960
$t_{0.05,n}$	1.833	1.812	1.753	1.725	1.708	1.697	1.684	1.645

Chi-square distribution

n	1	2	3	4	5	6	7	8	9
$\chi^2_{0.025,n}$	5.02	7.38	9.35	11.14	12.83	14.45	16.01	17.53	19.02
$\chi^2_{0.05,n}$	3.84	5.99	7.81	9.49	11.07	12.59	14.07	15.51	16.92
$\chi^2_{0.95,n}$	0^+	0.10	0.35	0.71	1.15	1.64	2.17	2.73	3.33
$\chi^2_{0.975,n}$	0^+	0.05	0.22	0.48	0.83	1.24	1.69	2.18	2.70

Chi-square distribution (continued)

n	10	15	20	25	30	40	50	100
$\chi^2_{0.025,n}$	20.48	27.49	34.17	40.65	46.98	59.34	71.42	129.56
$\chi^2_{0.05,n}$	18.31	25.00	31.41	37.65	43.77	55.76	67.50	124.34
$\chi^2_{0.95,n}$	3.94	7.26	10.85	14.61	18.49	26.51	34.76	77.93
$\chi^2_{0.975,n}$	3.25	6.27	9.59	13.12	16.79	24.43	32.36	74.22

Fisher distribution

n	1	2	3	4	5	6	7	8
$F_{0.025,n,n}$	647.8	39.00	15.44	9.60	7.15	5.82	4.99	4.43
$F_{0.05,n,n}$	161.4	19.00	9.28	6.39	5.05	4.28	3.79	3.44

n	9	10	15	20	30	60	120	∞
$F_{0.025,n,n}$	4.03	3.72	2.86	2.46	2.07	1.67	1.43	1.00
$F_{0.05,n,n}$	3.18	2.98	2.40	2.12	1.84	1.53	1.35	1.00

Values of $F_{0.025,n_1,n_2}$

$n_2 \backslash n_1$	1	2	3	4	5	6	7	8	9	10	11	12
1	648	800	864	900	922	937	948	957	963	969	973	977
2	38.5	39.0	39.2	39.2	39.3	39.3	39.4	39.4	39.4	39.4	39.4	39.4
3	17.4	16.0	15.4	15.1	14.9	14.7	14.6	14.5	14.5	14.4	14.4	14.3
4	12.2	10.6	9.98	9.60	9.36	9.20	9.07	8.98	8.90	8.84	8.79	8.75
5	10.0	8.43	7.76	7.39	7.15	6.98	6.85	6.76	6.68	6.62	6.57	6.52
6	8.81	7.26	6.60	6.23	5.99	5.82	5.70	5.60	5.52	5.46	5.41	5.37
7	8.07	6.54	5.89	5.52	5.29	5.12	4.99	4.90	4.82	4.76	4.71	4.67
8	7.57	6.06	5.42	5.05	4.82	4.65	4.53	4.43	4.36	4.30	4.24	4.20
9	7.21	5.71	5.08	4.72	4.48	4.32	4.20	4.10	4.03	3.96	3.91	3.87
10	6.94	5.46	4.83	4.47	4.24	4.07	3.95	3.85	3.78	3.72	3.66	3.62
11	6.72	5.26	4.63	4.28	4.04	3.88	3.76	3.66	3.59	3.53	3.47	3.43
12	6.55	5.10	4.47	4.12	3.89	3.73	3.61	3.51	3.44	3.37	3.32	3.28

Values of $F_{0.05, n_1, n_2}$

$n_2 \backslash n_1$	1	2	3	4	5	6	7	8	9	10	11	12
1	161	200	216	225	230	234	237	239	241	242	243	244
2	18.5	19.0	19.2	19.2	19.3	19.3	19.4	19.4	19.4	19.4	19.4	19.4
3	10.1	9.55	9.28	9.12	9.01	8.94	8.89	8.85	8.81	8.79	8.76	8.74
4	7.71	6.94	6.59	6.39	6.26	6.16	6.09	6.04	6.00	5.96	5.94	5.91
5	6.61	5.79	5.41	5.19	5.05	4.95	4.88	4.82	4.77	4.74	4.70	4.68
6	5.99	5.14	4.76	4.53	4.39	4.28	4.21	4.15	4.10	4.06	4.03	4.00
7	5.59	4.74	4.35	4.12	3.97	3.87	3.79	3.73	3.68	3.64	3.60	3.57
8	5.32	4.46	4.07	3.84	3.69	3.58	3.50	3.44	3.39	3.35	3.31	3.28
9	5.12	4.26	3.86	3.63	3.48	3.37	3.29	3.23	3.18	3.14	3.10	3.07
10	4.96	4.10	3.71	3.48	3.33	3.22	3.14	3.07	3.02	2.98	2.94	2.91
11	4.84	3.98	3.59	3.36	3.20	3.09	3.01	2.95	2.90	2.85	2.82	2.79
12	4.75	3.89	3.49	3.26	3.11	3.00	2.91	2.85	2.80	2.75	2.72	2.69

Appendix C: Classification of the Exercises

The list of exercises (among the multiple choice questions, problems or supplementary exercises) that can be done after having read each section is given below. The symbol 1m (respectively 1s), for example, denotes the multiple choice question (resp. supplementary exercise) no. 1 of the chapter in question.

Chapter 2

2.1: 1m, 6m, 12m, 19m, 32m
2.2: 2; 8m, 13m, 36m, 46m
2.3: 3, 8, 11, 18, 20; 2m, 7m, 11m, 14m, 15m, 21m, 24m, 29m, 34m
2.4: 1, 5, 24; 17m, 23m, 31m, 39m, 40m, 44m, 48m, 50m
2.5: all the others

Chapter 3

3.1: —
3.2: 3m, 15m, 52m
3.3: 1, 2, 10, 11, 42; 4m, 16m, 28m, 34m, 35m, 36m, 53m
3.4: 12, 19, 31, 34; 1m, 12m, 17m, 18m, 29m, 37m, 56m, 79m
3.5: 14, 27, 38; 6m, 19m, 20m, 41m, 58m, 64m, 80m
3.6: 6, 35; 21m, 40m, 47m
3.7: 3, 4, 5, 7, 9, 15, 16, 17, 20, 22, 23, 25, 28, 29, 32, 37, 39, 43, 46, 47; 2m, 5m, 22m, 23m, 30m, 32m, 38m, 39m, 45m, 46m, 48m, 49m, 54m, 55m, 57m, 59m, 66m, 67m, 68m, 72m, 73m, 74m, 76m, 77m, 78m
3.8: 8, 13, 18, 21, 24, 26, 30, 33, 36, 40, 41, 44, 45; 7m, 8m, 9m, 10m, 11m, 13m, 14m, 24m, 25m, 26m, 27m, 31m, 33m, 42m, 50m, 60m, 62m, 63m, 70m
3.9: 43m, 44m, 51m, 61m, 65m, 69m, 71m, 75m, 81m

Chapter 4

4.1: —
4.2: 1, 2, 12; 3m, 4m, 9m, 17m, 20m, 21m, 26m, 31m, 32m
4.3: 3, 14; 10m, 22m, 33m, 40m

4.4: 11m
4.5: 4, 5, 19; 12m, 28m, 34m
4.6: 6, 16, 22, 24, 25, 27, 29; 1m, 5m, 23m, 27m, 35m, 42m
4.7: 15, 17; 6m, 24m
4.8: 7, 20, 21; 13m, 30m, 47m
4.9: 8, 9, 28; 7m, 18m, 29m, 36m
4.10: 16m
4.11: 10, 11, 12, 13, 18, 23, 26; 2m, 8m, 14m, 15m, 19m, 25m, 37m, 38m, 39m, 41m, 43m, 44m, 45m, 46m, 48m

Chapter 5

5.1: —
5.2: 3; 2m, 6m, 16m, 19m
5.3: 8; 3m, 11m, 14m, 23m, 27m, 29m
5.4: 1, 4; 1m, 4m, 7m, 17m, 20m, 24m, 25m, 26m, 28m, 30m
5.5: 7; 5m, 12m, 18m
5.6: 8m, 21m
5.7: 2, 5, 6; 9m, 10m, 13m, 15m, 22m

Chapter 6

6.1: 1, 6, 8; 2m, 7m, 11m, 13m, 15m
6.2: 2, 5, 7; 1m, 3m, 4m, 9m, 12m, 14m, 23m; 1s, 2s, 4s
6.3: 3, 4, 9; 5m, 6m, 8m, 10m, 18m, 21m, 25m, 27m
6.4: 10–19; 16m, 17m, 19m, 20m, 22m, 24m, 26m, 28m; 3s, 5s–11s

Chapter 7

7.1: 2m
7.2: 1s
7.3: 2s, 17s
7.4: 1; 14s
7.5: 3s, 5s
7.6: 2, 3, 4, 5; 1m, 4m; 4s, 6s, 7s, 9s, 11s, 12s, 13s, 15s, 16s
7.7: 3m; 8s, 10s

Appendix D: Answers to the Multiple Choice Questions

Chapter 2

1 a; 2 d; 3 b; 4 d; 5 c; 6 c; 7 a; 8 c; 9 e; 10 c; 11 d; 12 c; 13 b; 14 a; 15 d; 16 c; 17 d; 18 c,e,b,b,c,d; 19 b,e; 20 d,c; 21 d,e; 22 d,a; 23 e,c; 24 a,b; 25 d,a; 26 b,c; 27 b,c,c,e; 28 d,a,a,c; 29 a,c,b; 30 e,b,a; 31 e,d,c; 32 b; 33 b; 34 d,c; 35 c,e,a,a; 36 c,a; 37 e,a,c,e; 38 b,d; 39 d,e; 40 c,b; 41 b,a,c,d; 42 c,e; 43 e,a; 44 d,d; 45 b,a; 46 e,c; 47 b,b; 48 d,d; 49 e,a; 50 b,d,a,e; 51 c,b,d,e.

Chapter 3

1 b,c,b,d; 2 a,e,c,d; 3 c; 4 e; 5 a; 6 d; 7 d; 8 b; 9 c; 10 e; 11 c; 12 e,d,c; 13 b,d,b; 14 d,e,d; 15 e; 16 c; 17 e; 18 b; 19 d; 20 b; 21 a; 22 d; 23 b; 24 e; 25 a; 26 e,c,e,b,a,c; 27 d,d,b,a,d,a; 28 b,d,c; 29 a,d,a; 30 b,e,a; 31 d,b,e,c; 32 c,d,d; 33 a,b,e; 34 c; 35 d; 36 b; 37 b; 38 e; 39 d; 40 a; 41 a; 42 e; 43 c; 44 b; 45 d,c; 46 a,b; 47 d,b; 48 c,e; 49 b,a; 50 e,a; 51 e,c; 52 b; 53 c; 54 e,c,e; 55 d,b; 56 a; 57 d; 58 a,e; 59 a; 60 b; 61 c; 62 c,d,a,e; 63 d,b,d; 64 e,c,b; 65 c,e,a,b; 66 c,b,d; 67 e,d,b,c; 68 e,a,e,b; 69 d,a,a; 70 d,a,d,c; 71 e,a,d,e; 72 b,b,c; 73 b,c,e; 74 c,a,c; 75 c,d,a,b; 76 c,a,d,e; 77 e,b,d; 78 d,a,d; 79 a,c,e; 80 c,b,e,e; 81 b,d,c,b.

Chapter 4

1 a,b,d,c,a; 2 d,a,e; 3 b; 4 c; 5 b; 6 d; 7 d; 8 e; 9 c; 10 b; 11 b; 12 d; 13 e; 14 b; 15 a; 16 b,c,c; 17 b; 18 c; 19 c; 20 d; 21 b; 22 a; 23 e; 24 c; 25 a; 26 c; 27 a; 28 a; 29 c; 30 e; 31 c; 32 e; 33 d; 34 a; 35 b; 36 a; 37 b; 38 a,e,b,b,c,e; 39 b,a,c,d,d,b; 40 d,d,b; 41 a,c; 42 d,a,a; 43 c,e; 44 e,a,a,d,e,a; 45 d,b,e,a,c; 46 a,c,d,e; 47 d,c,a; 48 b,e.

Chapter 5

1 d,e,d,c; 2 c; 3 e; 4 a; 5 d; 6 e; 7 e; 8 d; 9 a; 10 b,a,e; 11 d; 12 c; 13 a; 14 e; 15 d; 16 d; 17 b; 18 d; 19 c; 20 d; 21 d; 22 a; 23 a,d; 24 c,e; 25 e,c; 26 b,b; 27 c,b; 28 e,b; 29 c,a; 30 c,d.

Chapter 6

1 e,e,c; 2 e; 3 a; 4 c; 5 d; 6 e,b; 7 a; 8 b; 9 b; 10 e; 11 b; 12 e; 13 c; 14 e; 15 c,d; 16 c,b,a; 17 e,c,a,d,b; 18 e,a; 19 d,b,e; 20 e,c,c,a,d; 21 e,b; 22 d,c,b; 23 c,c; 24 a,d,b; 25 c,b,e; 26 b,d,e; 27 e,c; 28 b,d,a.

Chapter 7

1 e,c,e; 2 e; 3 d,c,d; 4 b,c,e.

Appendix E: Answers to Selected Supplementary Exercises

Chapter 6

5. a) H_1: $\mu < 240$; b) -1.27; c) 0.32; d) 82.

6. a) Test with paired observations; $t_0 \simeq 3.23$; b) 6.

7. a) 0.1; b) 0.6.

8. a) $w_0^2 \simeq 23.5 < \chi_{0.05,19}^2 \Rightarrow$ we do not reject H_0: $\sigma^2 > (1.5)^2$; b) 0.21; c) 40.

9. a) $f_0 \simeq 7.14 \Rightarrow$ we accept H_1: $\sigma_A^2 \neq \sigma_B^2$; $t_0^* \simeq -1.67 \Rightarrow$ we do not reject H_0; b) we do not reject H_0: $\sigma_A^2 = \sigma_B^2$; $t_0 \simeq -1.46 \Rightarrow$ we do not reject H_0.

10. a) $t_0 \simeq 1.81 < t_{0.01,6} \Rightarrow$ we do not reject H_0: $\mu = 0.07$; b) the formula with known σ gives 63 or 64 (exact answer = 66); c) the formula with known σ gives 0.9985 (more precise answer = 0.95).

11. a) X_i = diameter measured with calipers i; we assume that $X_i \sim N(\mu_i, \sigma_i^2)$; we perform a test with paired observations; $t_0 \simeq 0.43 \Rightarrow$ we do not reject H_0: $\mu_1 = \mu_2$; b) $w_0^2 \simeq 16.24 < \chi_{0.10,11}^2 \Rightarrow$ we do not reject H_0: $\sigma_1^2 = (0.001)^2$.

Chapter 7

2. a) $\hat{\beta}_0 \simeq 52.38$; $\hat{\beta}_1 \simeq -1.383$; $\hat{\sigma}^2 \simeq 1.89$; $f_0 \simeq 375.93$; b) $\hat{N}(40) \simeq -2.94 < 0 \Rightarrow 0$; c) $[0, 3.43]$ (approximately).

4. a) $\hat{\beta}_0 \simeq 8.31$; $\hat{\beta}_1 \simeq -1.08$; $\hat{\sigma}^2 \simeq 0.166$; b) $f_0 \simeq 113.55 > 5.32$; we reject H_0.

6. a) $\hat{\alpha} \simeq 14.76$; $\hat{\gamma} \simeq 2.65$; b) $f_0 \simeq 13.4$; $R^2 \simeq 0.81$; c) $[12.3, 44.3]$ (approximately).

8. a) $r_{X,Y} \simeq 0.83 \Rightarrow$ yes; b) 230; c) i) the observations of U and of V are the same, only the order of the observations is different; ii) $\hat{\beta}_0 \simeq 6.4$; $\hat{\beta}_1 \simeq 0.81$.

10. a) 0.94; b) 82.6.

12. a) $\bar{Y}$; b) $\bar{Y} - \hat{\beta}_1 \sum_{i=1}^{n} \dfrac{x_i^2}{n}$.

14. a) $0.8\bar{3}$; b) $1.\bar{6}$.

16. a) $Y' = \ln Y$; $x' = \ln x$; $\beta_0' = \ln \beta_0$; $\beta_1' = \beta_0 \beta_1$; b) $12 > F_{0.01,1,8} = t_{0.005,8}^2 \simeq 11.26$; we reject H_0.

Bibliography

1. Barnes, J. Wesley, *Statistical Analysis for Engineers: A Computer-Based Approach*, Prentice-Hall, Englewood Cliffs, New Jersey, 1988.
2. Bowker, Albert H., and Lieberman, Gerald J., *Engineering Statistics*, Prentice-Hall, Englewood Cliffs, New Jersey, 1959.
3. Breiman, Leo, *Probability and Stochastic Processes: With a View Toward Applications*, Houghton Mifflin, Boston, 1969.
4. Cartier, Jacques; Parent, Régis; and Picard, Jean-Marc, *Inférence Statistique*, Éditions Sciences et Culture, Montréal, 1976.
5. Chung, Kai Lai, *Elementary Probability Theory with Stochastic Processes*, Springer-Verlag, New York, 1975.
6. Clément, Bernard, *Analyse Statistique*, Tome I and II, Lectures notes, École Polytechnique de Montréal, 1988.
7. Dougherty, Edward R., *Probability and Statistics for the Engineering, Computing, and Physical Sciences*, Prentice-Hall, Englewood Cliffs, New Jersey, 1990.
8. Feller, William, *An Introduction to Probability Theory and Its Applications*, Volume I, 3rd Edition, Wiley, New York, 1968.
9. Feller, William, *An Introduction to Probability Theory and Its Applications*, Volume II, 2nd Edition, Wiley, New York, 1971.
10. Hastings, Kevin J., *Probability and Statistics*, Addison-Wesley, Reading, Massachusetts, 1997.
11. Hines, William W., and Montgomery, Douglas C., *Probability and Statistics in Engineering and Management Science*, 3rd Edition, Wiley, New York, 1990.
12. Hogg, Robert V., and Craig, Allen T., *Introduction to Mathematical Statistics*, 3rd Edition, Macmillan, New York, 1970.
13. Kallenberg, Olav, *Foundations of Modern Probability*, 2nd Edition, Springer-Verlag, New York, 2002.
14. Krée, Paul, *Introduction aux Mathématiques et à leurs Applications Fondamentales*, Dunod, Paris, 1969.
15. Leon-Garcia, Alberto, *Probability and Random Processes for Electrical Engineering*, 2nd Edition, Addison-Wesley, Reading, Massachusetts, 1994.
16. Lindgren, Bernard W., *Statistical Theory*, 3rd Edition, Macmillan, New York, 1976.
17. Maksoudian, Y. Leon, *Probability and Statistics with Applications*, International, Scranton, Pennsylvania, 1969.

18. Miller, Irwin; Freund, John E.; and Johnson, Richard A., *Probability and Statistics for Engineers*, 4th Edition, Prentice-Hall, Englewood Cliffs, New Jersey, 1990.
19. Papoulis, Athanasios, *Probability, Random Variables, and Stochastic Processes*, 3rd Edition, McGraw-Hill, New York, 1991.
20. Parzen, Emmanuel, *Stochastic Processes*, Holden-Day, San Francisco, 1962.
21. Roberts, Richard A., *An Introduction to Applied Probability*, Addison-Wesley, Reading, Massachusetts, 1992.
22. Ross, Sheldon M., *Introduction to Probability and Statistics for Engineers and Scientists*, Wiley, New York, 1987.
23. Ross, Sheldon M., *Introduction to Probability Models*, 7th Edition, Academic Press, San Diego, 2000.
24. Whittle, Peter, *Optimization Over Time*, Volume I, Wiley, Chichester, 1982.

Index